NEWLY REVISED FIFTH EDITION

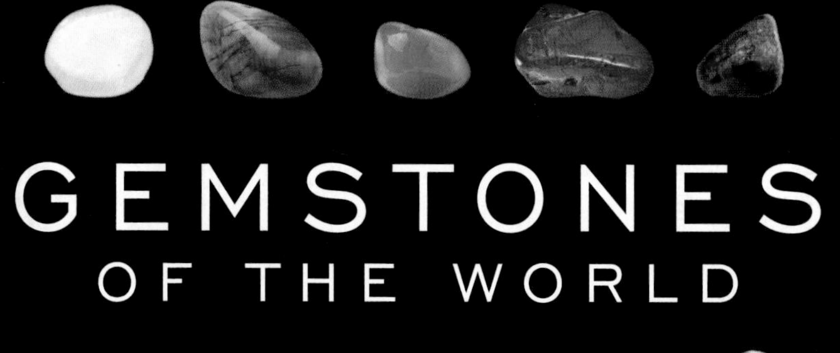

GEMSTONES OF THE WORLD

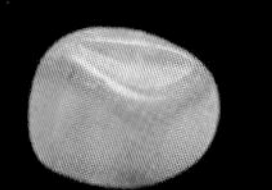
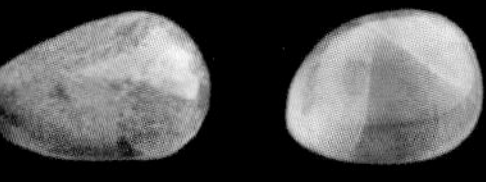
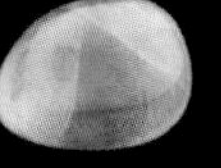
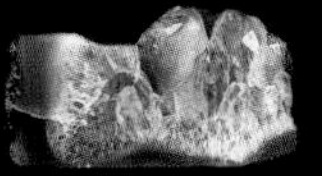
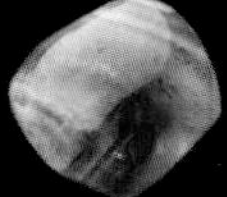

WALTER SCHUMANN

STERLING
New York

Contents

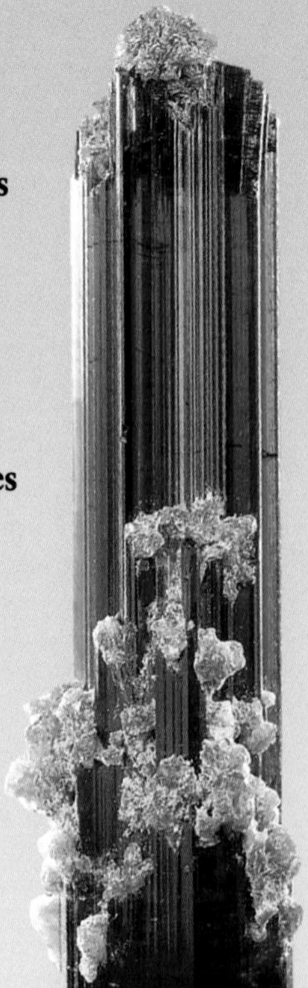

From the Preface to the First Edition

Gemstones have always fascinated mankind. In former centuries they were reserved for the ruling classes only. Today everybody can afford beautiful stones for jewelry and adornment. Precious stones for sale, especially if one includes those in so-called costume or fashion jewelry, are so numerous that it is hardly possible for the layman to survey or judge what is available. This little book has been written to help: it shows the many gems of the world in their many varieties, rough and cut, in true-to-nature color photographs. The accompanying text—always adjacent to the photographs—is designed to be of use to both the expert and the layman.

Introductory chapters on formation, properties, deposits, manufacture, synthesis, and imitations provide a survey of the world of beautiful stones. Unknown gemstones can be identified with the help of the tables at the end of the book.

I received valuable help from professional colleagues, friends, acquaintances, institutes, firms, and private persons who made stones available for illustration. My thanks are due to all of them with special thanks to Mr. Paul Ruppenthal, Idar-Oberstein. I also thank Mr. Karl Hartmann, Sobernheim, for taking the special photographs.

Preface to the 17th Edition

Gemstones of the World is distributed worldwide and has been translated into close to 20 languages. The number of copies has long exceeded the one million mark. Ever since the first edition in 1976, the scope of this work has been expanded by more than 60 pages of colorful illustrations.

Information has been updated and new scientific knowledge and economic conditions were taken into consideration. A separate chapter now critically examines the use of gemstones in cosmic and astral symbolism and for healing purposes.

Once again, I received valuable help and useful information, and I would like to express my sincere gratitude to Ms. Nicole Schiel, Company Groh + Ripp, Idar-Oberstein, as well as to Prof. Dr. Hermann Bank, Idar-Oberstein, Prof. Dr. Henry A. Hänni, Basel, Dieter Hahn, Idar-Oberstein, Dr. Gerhard Holzhey, Erfurt, Eckehard J. Petsch, Idar-Oberstein, Albert Ruppenthal, Idar-Oberstein, and Prof. Dr. Rainer Schultz-Güttler, Sao Paulo.

Walter Schumann

The Presentation of the Text

When describing gemstones, my objective has been to provide as much information as possible. Therefore, in part, smaller letter-types were used, short forms were chosen, and repetitions, which would take up space, were avoided.

The meaning of abbreviations, as well as the omission of the units of measurements, is clearly recognizable from the respective context.

Introduction

Gemstones and Their Influence

Gems have intrigued humans for at least 10,000 years. The first known, used for making jewelry, include amethyst, rock crystal, amber, garnet, jade, jasper, coral, lapis lazuli, pearl, serpentine, emerald, and turquoise. These stones were reserved for the wealthy, and served as status symbols. Rulers sealed documents with their jewel-encrusted seals. Such treasures can now be admired at many museums and treasure-vaults.

Today, gems are worn not so much to demonstrate wealth, but rather jewelry is bought increasingly for pleasure, in appreciation of its beauty.

Certainly, also today, when purchasing a gemstone, a certain love for a special stone is part of it. Formerly, when people were less scientifically knowledgeable, gems always had an aura of mystery, something almost spiritual. That's why they were worn as amulets and talismans. Up to the present day, gemstones have sometimes been used as remedies against illnesses. They could be used in three different ways: the mere presence of the stone was sufficient to effect a cure; the gem was placed on the afflicted part of the body; or the stone was powdered and eaten.

Presently, medical science worldwide is experiencing a revival of the ideas of the Middle Ages in the use of precious stones through the doctrines of the Esoterics. The chapter on Symbol and Beneficial Stones (p. 283) gives more information. Gems also have an assured place in modern religion. The breastplate of the high priests of Judea was studded with four rows of gemstones. Precious stones adorn the tiara and miter of the Pope and bishops as well as the monstrances, reliquaries, and icons found in Christian churches. All major religions use precious stones, be it as decoration for tools or adorning buildings.

As a capital investment, however, of all gemstones really only diamonds are suitable. In fact, these have proven to hold on to their value, despite the travails of war or depressions in the economy.

A truly modern problem is the imitation of gemstones that have become more and more sophisticated. See also the chapter "Imitation and Synthetic Gemstones" (p. 266). Although replicas of precious matter have always existed, it is only in our modern times that imitating has turned into a booming economic sector with an often disreputable background.

The English Imperial State Crown with rubies, emeralds, sapphires, pearls, and more than 3000 diamonds

In the center below "Cullinan II." (or "Lesser Star of Africa") with 317.40ct. It is the second-largest polished diamond in the world, having 66 facets. It was cut from the largest rough diamond ever found, the "Cullinan," besides 104 other stones. The company Asscher in Amsterdam polished it (compare also "Cullinan I.," page 94, No. 3).

The large red stone above "Cullinan II." is the so-called "Black Prince's Ruby." Once thought to be a ruby, it is, in fact, a spinel, uncut, only polished, 2in (5cm) high.

The State Crown is exhibited in the Tower of London.

Terminology

In the following, important terms of the trade, used throughout the book, are explained:

Gem/Gemstone There is no generally accepted definition for the term gem or gemstone, but they all have something special, something beautiful about them. Most gemstones are minerals (e.g., diamond), mineral aggregates (such as lapis lazuli), or rocks (such as onyxmarble). Some are organic formations (e.g., amber), and other gem materials are of synthetic origin (e.g., YAG).

There is no definite demarcation line, and woods, coal, bones, glass, and metals are all used for ornamentation. Some examples include jet (a form of coal), ivory (tusks of elephants as well as teeth of other large animals), moldavite (a glassy after-product of the striking of a meteorite), and gold nugget (more or less large gold-lumps). Even fossils are sometimes used as ornamental material.

For some gemstones the source of specialness and beauty is the color, an unusual optical phenomenon, or the shine that makes them stand out in comparison to other stones. For other stones it is the hardness or an interesting inclusion that makes them special. Rarity also plays a role in the classification as gemstone.

Since the valued characteristics usually come into effect only through cutting and polishing, gemstones are also normally considered to be the cut stones. Cutting and polishing means refinement of what might be an otherwise insignificant raw material.

There are several hundred distinct types of gems and gem materials. The number of the variations is about double that. From time to time, new gemstones are discovered or varieties with gemstone quality are found in minerals which have been already known.

Harder stones are suitable for jewelry, whereas softer stones are often sought after by amateur collectors as well as serious lapidaries.

Gemology Internationally, the science of gemstones is most commonly referred to as gemology.

Colored Stone Colored stone is a trade term for all gemstones except diamonds (even those that are colorless).

Rock With regard to gems, a rock is a natural aggregate of two or more minerals. Lapis lazuli is an example of a gem that is classified as a rock. It forms independent geological bodies of greater expansion and comprises both solid rock and accumulated unconsolidated rock.

Semi-Precious Stone The term semi-precious has generally fallen out of use because of its derogatory meaning. Formerly, one meant with this term the less valuable and not very hard gemstones, which one opposed to the "precious" stones. "Precious" and "semi-precious" are adjectives, however, that cannot be adequately defined to distinguish between gems.

Imitation Imitations are made to resemble natural or synthetic gem materials, completely or partially manmade. They imitate the look, color, and effect of the original substance, but they possess neither their chemical nor their physical characteristics. To these belong—strictly speaking—also those synthetic stones that do not have a counterpart in nature (for instance, fabulite or YAG). However, in the trade these are often counted as synthetic stones.

Jewel Every individual ornamental piece is a jewel. Generally, a jewel refers to a piece of jewelry containing one or more gems set in precious metal. Sometimes, it can also refer to cut, unset gemstones.

Crystal A crystal is a uniform body with an ordered structure, i.e., a strict order of the smallest components (the atoms, ions, or molecules) in a geometric crystal lattice. The varying structures of the lattice, together with the chemical components, are the causes of the varying physical properties of the crystals and therefore also of the gems.

Crystallography Crystallography is the science of crystals.

Matrix Finer-grained ground mass than the embedded crystals

Mineral A mineral is a naturally occurring, inorganic, solid constituent of the earth's (or other celestial body's) crust. Most minerals have definite chemical compositions and crystal structures.

Mineralogy Mineralogy is the science of minerals.

Petrography Petrography is the descriptive science of rocks, typically utilizing a petrographic microscope or other instruments. The term petrography is often used more comprehensively to mean the same as petrology.

Petrology Petrology is the science of the origin, history, occurrence, structure, chemical composition, and classification of rocks. Sometimes petrology is loosely used to mean the same as petrography.

Species A mineral species is distinguished by a specific combination of chemical composition and crystal structure (e.g., diamond is carbon in a cubic structure). This combination produces the mineral's distinctive optical, physical, and chemical properties.

Stone Popularly, stone is the collective name for all solid constituents of the earth's crust except for ice and coal. For jewelers and gem collectors, the word stone means only gemstones. For the architect, on the other hand, it means the material used for constructing buildings and streets. In the science of the earth, geology, one does not talk of stones, but of rocks and minerals.

Synthesis Short term for a synthetic stone.

Synthetic Stone Synthetic gemstones (in short called syntheses) are crystallized manmade products whose physical and chemical properties for the most part correspond to those of their natural gemstone counterparts.

This term is also used by the gemstone trade for those synthetic stones which do not have a counterpart in nature (for instance, fabulite or YAG). In fact, these stones are properly termed imitations.

Variety By a variety, one understands in the case of gemstones a modification which distinguishes itself through its look, color, or other characteristics from the actual gemstone species.

The Nomenclature of Gemstones

The oldest names for gemstones can be traced back to Oriental languages, Greek and Latin. Greek names especially have left their stamp on modern gem nomenclature. The meaning of old names is not always certain, especially where the first meaning of the word has been changed. Also, in antiquity, totally different stones were given the same name simply because they may have had the same color.

Original names referred to special characteristics of the stones, such as color (for instance "prase" for its green color), their place of discovery ("agate" for a river in Sicily), or mysterious powers ("amethyst" was thought to protect against drunkenness). Many mineral names, which were later also used to name gemstones, have their origin in the miners' language of the Middle Ages.

Nomenclature has been viewed scientifically only since the beginning of the modern age. Because of the discovery of many hitherto unknown minerals, new names had to be found. A principle for naming new minerals and gemstones was established which is still adhered to today. A new name is devised to refer to some special characteristic of the mineral, based on Greek or Latin, the chemical constituents, the place of occurrence, or a person's name.

Through such name-giving, not only experts are honored but also patrons or others who may or may not have any connection with mineralogy or gemology. But since everyone did not always agree when a name was given, various names for the same mineral have come into use and persist to this day.

The gemstone and jewelry trade added more of their own names, mainly to stimulate sales, producing a large number of synonyms and variety names for gemstones.

In order to correct this state of affairs, all newly discovered minerals, as well as the intended new names, must now be presented for evaluation to the Commission on New Mineral Names of the IMA (International Mineralogical Association), to which experts from all over the world belong.

Anyone who believes that he has found a new mineral or an important gemstone variety must have his first right and other legalities as well as the name-giving checked. Only then is the name of the mineral and/or the gemstone sanctioned and standardized.

Because the danger of purposefully giving a wrong name and improper evaluation of goods is especially high in the gemstone trade, the Commission for Delivery Conditions and Quality Securing at the German Norm Commission has published in 1963/70 in the RAL A5/A5E *Guidelines for Precious Stones, Gemstones, Pearls and Corals* (for Germany) to protect against unfair competition.

Permitted definitions and the trade customs for gemstones are internationally regulated through the International Association of Jewelry, Silverwares, Diamonds, Pearls, and Stones, in short CIBJO (Confédération Internationale de la Bijouterie, Joaillerie, Orfevrerie des Diamants, Perles, et Pierres).

In the United States, the Federal Trade Commission (FTC) *Guides for the Jewelry, Precious Metals, and Pewter Industries* serve a similar purpose.

Certainly, the aforementioned institutions have led to better communication and to more security for the gemstone buyer and seller, but such measures, naturally, cannot ensure the enforcement of an absolute guarantee for genuineness.

False and Misleading Names of Some Gemstones

False Gemstone Name	Preferred Gemnological Name
Adelaide ruby	Almandite
African emerald	Green fluorite
Alaska diamond	Rock crystal (quartz)
American jade	Green idocrase
American ruby	Pyrope or almandite (garnet) or rose quartz
Arizona ruby	Pyrope (garnet)
Arizona spinel	Red or green garnet
Arkansas diamond	Rock crystal (quartz)
Balas ruby	Red spinel
Blue alexandrite	Color-change sapphire
Blue moonstone	Artificially blue-tinted chalcedony
Bohemian chrysolite	Moldavite (natural glass)
Bohemian diamond	Rock crystal (quartz)
Bohemian ruby	Pyrope (garnet) or rose quartz
Brazilian aquamarine	Bluegreen topaz
Brazilian ruby	Red or pink topaz
Brazilian sapphire	Blue tourmaline
Californian ruby	Hessonite (grossular garnet)
Candy spinel	Almandite (garnet)
Cape-chrysolite	Green prehnite
Cape-ruby	Pyrope (garnet)
Ceylon diamond	Colorless zircon
Ceylon opal	Opal-like glimmery moonstone
Copper lapis	Azurite
German diamond	Rock crystal (quartz)
German lapis	Artificially blue-tinted jasper (chalcedony)
Gold topaz	Citrine (quartz)
Indian jade	Aventurine (quartz)
King's topaz	Yellow sapphire
Korean jade	Serpentine
Lithia amethyst	Kunzite (spodumene)
Lithia emerald	Hiddenite (spodumene)
Madeira topaz	Citrine (quartz)
Marmarosch diamond	Rock crystal (quartz)
Matura diamond	Colorless fired zircon
Mexican diamond	Rock crystal (quartz)
Mexican jade	Artificially tinted green marble
Montana ruby	Red garnet
Oriental amethyst	Violet sapphire
Oriental hyacinth	Pink sapphire
Oriental topaz	Yellow sapphire
Palmyra topaz	Brown synthetic sapphire
Salmanca topaz	Citrine (quartz)
Saxon chrysolite	Greenish-yellow topaz
Saxon diamond	Colorless topaz
Serra topaz	Citrine (quartz)
Siamese aquamarine	Blue zircon
Siberian chrysolite	Demantoid (garnet)
Siberian ruby	Red tourmaline
Simili diamond	Glass imitation
Slave-diamond	Colorless topaz
Smoky topaz	Smoky quartz
Spanish topaz	Citrine (quartz)
Strass diamond	Glass imitation
Transvaal jade	Green hydrogrossular garnet
Ural sapphire	Blue tourmaline
Viennese turquoise	Artificially blue-tinted argillaceous earth

Origin and Structure of Gemstones

Since, with few exceptions, most gemstones are minerals, it is the origin and structure of minerals that concerns the gemologist. The formation of the non-mineral gemstones (for instance, amber, coral, and pearl) will be dealt with in more detail when they are described.

Origin Minerals can be formed in various ways. Some crystallize from molten magma and gases of the earth's interior or from volcanic lava streams that reach the earth's surface (igneous or magmatic minerals). Others crystallize from hydrous solutions or grow with the help of organisms on or near the earth's surface (sedimentary minerals). Lastly, new minerals are formed by recrystallization of existing minerals under great pressure and high temperatures in the lower regions of the earth's crust (metamorphic minerals).

The chemical composition of the minerals is shown by a formula. Impurities are not included in this formula, even where they cause the color of the stone.

Formations of Crystals Nearly all minerals grow in certain crystal forms; i.e., they are homogeneous bodies with a regular lattice of atoms, ions, or molecules. They are geometrically arranged and their outer shapes are limited by flat surfaces (in the ideal case), resulting in crystal faces.

Most crystals are small, sometimes even microscopically small, but there are also some giant specimens. In general, the smallest minerals (due to their tiny size) as well as the giant minerals (due to their inclusions, impurities, or uneven growth marks) are unsuitable as gems.

The mineral's chemical composition and inner structure, the lattice, determine the physical properties of the crystal (its outer shape, hardness, cleavage, type of fracture, specific gravity) and also its optical properties.

Most crystals are not regularly shaped, but have an irregular form, because some crystal faces have developed better at the cost of others; however, the angle between the faces always remains constant. When individual single crystals occur in combination with other crystal forms, e.g., hexahedron with

Crystal lattice of sphalerite.

Crystal lattice of quartz.

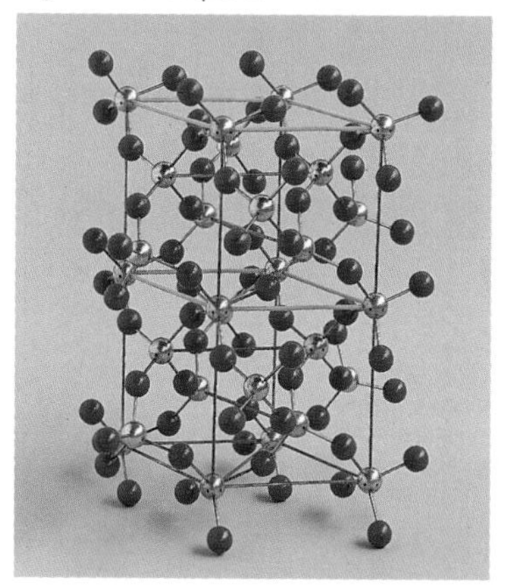

Rubellite, Madagascar (⅔ natural size).

octahedron, the identification of a mineral on the basis of crystal shape can be extraordinarily complicated.

The arrangement of faces preferred by a mineral is called the "habit"; for instance, pyrite is often found in the shape of a pentagon dodecahedron, garnet, as a rhomb-dodecahedron. The habit of a crystal also refers to its type and can be tabular, acicular, foliated, columnar, or compact. For the benefit of laymen, the technical terms *habit* and *form* are sometimes called structure. Commonly minerals adopt the crystal form of another—usually through replacing it. These occurrences are called pseudomorphs.

Where two or more crystals are intergrown according to certain laws, one speaks of twins, triplets, or quadruplets. Depending on whether the individual crystals are grown together or intergrown, one speaks of contact twins or penetration twins.

Apart from twinning, which adheres to certain laws, many crystals are irregularly intergrown into aggregates. Depending on the growth process, filiform (wire-like), fibrous, radial-shaped, leaf-like, shell-like, scaly, or grainy aggregates are formed. According to miners' lingo, a mineral aggregate with free-standing individual crystals is called a "step."

Well-developed, characteristic minerals are formed as druses on the inner walls of rock openings (geodes); these are mainly round hollows created by gas bubbles in magmatic rocks or spaces where organic material has been removed in sedimentary rocks.

Crystal Systems In crystallography, crystals are divided into seven systems. The distinction is made according to crystal axes and the angles at which the axes intersect. On page 17, the crystal systems are depicted with some typical crystal formations.

Cubic System (also called the isometric system) All three axes have the same length and intersect at right angles. Typical crystal shapes are the cube, octahedron, rhombic dodecahedron, pentagonal dodecahedron, icosi-tetrahedron, and hexacisochedron.

Tetragonal System The three axes intersect at right angles, two are of the same length and are in the same plane, while the main axis is either longer or shorter. Typical crystal shapes are four-sided prisms and pyramids, trapezohedrons and eight-sided pyramids as well as double pyramids.

Hexagonal system Three of the four axes are in one plane, are of the same length, and intersect each other at angles of 120 degrees. The fourth axis, which is a different length, is at right angles to the others. Typical crystal shapes are hexagonal prisms and pyramids, as well as twelve-sided pyramids and double pyramids.

Trigonal system (rhombohedral system) Three of four axes are in the same plane, are of equal length, and intersect each other at angles of 120 degrees. The fourth axis, which is of different length, is at right angles to the others. The difference is one of symmetry. In the case of the hexagonal system, the cross section of the prism base is six-sided; in the trigonal system, it is three-sided. The six-sided hexagonal shape is formed by a cutting-off process of the corners of the triangles. Typical crystal forms of the trigonal system are three-sided prisms and pyramids, rhombohedra, and scalenohedra.

Orthorhombic system (rhombic system) Three axes of different lengths are at right angles to each other. Typical crystal shapes are basal pinacoids, rhombic prisms, and pyramids as well as rhombic double pyramids.

Monoclinic system The three axes are each of different lengths; two are at right angles to each other, and the third one is inclined. Typical crystal forms are basal pinacoids and prisms with inclined end faces.

Triclinic system All three axes are of different lengths and inclined to each other. Typical crystal forms are paired faces.

In the table on page 18, minerals from which many gemstone varieties come are allocated to their respective crystal systems.

Horizontal gemstone microscope. In the center of the apparatus is a box with a red gemstone, which is held by a movable claw holder.

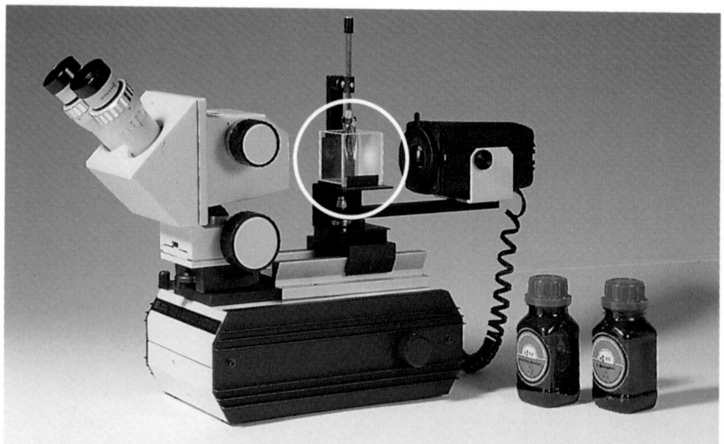

Crystal Systems

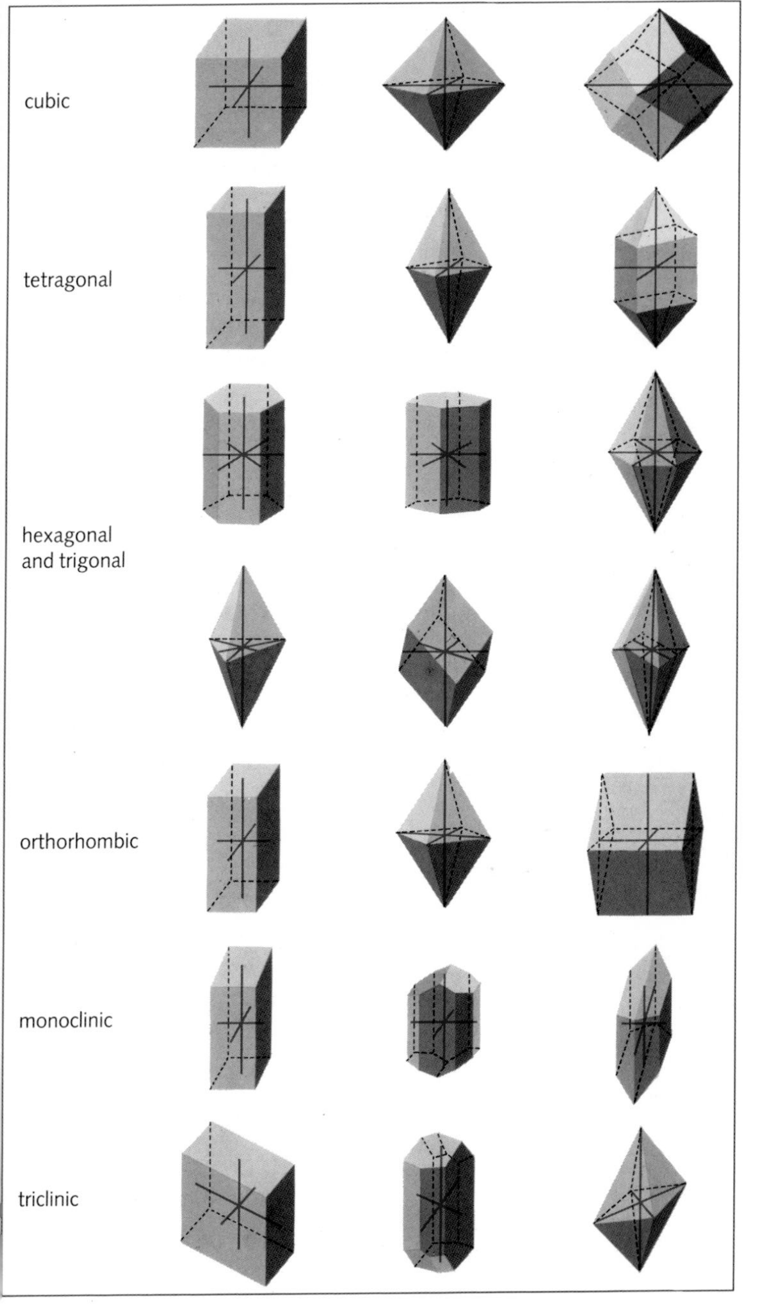

Selected Gemstones Ordered by Crystal System

cubic

Almandine
Analcite
Andradite
Bixbyite
Boleite
Chromite
Cuprite
Demantoid
Diamond
Fabulite
Fluorite
Gahnite
Gahnospinel
Galaxite
GGG
Gold
Garnet
Grossularite
Hackmanite
Haüynite
Helvite
Hercynite
Hessonite
Katoite
Langbeinite
Lapis Lazuli
Lazurite
Leucogranite
Magnesiochromite
Magnetite
Melanite
Microlite
Pentlandite
Periclase
Picotite
Pleonaste
Pollucite
Pyrite
Pyrope
Rhodozite
Rhodolite
Schorlomite
Senarmontite
Silver
Sodalite
Spessartine
Sphalerite
Spinel
Synth. Spinel
Thorianite
Topazolite
Tsavorite
Uvarovite
Villiaumite
YAG
Zirconia
Zunyite

tetragonal

Anatase
Apophyllite
Carletonite
Cassiterite
Chalcopyrite
Chiolite
Ekanite
Fergusonite
Hyacinth
Leucite
Marialite
Meionite
Melinophane
Mellite
Phosgenite
Powellite
Pyrolusite
Rutile
Sarcolite
Scapolite
Scheelite
Sellaite
Stolzite
Synth. Rutile
Thorite
Tugtupite
Vesuvianite
Wardite
Wulfenite
Zircon

hexagonal

Algodonite
Apatite
Aquamarine
Bastnasite
Benitoite
Beryl
Bixbite
Breithauptite
Cacoxenite
Cancrinite
Catapleiite
Chlorapatite
Covellite
Emerald
Ettringite
Fluorapatite
Gold Beryl
Goshenite
Greenockite
Heliodor
Hydroxylapatite
Jeremejevite
Lizardite
Manganapatite
Milarite
Mimetite Moissan-
ite
Morganite
Nepheline
Nickeline
Painite
Poudretteite
Precious Beryl
Simpsonite
Sogdianite
Sturmanite
Sugilite
Taaffeite
Thaumasite
Vanadinite
Wurtzite
Zincite

trigonal

Agate
Achroite
Amethyst
Amethyst Quartz
Ametrine
Ankerite
Aventurine
Bloodstone
Blue Quartz
Brucite
Buergerite
Calcite
Carnelian
Cat's-Eye Quartz
Chalcedony
Chromdravite
Chrysoprase
Citrine
Cinnabar
Corundum
Davidite
Dendritic Agate
Dioptase
Dolomite
Dravite
Elbaite
Eudialyte
Gaspeite
Hawk's-Eye
Hematite
Ilmenite
Indicolite
Jasper
Liddicoatite
Linobate
Lizardite
Magnesite
Melonite
Millerite
Moss Agate
Parisite
Pezzottaite
Petrified Wood
Phenakite
Povondraite
Prase
Prasiolite
Proustite
Pyrargyrite
Quartz
Rhodochrosite
Rock Crystal
Rose Quartz
Rubellite
Ruby
Sapphire
Sardonyx
Schlossmacherite
Schorl
Siberite
Siderite
Smithsonite
Sphaerocobaltite
Smoky Quartz
Stichtite
Tiger's-Eye
Tsilaisite
Tourmaline
Uvite
Verdelite
Willemite

orthorhombic

Adamite
Aeschynite
Alexandrite
Andalusite
Anglesite
Anhydrite
Aragonite
Baryte
Bastite
Bismutotantalite

Boracite
Bornite
Bronzite
Brookite
Celestine
Cerussite
Chambersite
Childrenite
Chrysoberyl
Cobaltite
Cordierite
Danburite
Descloizite
Diaspore
Dumortierite
Enstatite
Eosphorite
Euchroite
Euxenite
Ferrosilite
Forsterite
Goethite
Grandidierite
Hambergite
Hemimorphite
Heterosite
Holtite
Humite
Hypersthene
Kornerupine
Lawsonite
Libethenite
Lithiophilite
Manganotantalite
Marcasite
Mordenite
Natrolite
Norbergite
Pearl
Peridot
Prehnite
Purpurite
Samarskite
Scorodite
Sekaninaite
Sepiolite
Shattuckite
Shomiokite
Shortite
Sillimanite
Sinhalite
Stibiotantalite
Strontianite
Sulfur
Tanzanite
Tantalite
Tephroite
Thomsonite
Thulite
Topaz
Triphylite
Variscite
Wavellite
Witherite
Yttrotantalite
Zektzerite
Zoisite

monoclinic

Aegirine
Aegirine-Augite
Actinolite
Allanite
Antigorite
Augelite
Azurite
Barytocalcite
Bayldonite
Beryllonite
Bikitaite
Bowenite
Brazilianite
Canasite
Chalcocite
Charoite
Childrenite
Chondrodite
Chromian Diopside
Chrysocolla
Clinochlore
Clinochrysotile
Clinoenstatite
Clinohumite
Clinozoisite
Colemanite
Creedite
Crocoite
Cryolite
Datolite
Dickinsonite
Diopside
Durangite
Epidote
Euclase
Friedelite
Fuchsite
Gadolinite
Gaylussite
Gypsum
Hancockite
Hedenbergite
Herderite
Hiddenite
Hodgkinsonite
Hornblende
Howlite
Hübnerite
Huréaulite
Hurlbutite
Hyalophane
Hydroxylherderite
Inderite
Jadeite
Kämmererite
Kunzite
Lazulite
Legrandite
Lepidolite
Linarite
Ludlamite
Malachite
Mesolite
Monazite
Moonstone
Muscovite
Nephrite
Neptunite
Orthoclase
Palygorskite
Papagoite
Pargasite
Petalite
Phosphophyllite
Piemontite
Prosopite
Psilomelane
Pumpellyite
Pyrophyllite
Realgar
Richterite
Rinkite
Sanidine
Sapphirine
Scolecite
Scorzalite
Serpentine
Smaragdite
Spodumene
Spurrite
Staurolite
Talc
Tawmawite
Titanite
Tremolite
Väyrynenite
Violane
Vivianite
Vlasovite
Whewellite
Williamsite
Wolframite
Wollastonite
Xonotlite
Yugawaralite

triclinic

Albite
Amazonite
Amblygonite
Andesine
Anorthite
Anorthoclase
Axinite
Bustamite
Bytownite
Ceruleite
Chabazite
Ferro-Axinite
Fowlerite
Kurnakovite
Kyanite
Labradorite
Leucophanite
Magnesio-Axinite
Manganaxinite
Microline
Montebrasite
Nambulite
Natromontebrasite
Oligoclase
Pectolite
Peristerite
Pyrophyllite
Pyroxmangite
Rhodonite
Sérandite
Serendibite
Sunstone
Talc
Tinzenite
Turquoise
Ulexite
Ussingite
Weloganite
Wollastonite
Xonotlite

amorphous

Amber
Glass
Moldavite
Obsidian
Opal
Petrified Wood
Strass

Properties of Gemstones

Specific knowledge about the most important properties of gemstones are of inestimable value to the gemstone cutter and the setter, as well as to the wearer of the jewelry and the collector. Only with proper knowledge can one correctly work the gemstone, make use of it, and take care of it.

Hardness

In the case of minerals and gemstones, hardness refers first to scratch hardness, then to cutting resistance.

Scratch Hardness

The Viennese mineralogist Friedrich Mohs (1773–1839) introduced the term *scratch hardness*. He defined the scratch hardness as the resistance of a mineral when scratched with a pointed testing object.

Mohs set up a comparison scale using ten minerals of different degrees of hardness (Mohs' hardness scale), which is still widely in use. Number 1 is the softest, number 10 the hardest degree. Each mineral in the series scratches the previous one with the lesser hardness and is scratched by the one which follows after. Minerals of the same hardness will not scratch each other. In practice, the hardness grades have also been subdivided into half degrees. All minerals and gemstones known to us today are allocated to Mohs' hardness scale. (See table, pages 22 and 23.)

Gemstones of the scratch hardness (Mohs' hardness) 1 and 2 are considered soft, those of the degrees 3 to 5 medium hard, and those over 6 hard. Formerly, one spoke also of gemstone hardness in regard to the steps 8 to 10. This is no longer in use, since there are valuable gemstones that do not have these high Mohs' hardness values.

The luster and polish of gemstones of the Mohs' hardness below 7 can be damaged by dust, as this may contain small particles of quartz (Mohs' hardness 7). Through the scratching of this quartz dust, the stones become dull in the course of time. Such stones must be carefully handled when being worn and stored.

Relative and Absolute Hardness Scale

Scratch Hardness (Mohs)	Mineral used for resistance comparison (Rosiwal)	Simple hardness tester	Cutting
1	Talc	Can be scratched with fingernail	0.03
2	Gypsum	Can be scratched with fingernail	1.25
3	Calcite	Can be scratched with copper coin	4.5
4	Fluorite	Easily scratched with knife	5.0
5	Apatite	Can be scratched with knife	6.5
6	Orthoclase	Can be scratched with steel file	37
7	Quartz	Scratches window glass	120
8	Topaz		175
9	Corundum		1,000
10	Diamond		140,000

The Mohs' hardness scale is a relative hardness scale. It only shows which gemstone is harder than another one. Nothing is said about increase of hardness within the scale. That is possible only with absolute hardness scales, for example, cutting resistance. (See table, page 20.)

Hardness Test Formerly, when the optical examination methods had not been developed to such a degree as they are today, the scratch-hardness test played a larger role for the determination of gemstones. Now, the scratch-hardness test is used only very rarely by gemologists. It is too inaccurate for a precise testing of hardness, and there is the danger of hurting the gemstone.

In the trade, test pieces and scratch tools for the testing of hardness can be bought. When doing the scratch test, one must make sure that the examination is done only with sharp-edged objects on fresh, nondecomposed crystal or cut surfaces. Corrugated or foliated formations feign a lesser hardness.

Always begin with softer test materials in order not to damage the gemstone unnecessarily. If possible, do not scratch transparent cut stones at all. With translucent or opaque gemstones test only at an unnoticeable spot, or on the underside.

Cutting Resistance

For the gemstone cutter, naturally the hardness of a stone, when cutting, plays an important role. There are also gemstones which are of a different hardness on different crystal faces and in different directions. For the stone collector, small differences in hardness are of lesser importance. In kyanite (compare drawing below and page 212), for example, the Mohs' hardness along the stem-like crystals is 4-4½, but across it is 6 to 7. There are also important differences in hardness in diamond on the crystal faces (see drawing below). (For more about the cutting of diamond, see page 77.)

For the gemstone cutter, it would be helpful to have absolute values regarding the cutting hardness of gemstones. Unfortunately, there are hardly any useful numbers available. The cutter must rather discover for himself in practice and rely on his experience.

It is real art to cut softer gemstones, which only a few specialists master. If the crystal faces of a stone, in addition, are of different degrees of hardness, it takes a lot of skill to form sharp and even edges on these stones.

When polishing gemstones, the hardness is of utmost importance, because harder gemstones take a polish better than softer stones.

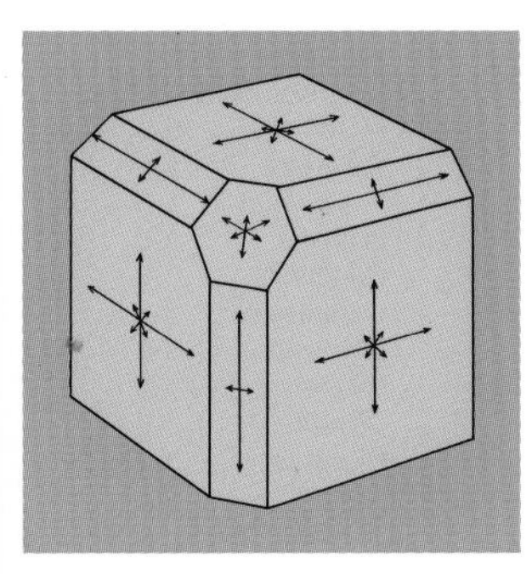

Left: Differences in hardness in diamond. The shorter the arrow, the larger the cutting hardness in this direction (according to E.M. and J. Wilks).

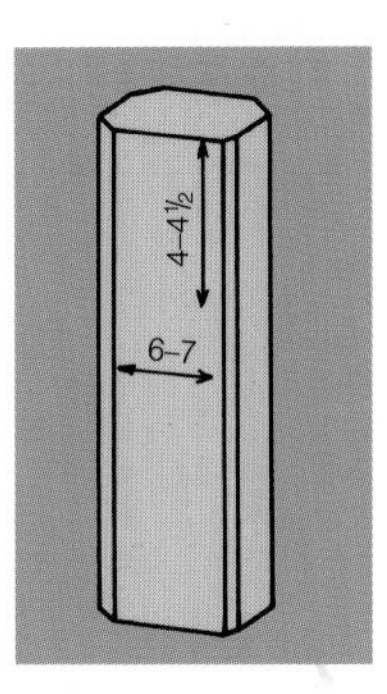

Right: Differences in hardness in kyanite. Clearly different Mohs' hardnesses in longitudinal and transverse directions.

Selected Gemstones Ordered by Mohs' Hardness

Diamond	10
Synth. Moissanit	9½
Ruby	9
Sapphire	9
Alexandrite	8½
Chrysoberyll	8½
Holtite	8½
YAG	8½
Zirconia	8½
Rhodizite	8-8½
Taaffeite	8-8½
Spinel	8
Topaz	8
Aquamarine	7½-8
Red Beryl	7½-8
Precious Beryl	7½-8
Gahnite	7½-8
Galaxite	7½-8
Painite	7½-8
Phenakite	7½-8
Emerald	7½-8
Andalusite	7½
Euclase	7½
Hambergite	7½
Sapphirine	7½
Dumortierite	7-8½
Almandine	7-7½
Boracite	7-7½
Cordierite	7-7½
Danburite	7-7½
Grandidierite	7-7½
Pyrope	7-7½
Schorlomite	7-7½
Sekaninaite	7-7½
Simpsonite	7-7½
Spessartine	7-7½
Staurolite	7-7½
Turmaline	7-7½
Uvarovite	7-7½
Amethyst	7
Aventurine	7
Rock Crystal	7
Chambersite	7
Chromdravite	7
Citrine	7
Forsterite	7
Povondraite	7
Quartz	7
Smoky Quartz	7
Rose Quartz	7
Zunyite	7
Garnet	6½-7½
Jeremejevite	6½-7½
Sillimanite	6½-7½
Zircon	6½-7½
Axinite	6½-7
Chalcedony	6½-7
Chrysoprase	6½-7
Diaspore	6½-7
Ferro-Axinite	6½-7
Gadolinite	6½-7
Grossular	6½-7
Hiddenite	6½-7
Jadeite	6½-7
Jasper	6½-7
Kornerupine	6½-7
Kunzite	6½-7
Mangan-Axinite	6½-7
Peridot	6½-7
Pollucite	6½-7
Serendibite	6½-7
Sinhalite	6½-7
Spodumene	6½-7
Tanzanite	6½-7
Thorianite	6½-7
Tinzenite	6½-7
GGG	6½
Magnesio-Axinite	6½
Nambulite	6½
Vesuvianite	6½
Cassiterite	6-7
Clinozoisite	6-7
Epidote	6-7
Hancockite	6-7
Pyrolusite	6-7
Sogdianite	6-7
Amazonite	6-6½
Andesine	6-6½
Anorthoclase	6-6½
Benitoite	6-6½
Bixbyite	6-6½
Bytownite	6-6½
Chondroite	6-6½
Helvite	6-6½
Hyalophane	6-6½
Labradorite	6-6½
Marcasite	6-6½
Microcline	6-6½
Nephrite	6-6½
Norbergite	6-6½
Oligoclase	6-6½
Orthoclase	6-6½
Petalite	6-6½
Prehnite	6-6½
Pyrite	6-6½
Rutile	6-6½
Sanidine	6-6½
Smaragdite	6-6½
Sugilite	6-6½
Tantalite	6-6½
Xonotlite	6-6½
Zoisite	6-6½
Aegirine	6
Amblygonite	6
Clinohumite	6
Humite	6
Hurlbutite	6
Lawsonite	6
Pumpellyite	6
Tephroite	6
Vlasovite	6
Zektzerite	6
Hematite	5½-6½
Hedenbergite	5½-6½
Magnetite	5½-6½
Manganotantalite	5½-6½
Opal	5½-6½
Rhodonite	5½-6½
Actinolite	5½-6
Allanite	5½-6
Anatase	5½-6
Beryllonite	5½-6
Brookite	5½-6
Bustamite	5½-6
Canasite	5½-6
Cobaltite	5½-6
Euxenite	5½-6
Fabulite	5½-6
Fergusonite	5½-6
Haüyne	5½-6
Leucite	5½-6
Marialite	5½-6
Meionite	5½-6
Milarite	5½-6
Montebrasite	5½-6
Natromontebrasite	5½-6
Periclase	5½-6
Pyroxmangite	5½-6
Sarcolite	5½-6
Scorzalite	5½-6
Scapolite	5½-6
Sodalite	5½-6
Tugtupite	5½-6
Brazilianite	5½
Breithauptite	5½
Chromite	5½
Enstatite	5½
Linobate	5½
Magnesiochromite	5½
Moldavite	5½
Willemite	5½
Aeschynite	5-6
Bronzite	5-6
Cancrinite	5-6
Catapleiite	5-6
Ceruléite	5-6
Clinoenstatite	5-6
Davidite	5-6
Diopside	5-6
Ferrosilite	5-6
Hornblende	5-6

Hypersthene 5-6
Ilmenite 5-6
Lapis Lazuli 5-6
Lazulite 5-6
Nepheline 5-6
Neptunite 5-6
Pargasite 5-6
Richterite 5-6
Samarskite 5-6
Stibiotantalite 5-6
Tremolite 5-6
Turquoise 5-6
Analcime 5-5½
Datolite 5-5½
Durangite 5-5½
Eudialyte 5-5½
Goethite 5-5½
Herderite 5-5½
Hydroxylherderite 5-5½
Meliphanite 5-5½
Mesolite 5-5½
Microlite 5-5½
Monazite 5-5½
Natrolite 5-5½
Nickeline 5-5½
Papagoite 5-5½
Psilomelane 5-5½
Scolecite 5-5½
Sellaite 5-5½
Thomsonite 5-5½
Titanite 5-5½
Wolframite 5-5½
Yttrotantalite 5-5½
Apatite 5
Bismutotantalite 5
Childrenite 5
Chlorapatite 5
Dioptase 5
Eosphorite 5
Fluorapatite 5
Hemimorphite 5
Hydroxylapatite 5
Mangan-Apatite 5
Odontolite 5
Rinkite 5
Schlossmacherite 5
Smithsonite 5
Spurrite 5
Strass 5
Vayrynenite 5
Ekanite 4½-6½
Apophyllite 4½-5
Augelite 4½-5
Charoite 4½-5
Gaspéite 4½-5
Hodgkinsonite 4½-5
Legrandite 4½-5
Pectolite 4½-5
Scheelite 4½-5
Wardite 4½-5
Wollastonite 4½-5
Bayldonite 4½
Colemanite 4½
Parisite 4½
Prosopite 4½
Yugawaralite 4½
Kyanite 4-7
Sérandite 4-5½
Chabazite 4-5
Friedelite 4-5
Lithiophilite 4-5
Mordenite 4-5
Triphylite 4-5
Variscite 4-5
Zincite 4-5
Carletonite 4-4½
Hübnerite 4-4½
Purpurite 4-4½
Algodonite 4
Ammonite 4
Barytocalcite 4
Fluorite 4
Leucophanite 4
Libethenite 4
Rhodochrosite 4
Magnesite 3½-4½
Siderite 3½-4½
Ankerite 3½-4
Aragonite 3½-4
Azurite 3½-4
Chalcopyrite 3½-4
Creedite 3½-4
Cuprite 3½-4
Dickinsonite 3½-4
Dolomite 3½-4
Euchroite 3½-4
Langbeinite 3½-4
Malachite 3½-4
Mimetite 3½-4
Pentlandite 3½-4
Powellite 3½-4
Scorodite 3½-4
Shungite 3½-4
Shattuckite 3½-4
Sphalerite 3½-4
Wavellite 3½-4
Wurtzite 3½-4
Adamite 3½
Anhydrite 3½
Chiolite 3½
Huréaulite 3½
Strontianite 3½
Thaumasite 3½
Weloganite 3½
Cacoxenite 3-4
Coral 3-4
Ludlamite 3-4
Anglesite 3-3½
Baryte 3-3½
Boleite 3-3½
Cerussite 3-3½
Celestine 3-3½
Descloizite 3-3½
Greenockite 3-3½
Howlite 3-3½
Millerite 3-3½
Phosphophyllite 3-3½
Witherite 3-3½
Bornite 3
Calcite 3
Kurnakovite 3
Shortite 3
Wulfenite 3
Serpentine 2½-5½
Pearl 2½-4½
Jet 2½-4
Chalcocite 2½-3
Crocoite 2½-3
Gaylussite 2½-3
Gold 2½-3
Inderite 2½-3
Lepidolite 2½-3
Pyrargyrite 2½-3
Silver 2½-3
Stolzite 2½-3
Vanadinite 2½-3
Whewellite 2½-3
Brucite 2½
Cryolite 2½
Linarite 2½
Lizardite 2½
Proustite 2½
Sturmanite 2½
Chrysocolla 2-4
Clinochrysotile 2-3
Fuchsite 2-3
Muscovite 2-3
Phosgenite 2-3
Shomiokite 2-3
Amber 2-2½
Cinnabar 2-2½
Ettringite 2-2½
Kämmererite 2-2½
Mellite 2-2½
Senarmontite 2-2½
Ulexite 2-2½
Villiaumite 2-2½
Gypsum 2
Stichtite 1½-2½
Sulphur 1½-2½
Covellite 1½-2
Melonite 1½-2
Realgar 1½-2
Vivianite 1½-2
Palygorskite 1-2
Pyrophyllite 1-2
Talc 1

Cleavage and Fracture

Many gemstones can be split along certain flat planes, which experts call *cleavage*. Cleavage is related to the lattice of the crystal—the cohesive property of the atoms. Unfortunately, there is no unified definition for the denomination of cleavages, so completely different names are used for individual parts.

Depending on the ease with which a crystal can be cleaved, one differentiates between a perfect (euclase), a good (sphene), and an imperfect cleavage (peridot). Some gemstones cannot be cleaved at all (quartz). One then says they do not have cleavage. A loosening of contact twins is not called cleavage, but separation or parting.

Lapidaries and stone setters must take account of the cleavage. Often a small tap or too much pressure when testing for Mohs' hardness is sufficient to split the stone. When soldering, the temperature can cause fissures along the cleavage planes, where eventually the gem may break completely along this line. In gemstones with perfect cleavage the facets must be transverse to the cleavage planes, otherwise the stone will be more vunerable to breaking. Piercings should, if possible, be done vertically to the cleavage surfaces.

Cleavage is used to divide large gem crystals or remove faulty pieces. The largest diamond of gem quality ever found, the Cullinan of 3106cts, was cleaved in 1908 into three large pieces which were then cleaved into numerous smaller pieces. Today, small pieces are sawn or divided by means of laser beam in order to avoid unwanted cleavages, and to make the best use of the shape of the stone.

The breaking of a gemstone with a blow producing irregular surfaces is called *fracture*. It can be conchoidal (shell-like), uneven, smooth, fibrous, splintery, or grainy. Sometimes, the type of fracture helps to identify a mineral. Conchoidal fracture is, for instance, characteristic for all quartz and glass-like minerals.

Large conchoidal fracture with flat cavities (obsidian).

Density and Specific Gravity

In general scientific usage the measurement *specific gravity,* which indicates the ratio of the weight of a specific material to the weight of the same volume of water, is now replaced by the term *density,* which is expressed typically as grams per cubic centimeter (g/cm^3). Its numerical value indicates the relation of the measured gemstone to an equal amount of water.

Weight is in fact not a constant attribute. It depends on the magnitude of the gravity at the respective location where it is measured. But for the determination of the weight of gemstones, this does not matter, since the measurement is always done under the same gravitation conditions. A more obvious factor affecting weight is an object's size.

The density is a property independent of location and size. It is defined as weight per volume represented in g/cm^3 and/or kg/m^3.

In practice, there is not a significant distinction between using the term density and specific gravity, since nothing is really changed in actual measurement. Nevertheless, in the following, the term *density* is used instead of *specific gravity.*

In regard to the "heaviness" of a gemstone, we will continue to speak of "weight" and not of the "density" of a gemstone, since that corresponds to the practice of the gemstone trade.

The density of gemstones varies between 1 and 8. Values under 2 are considered light (e.g., amber about 1), those from 2 to 4 normal (quartz 2.6), and those over 4 are considered heavy (casserite around 7). The more valuable gemstones (such as diamond, ruby, and sapphire) have densities that are much greater than the common rock-forming minerals, especially quartz and feldspar. Such heavy minerals are deposited in flowing waters before the sands, which are rich in quartz, and they form the so-called placer deposits (compare with page 62).

Determination of Density

In identifying gemstones, a determination of the density can be useful. But for specialists, optical procedures for the determination of gemstones are more common. These procedures require expensive instruments, however.

Two methods have proven a success for the determination of the density of gemstones: the buoyancy method with the help of a hydrostatic balance and the suspension, or heavy liquid, method. The first is time consuming, but inexpensive. It is also inaccurate with small stones. Even though the second method is a little bit more expensive, it produces good results in a short time, especially with lots of unknown gems. However, the immersion liquids are quite hazardous.

Hydrostatic Balance The measuring procedure with a hydrostatic scale works on Archimedes' Principle of buoyancy. The volume of the unknown gem is determined. The density is then easily worked out. A hydrostatic balance can be constructed by anybody (see illustration on the next page). The beginner can adapt letter scales. Anyone more advanced should use a precision balance as used by a chemist or pharmacist. The object to be tested is first weighed in air (in the pan under the bridge) and then in water (in the net in the beaker).

The difference in weight corresponds to the weight of the water that is displaced, and therefore equal to the volume of the gemstone. It is possible to determine the density with this method to one, and with practice, to two decimal places. It is important to ensure that the stone is not in contact with a

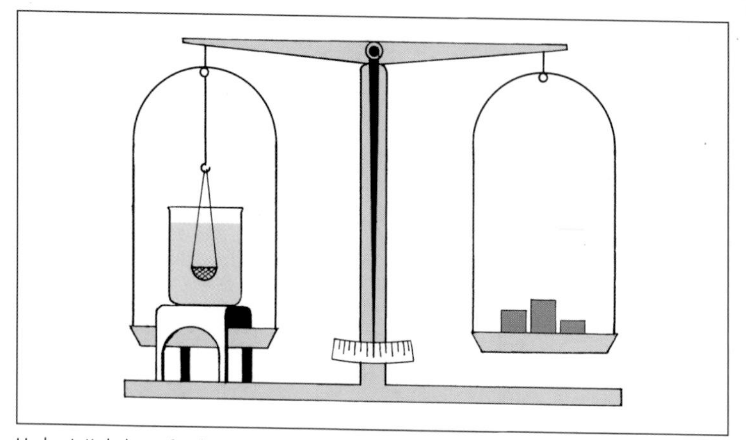
Hydrostatic balance for determining gemstone volume.

foreign substance, that it is not set, and that, when weighed in air, it is dry. It is also necessary to take into account the weight of the net used to suspend the gem in the beaker.

Example:

Substance (weight) in air	5.2 g
Substance (weight) in water	3.3 g
Difference = Volume	1.9 cm^3

$$\text{Density} = \frac{\text{Weight in air}}{\text{Volume}} = \frac{5.2}{1.9} = 2.7\ g/cm^3$$

Suspension Method This method rests on the idea that an object will float in a liquid of higher density, sink in a liquid of lower density, and remain suspended in a liquid of the same density. To perform the test, a set of liquids with known high densities (heavy liquids) is normally used.

When a gemstone is examined, the unknown stone is placed into a high density liquid (heavy liquid), which becomes lighter through dilution and finally takes on the same density as the object examined. The density of the object can thus be determined through the suspension in the high density liquid.

Solely for the purpose of determining the density of the diluted heavy liquid, experts have constructed a special scale (the Westphal Balance, page 27); amateurs, on the other hand, are better off using indicators. These are pieces of glass or minerals of varying but known density. If such an indicator remains suspended in the liquid, its density equals that of the liquid and is consequently equal to that of the object examined.

There are liquids of varying heaviness. Those liquids that can be diluted with distilled water are particularly well suited. Thoulet's solution is one of them (a potassium-mercuric-iodide solution) with a density of 3.2. One can identify about one half of all gemstones with this liquid.

Clerici's solution (thallium formate and thallium malonate solution) can be used for heavier stones, which has a maximum density of 4.25. Its density encompasses the entire range of gemstones, with the exception of about two dozen. However, this solution is expensive, very toxic, and vitriolic. Therefore, amateurs should refrain from using it, if possible.

If a density of up to 3.5 is needed, one can also make use of the Rohrbach solution (a barium-mercuric-iodide solution; density 3.59).

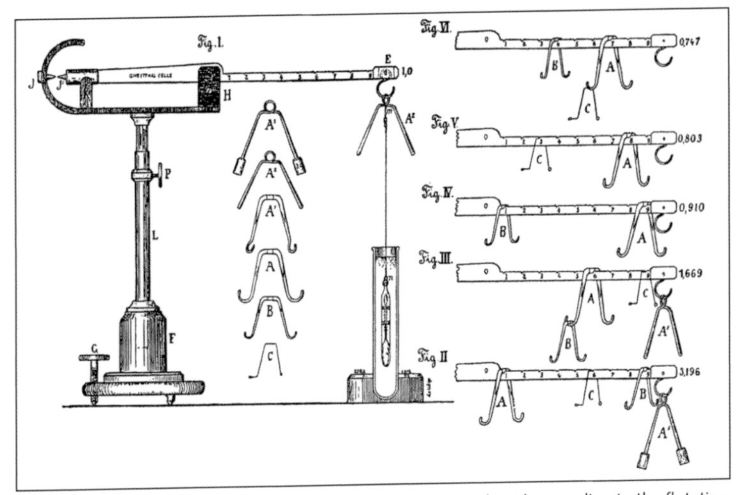

The Westphal Balance is used to measure the density of heavy liquids according to the flotation method. Original illustration of the mechanical workshop of G. Westphal, Celle, around 1890. On the top, beam with numbered notches, on which a series of riders can be placed (as illustrated to the right of the tripod) and a plummet made of glass with a built-in mercury thermometer. Figures VI to II, on the very right, show some balance examples.

The simplest and fastest work method is to employ a whole series of standardized heavy liquids. In practice, this method is used most frequently; however, it is too costly for the amateur who wishes only to identify individual stones.

It is not necessary to pour out diluted liquids since their original density can be regained through vaporization in a water bath. All named heavy liquids are toxic. Therefore, one must follow proper precautions when using them. Do not breathe in the vapors, and do not eat while you are working.

The suspension method is particularly recommended if you wish to sort out certain gemstones from among a number of unknown ones. It can also be recommended when you wish to distinguish real gemstones from synthetics and other imitations.

However, in recent years, optical methods have become more popular than density for the identification of gemstones.

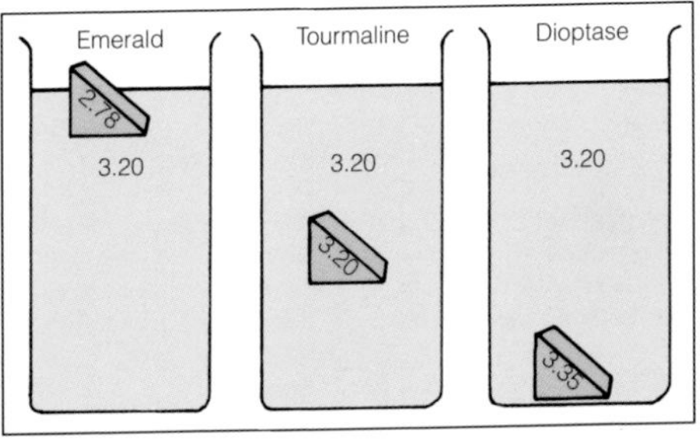

Determining the density of a gemstone with the help of heavy liquids. (e.g., with Thoulet solution with the density of 3.2). Lighter stones float at the surface; heavier ones sink to the bottom. Gemstones with the same density as that of the liquid are suspended in the liquid.

Selected Gemstones Ordered by Density

Gold	15.5-19.3
Thorianite	9.7-9.8
Silver	9.6-12.0
Algondonite	8.38
Bismutotantalite	8.15-8.89
Cinnabar	8.0-8.2
Stolzite	7.9-8.34
Nickeline	7.78
Manganotantalite	7.73-7.97
Melonite	7.72
Breithauptite	7.59-8.23
Stibiotantalite	7.53
Mimetesite	7.24
Hübnerite	7.12-7.18
Wolframite	7.1-7.6
GGG	7.00-7.09
Cassiterite	6.7-7.1
Wulfenite	6.5-7.00
Vanadinite	6.5-7.1
Cerussite	6.46-6.57
Cobaltite	6.33
Anglesite	6.30-6.39
Phosgenite	6.13
Simpsonite	5.92-6.84
Scheelite	5.9-6.3
Crocoite	5.9-6.1
Cuprite	5.85-6.15
Pyrargyrite	5.85
Yttrotantalite	5.7
Zincite	5.66
Proustite	5.51-5.64
Descloizite	5.5-6.2
Chalcosite	5.5-5.8
Zirconia	5.50-6.00
Millerite	5.5
Fergusonite	5.35-5.44
Euxenite	5.3-5.9
Linarite	5.30
Senarmontite	5.2-5.5
Magnetite	5.2
Aeschynite	5.19
Tantalite	5.18-8.20
Hematite	5.12-5.28
Fabulite	5.11-5.15
Bornite	5.06-5.08
Boleite	5.05
Samarskite	5.0-5.69
Pyrite	5.00-5.20
Monazite	4.98-5.43
Bixbyite	4.93
Marcasite	4.83-4.92
Greenockite	4.73-4.79
Psilomelane	4.70-4.74
Linobate	4.64-4.66
Pentlandite	4.6-5.0
Covellite	4.6-4.76
YAG	4.55-4.65
Illmenite	4.5-5.0
Pyrolusite	4.5-5.0
Chromite	4.5-4.8
Davidite	4.5
Baryte	4.43-4.46
Magnesiochromite	4.39-4.67
Bayldonite	4.35
Parisite	4.33-4.42
Microlite	4.3-5.7
Adamine	4.30-4.68
Witherite	4.27-4.79
Galaxite	4.23
Powellite	4.23
Rutile	4.20-4.30
Spessartine	4.12-4.18
Shattuckite	4.11
Chalycopyrite	4.10-4.30
Wurtzite	4.09
Brookite	4.08-4.18
Hancockite	4.03
Painite	4.01
Gadolinite	4.00-4.65
Smithsonite	4.00-4.65
Gahnite	4.00-4.62
Legrandite	3.98-4.04
Ruby	3.97-4.05
Celestine	3.97-4.00
Libethenite	3.97
Ferrosilite	3.96
Sapphire	3.95-4.03
Durangite	3.94-4.07
Zircon	3.93-4.73
Almandine	3.93-4.30
Hodgkinsonite	3.91-3.99
Sphalerite	3.90-4.10
Willemite	3.89-4.18
Tephroite	3.87-4.12
Siderite	3.83-3.96
Anatase	3.82-3.97
Goethite	3.8-4.3
Gaspéite	3.71
Andradite	3.7-4.1
Azurite	3.7-3.9
Periclase	3.7-3.9
Chrysoberyl	3.70-3.78
Schorlomite	3.69-3.88
Barytocalcite	3.66
Staurolite	3.65-3.77
Benitoite	3.64-3.68
Strontianite	3.63-3.79
Pyrope	3.62-3.87
Pyroxmangite	3.61-3.80
Holtite	3.60-3.90
Taaffeite	3.60-3.62
Grossular	3.56-3.73
Realgar	3.56
Spinel	3.54-3.63
Kyanite	3.53-3.70
Nambulite	3.53
Titanite	3.52-3.54
Allanite	3.5-4.2
Aegirine	3.50-3.60
Hedenbergite	3.50-3.56
Diamond	3.50-3.53
Topaz	3.49-3.57
Chambersite	3.49
Sinhalite	3.46-3.50
Rhodochrosite	3.45-3.70
Piemontite	3.45-3.52
Euchroite	3.44
Rhodizite	3.44
Serendibite	3.42-3.52
Uvarovite	3.41-3.52
Rhodonite	3.40-3.74
Lithiophilite	3.4-3.6
Sapphirine	3.40-3.58
Aegirine-augite	3.40-3.55
Hypersthene	3.4-3.5
Chromdravite	3.40
Scorzalite	3.38
Tinzenite	3.36-3.43
Tanzanite	3.35
Triphylite	3.34-3.58
Dickinsonite	3.34-3.41
Vesuvianite	3.32-3.47
Bustamite	3.32-3.42
Sérandite	3.32
Manganaxinite	3.31
Epidote	3.3-3.5
Hemimorphite	3.30-3.50
Diaspore	3.30-3.39
Jadeite	3.30-3.38
Peridotite	3.28-3.48
Dioptase	3.28-3.35
Ekanite	3.28-3.32
Jeremejevite	3.28-3.31
Parascorodite	3.28-3.29
Forsterite	3.28
Kornerupine	3.27-3.45
Dumortierite	3.26-3.41
Axinite	3.26-3.36
Povondraite	3.28
Malachite	3.25-4.10
Ferro-axinite	3.25-3.28
Papagoite	3.25
Smaragdite	3.24-3.50
Sillimanite	2.23-3.27
Diopside	3.22-3.38
Vayrynenite	3.22
Weloganite	3.22
Clinozoisite	3.21-3.28
Helvite	3.20-3.44
Purpurite	3.2-3.4
Humite	3.20-3.32
Enstatite	3.20-3.30
Neptunite	3.19-3.23

Clinoenstatite 3.19
Rinkite 3.18-3.44
Pumpellyite 3.18-3.33
Magnesio-axinite 3.18
Norbergite 3.18
Chlorapatite 3.17-3.18
Chondrodite 3.16-3.26
Apatite 3.16-3.23
Strass 3.15-4.20
Zoisite 3.15-3.36
Hiddenite 3.15-3.21
Kunzite 3.15-3.21
Spodumene 3.15-3.21
Hureaulite 3.15-3.19
Sellaite 3.15
Clinohumite 3.13-3.75
Fluorapatite 3.1-3.25
Carborundum 3.10-3.22
Ludlamite 3.1-3.2
Euclase 3.10
Lawsonite 3.08-3.09
Phosphophyllite 3.07-3.13
Friedelite 3.06-3.19
Andalusite 3.05-3.20
Eosphorite 3.05-3.08
Pargasite 3.04-3.17
Lazulite 3.04-3.14
Actinolite 3.03-3.07
Spurrite 3.02
Amblygonite 3.01-3.11
Natromontebrasite 3.01-3.11
Fluorite 3.00-3.25
Melinophane 3.00-3.03
Leucophanite 3.0
Schlossmacherite 3.00
Chiolite 2.99-3.01
Montebrasite 2.98-3.11
Brazilianite 2.98-2.99
Richterite 2.97-3.45
Danburite 2.97-3.03
Ankerite 2.97
Cryolite 2.97
Magnesite 2.96-3.12
Tremolite 2.95-3.07
Herderite 2.95-3.02
Phenakite 2.95-2.97
Boracite 2.95-2.96
Aragonite 2.94
Vlasovite 2.92-2.96
Sarcolite 2.91-2.96
Hornblende 2.9-3.4
Nephrite 2.90-3.03
Datolite 2.90-3.00
Anhydritspate 2.90-2.98
Prosopite 2.88-2.89
Hurlbutite 2.88
Zunyite 2.87
Wollastonite 2.86-3.09
Grandidierite 2.85-3.00
Pollucite 2.85-2.94
Langbeinite 2.83
Turmaline 2.82-3.32
Prehnite 2.82-2.94
Wardite 2.81-2.87
Dolomite 2.80-2.95
Lepidolite 2.80-2.90
Beryllonite 2.80-2.87
Zektzerite 2.79
Muscovite 2.78-2.88
Sogdianite 2.76-2.90
Sugilite 2.76-2.80
Sekaninaite 2.76-2.77
Ammonite 2.75-2.80
Eudialyte 2.74-2.98
Pectolite 2.74-2.88
Meionite 2.74-2.78
Bytownite 2.72-2.74
Catapleiite 2.72
Creedite 2.72
Xonotlite 2.71-2.72
Canasite 2.71
Ceruléite 2.70-2.80
Augelite 2.70-2.75
Calcite 2.69-2.71
Aquamarine 2.68-2.74
Emerald 2.67-2.78
Noble beryl 2.66-2.87
Pyrophyllite 2.65-2.90
Labradorite 2.65-2.75
Andesine 2.65-2.69
Amethyst 2.65
Citrine 2.65
Prasiolite 2.65
Quartz 2.65
Rock crystal 2.65
Rose quartz 2.65
Smoky quartz 2.65
Vivianite 2.64-2.70
Aventurine 2.64-2.69
Kämmererite 2.64
Oligoclase 2.62-2.67
Sunstone 2.62-2.65
Pearl 2.60-2.85
Coral 2.60-2.70
Agate 2.60-2.64
Schaurteite 2.60
Peristerite 2.59-2.68
Petrified wood 2.58-2.91
Jasper 2.58-2.91
Cordierite 2.58-2.66
Chalcedony 2.58-2.64
Chrysoprase 2.58-2.64
Moss agate 2.58-2.64
Tigers eye 2.58-2.64
Scapolite 2.57-2.74
Anorthoclase 2.57-2.60
Moonstone 2.56-2.59
Amazonite 2.56-2.58
Orthoclase 2.56-2.58
Sanidine 2.56-2.58
Talc 2.55-2.80
Nepheline 2.55-2.65
Lizardite 2.55
Charoite 2.54-2.78
Microline 2.54-2.57
Clinochrysotile 2.53
Lapis lazuli 2.50-3.00
Marialite 2.50-2.62
Milarite 2.46-2.61
Howlite 2.45-2.58
Leucite 2.45-2.50
Carletonite 2.45
Serpentine 2.44-2.62
Variscite 2.42-2.58
Cancrinite 2.42-2.51
Haüyne 2.4-2.5
Colemanite 2.40-2.42
Petalite 2.40
Brucite 2.39
Tugtupite 2.36-2.57
Wavellite 2.36
Obsidian 2.35-2.60
Hambergite 2.35
Moldavite 2.32-2.38
Turquoise 2.31-2.84
Apophyllite 2.30-2.50
Mesolite 2.26-2.40
Thomsonite 2.23-2.39
Analcime 2.22-2.29
Scolecite 2.21-2.29
Palygorskite 2.21
Cacoxenite 2.2-2.6
Gypsum 2.20-2.40
Natrolite 2.20-2.26
Whewellite 2.19-2.25
Yugawaralite 2.19-2.23
Stichtite 2.16-2.18
Sodalite 2.14-2.40
Mordenite 2.12-2.15
Chabasite 2.05-2.20
Sulphur 2.05-2.08
Glass 2.0-4.5
Chrysocolla 2.00-2.40
Sepiolite 2.0-2.1
Gaylussite 1.99
Opal 1.88-2.50
Thaumasite 1.91
Kurnakovite 1.86
Sturmanite 1.85
Inderite 1.78-1.86
Ettringite 1.77
Ulexite 1.65-1.95
Mellite 1.58-1.60
Jet 1.19-1.35
Amber 1.05-1.09

Weights Used in the Gem Trade

In the international gem trade, the carat, gram, grain, and momme are used as units of weight.

Carat The weight used in the gem trade since antiquity. The name is derived from the seed *kuara* of the African Coraltree or from the kernel (Greek *keration*) of the Carob bean.

Since 1907 Europe and America, followed by other countries, have adopted the metric carat (mct) of 200 mg or 0.2 g. Local carat weights have differed historically from location to location, varying between 188 and 213 mg.

The carat is subdivided into fractions (e.g., 1/10ct) or decimals (e.g., 1.25ct) up to two decimal places. Small diamonds are weighed in *points*, which are 1/100 carat (0.01ct). Small brilliants with a weight of 0.07-0.15 are internationally called *Mêlé* (French for mixed), heavy ones with about 0.12-0.15 carat are called coarse *Mêlé*.

The table below illustrates diameter and corresponding carat weight for diamonds cut in the modern brilliant cut (see page 97). Gems with different density and different cuts obviously have different diameters.

The price of a gemstone is usually indicated in the gem trade "per carat." By calculating the actual weight, one receives the price per piece. When selling to the final buyer, though, usually the total price is given. The carat price often progressively increases with the size of the gemstone in very small increments. When a one-carat piece, for example, costs $750, then a two-carat-piece is not necessarily worth $1500, but maybe $3000 or even more.

The carat weight of gems is not to be confused with the karat used by the goldsmith. In the case of gold the karat is not a weight measure at all but rather a measure of quality. The higher the karatage, the higher the content of gold in the piece of jewelry. The weight can be variable.

Gram The weight measure used in the trade for less precious gemstones and especially for rough stones.

Grain Formerly, the weight measure for pearls. It corresponds to 0.05g or 1/4ct. Today, it is increasingly substituted by use of the carat.

Momme The old Japanese measure of *momme* (=3.75g=18.75ct) for cultured pearls is hardly used anymore except by wholesale distributors.

Diameters and Weights of Brilliant Diamonds

Diameter in mm	2.2	3.0	4.1	5.2	6.5
Weight in carat	.04	.10	.25	.50	1.00

7.4	8.2	9.0	9.3	11.0
1.50	2.00	2.50	3.00	5.00

Optical Properties

Of all the various properties of gemstones, the optical characteristics are of unsurpassed importance. They produce color and luster, fire and luminescence, play of light, and schiller (iridescence). In the examination of gems, nowadays, there is more and more concentration on the optical effects.

Color

Color is the most important characteristic of gems. In the case of most stones, it is not diagnostic in identification, because many have the same color and numerous stones occur in many colors. Color is produced by light; light is an electromagnetic vibration at certain wavelengths. The human eye can only perceive wavelengths between 750 and 380 nm (see page 41). This visible field is divided into several sectors of certain wavelengths, each of a particular color (spectral colors: red, orange, yellow, green, blue, violet).

The mixture of all these colors produces white light. If, however, a certain wavelength (i.e., also the corresponding color) is absorbed out of the entire spectrum, the remaining mixture produces a certain color, but not white. If all wavelengths pass through the stone, it appears colorless. If all light is

Crystal with scepter amethyst; Mexico (somewhat reduced).

absorbed, the stone appears black. If all wavelengths are absorbed to the same degree, the stone is dull white or gray.

In the case of gemstones, the metals and their combinations, especially chrome, iron, cobalt, copper, manganese, nickel, and vanadium, absorb certain wavelengths of light and so cause coloration. In the case of zircon and smoky quartz, no impurity, or foreign substance, is responsible for the color, but rather a deformation of the internal crystal structure (lattice) results in the selective absorption of light, giving a change in the original color.

The distance the light ray travels through the stone can also influence absorption and thus color. The cutter must therefore use this fact to his advantage. Lightcolored stones are made thicker and/or are given such an arrangement of facets that the absorption path lengthens, giving a deeper color. Materials with colors that are too dark are cut thinly. The dark red almandite garnet, for example, is therefore often hollowed out on the underside.

Artificial light has an influence on the color of gemstones, as it is usually differently composed than daylight. There are gemstones whose color is influenced unfavorably by artificial incandescent light (e.g., sapphire), and those which have an especially radiating effect in such light (e.g., ruby and emerald). The most obvious change in color occurs in alexandrite, which is green in daylight and red in artificial incandescent light.

Although color is of great importance in gems, with the exception of diamonds, no practical method of objective color determination is widely accepted. Color comparison charts are poor substitutes because there is too much room for subjective consideration. The measuring methods used in science for color determination are too complicated for the trade.

Color of Streak

The color appearance of gems, even in the same species, can vary greatly. For instance, beryl can have all the colors of the spectrum, but can also be colorless. This colorlessness is, in fact, the true color. It is called *inherent color.* All other colors are produced by impurities.

The inherent color, as it is constant, can help to identify a rough stone. This color can be seen by streaking the mineral on a rough porcelain plate, called the *streak plate.* The finely ground powder has the same effect as thin transparent platelets, from which the color-producing impurities have been abstracted. Steely hematite, for instance, has a streak color (called *streak*) which is red.

When no color line can be noticed when streaking the mineral on a streak plate, then the streak is said to be colorless or white. With minerals that are harder than the streak plate, i.e., Mohs' hardness above 6, a small part of the gem to be examined must be pulverized and must then be rubbed on the streak plate.

Brass-colored pyrite's streak is black, and blue sodalite's is white. In the case of very hard gemstones, it is advisable first to remove a little powder with a steel file, and then rub it on the streak plate. This method of determination is of special interest to collectors. Because of the danger of damage, a cut gem should never be tested for streak. Refer to the table of streaks on pages 34 and 35.

Streak test on a streak plate. The brassy yellow pyrite leaves a greenish-black streak of color.

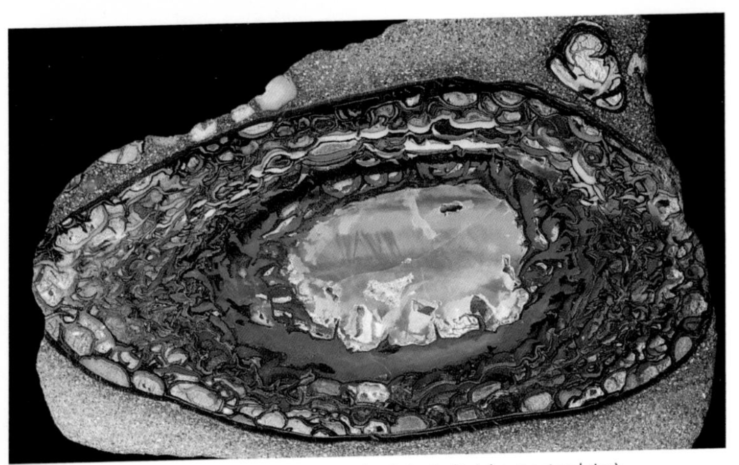
Opal in matrix, so-called boulder opal; Queensland, Australia (about natural size).

Changes in Color

The color of gemstones and decorative stones can change. If this alteration is only temporary, then it is called a change in color. With permanently altered colors, however, one should correctly speak of an artificial color change. Temporarily altered colors exist, for instance, with color play and with a Schiller effect on the surface of some gemstones. The color change from daylight to artificial light, clearly seen with alexandrite, has already been explained (page 114).

Tourists experience changes in color (or perhaps only a trick of the senses) again and again when traveling to the tropics. In regions with a lot of sunshine, some of the gemstones that were purchased in the countries of origin sometimes appear more colorful and radiant than at home. Whether there are indeed differences in quality or whether simply "felt" vacation opinions are involved cannot be proven scientifically. In any case, several gemstone buyers from Western industrial countries use arranged sample collections in their home countries to compare colors of gemstones that were purchased in tropic and sub-tropic countries.

The color of some gemstones and decorative stones is permanently altered by time. Amethyst, rose quartz, and kunzite, for instance, can become paler even to the extent of becoming colorless when exposed to direct sunlight. Generally, however, color changes effected by nature are rare.

Man-made influences are much more common in the color alteration of gemstones and decorative stones in such a way that they appear as perfect as possible (according to the standards of the time). Today, a wide range of technical treatment methods enables color changes in gemstones and decorative stones in all color shades.

Further explanations regarding the technical alteration of the color of gemstones through these treatments can be found on page 271.

Selected Gemstones Ordered by Streak

White, Colorless and Gray

Amber
Ametrine
Analcime
Anatase
Ancerite
Andalusite
Andesine
Andradite
Anglesite
Anhydrite
Anorthite
Anorthoclase
Apatite
Apophyllite
Aquamarine
Aragonite
Augelite
Aventurine
Axinite
Baryte
Barytocalcite
Benitoite
Beryl
Beryllonite
Bixbite
Blue quartz
Boracite
Brazilianite
Bronzite
Brookite
Brucite
Buergerite
Bustamite
Bytownite
Calcite
Canasite
Cancrinite
Carletonite
Carnelian
Cassiterite
Catapleiite
Cat's-eye quartz
Celestine
Cerussite
Chabasite
Chalcedony
Chambersite
Charoite
Childrenite
Chiolite
Chlorapatite
Chrysoberyl
Chrysoprase
Citrine
Clinochrysotile
Clinoenstatite
Clinohumite
Clinozoisite
Colemanite
Coral
Cordierite
Corundum
Creedite
Cryolite
Danburite
Datolite
Demantoid
Diamond
Diaspore
Dickinsonite
Diopside
Dolomite
Dravite
Dumortierite
Ekanite
Elbaite
Emerald
Enstatite
Eosphorite
Epidote
Ettringite
Eudialyte
Euclase
Fabulite
Ferro-axinite
Fire Opal
Fluorapatite
Fluorite
Forsterite
Fuchsite
Gadolinite
Gahnite
Garnet
Gaylussite
GGG
Glass
Golden beryl
Goshenite
Grandidierite
Grossular
Gypsum
Hambergite
Haüyne
Hedenbergite
Heliodor
Helvite
Hemimorphite
Herderite
Hessonite
Hiddenite
Hodgkinsonite
Holtite
Howlite
Humite
Hureaulite
Hurlbutite
Hyalophane
Hyacinth
Hydroxylapatite
Hydroxylherderite
Hypersthene
Inderite
Indicolite
Ivory
Jade
Jadeite
Jasper
Jeremejevite
Kämmererite
Katoite
Kornerupine
Kunzite
Kurnakovite
Kyanite
Labradorite
Langbeinite
Lawsonite
Lazulite
Legrandite
Lepidolite
Leucite
Leucogarnet
Leucophanite
Liddicoatite
Linobate
Lithiophilite
Lizardite
Ludlamite
Magnesio-axinite
Magnesite
Mangan-axinite
Marialite
Meionite
Melanite
Meliphanite
Mellite
Mesolite
Microline
Microlite
Milarite
Mimetite
Moissanite
Moldavite
Monazite
Montebrasite
Moonstone
Moss Agate
Mordenite
Morganite
Morion
Muscovite
Natrolite
Natromontebrasite
Nepheline
Nephrite
Norbergite
Obsidian
Oligoclase
Opal
Orthoclase
Painite
Palygorskite
Pargasite
Parisite
Pearl
Pectolite
Periclase
Peridot
Peristerite
Petalite
Petrified wood
Phenakite
Phosgenite
Phosphophyllite
Pollucite
Powellite
Prase
Prasiolite
Precious beryl
Prehnite
Prosopite
Pyrope
Pyrophyllite
Pyroxmangite
Red beryl
Rhodizite
Rhodochrosite
Rhodolite
Rhodonite
Richterite
Rock crystal
Rose quartz
Rubellite
Ruby
Sanidine
Sapphire
Sapphirine
Sarcolite
Scapolite
Scheelite
Schlossmacherite
Schorl
Scolecite
Scorzalite
Scorodite
Sekaninaite
Sellaite
Senarmontite
Sepiolite

Sérandite
Serendibite
Serpentine
Shomiokite
Shortite
Siberite
Siderite
Silver
Sillimanite
Simpsonite
Sinhalite
Smaradgite
Smithsonite
Smoky quartz
Sodalite
Spessartine
Spinel
Spodumene
Spurrite
Staurolite
Stichite
Stolzite
Strass
Strontianite
Sugilite
Sunstone
Sulphur
Synthetic
Synthetic rutile
Synthetic spinel
Taaffeite
Talc
Tanzanite
Tephroite
Thaumasite
Thomsonite
Thorianite
Tiger's-eye
Titanite
Topaz
Topazolite
Tremolite
Triphylite
Tsavorite
Tsilaisite
Turquoise
Tugtupite
Tourmaline
Ulexite
Uvite
Uvavorite
Vanadinite
Variscite
Verdelite
Vesuvianite
Villiaumite
Vivianite
Vlasovite
Wardite
Wavellite
Weloganite
Whewellite
Willemite
Witherite
Wollastonite
Wulfenite
Xonotlite
YAG
Yugawaralite
Zektzerite
Zircon
Zirconia
Zoisite
Zunyite

Red, Pink and Orange

Chambersite
Cinnabar
Crocoite
Cuprite
Friedelite
Greenockite
Hematite
Hübnerite
Jasper
Manganotantalite
Piemontite
Proustite
Purpurite
Pyrargyrite
Realgar
Samarskite

Yellow, Orange and Brown

Aegirine
Aeschynite
Allanite
Apatite
Bismutotantalite
Breithauptite
Bronzite
Cacoxenite
Cassiterite
Chondrodite
Chromite
Crocoite
Cuprite
Descloizite
Durangite
Euxenite
Fergusonite
Galaxite
Goethite
Gold
Greenockite
Hematite
Hancockite
Hornblende
Hübnerite
Jasper
Jet
Nambulite
Neptunite
Nickeline
Pentlandite
Petrified wood
Povondraite
Psilomelane
Realgar
Rinkite
Rutile
Samarskite
Schorlomite
Sphalerite
Stibiotantalite
Sturmanite
Tantalite
Thorianite
Vanadinite
Wolframite
Wurtzite
Zincite

Green, Yellow-Green and Blue-Green

Aegirine-augite
Bayldonite
Chalcopyrite
Chromdravite
Chrysocolla
Dioptase
Euchroite
Fuchsite
Gadolinite
Gaspéite
Hornblende
Libethenite
Malachite
Marcasite
Millerite
Pumpellyite
Pyrite

Blue, Blue-Green and Blue-Red

Azurite
Boleite
Ceruléite
Haüyne
Lapis lazuli
Lazurite
Linarite
Papagoite
Pumpellyite
Purpurite
Shattuckite
Vivianite

Black and Gray

Aegirine-augite
Bismutotantalite
Bixbyite
Bornite
Chalcocite
Chalcopyrite
Clinozoisite
Cobaltite
Covellite
Davidite
Epidote
Hedenbergite
Ilmenite
Jet
Magnesiochromite
Magnetite
Marcasite
Melonite
Millerite
Nickeline
Pentlandite
Powellite
Psilomelane
Pyrite
Pyrolusite
Shungite
Tantalite
Wolframite
Yttrotantalite

Refraction of Light

Most of us, when children, noticed that when a stick was partially immersed in water at a slant, it appeared to "break" at water level. The lower part of the stick appeared to be at a different angle from the upper part. What we observed was caused by the refraction of light. It always occurs when a ray of light leaves one medium (for instance, air) and enters obliquely into another (for instance, water and/or a gem crystal) at the interface between the two media.

The amount of refraction in the crystals is constant for each specific gemstone. It can therefore be used in the identification of the type of stone. The amount of the refraction is called the refractive index and is defined as the proportional relation between the speed of light in air and that in the stone. A decrease in the velocity of light in the stone causes a deviation of the light rays.

Example:		
	Speed of light in air (V1)	300,000 km/sec
	Speed of light in diamond (V2)	124,120 km/sec

$$\text{Refractive index} = \frac{\text{V1 (air)}}{\text{V2 (diamond)}} = \frac{300{,}000}{124{,}120} = 2.415$$

This means that the speed of light in air is 2.4 times faster than the speed of light in diamond. The refractive indices of gems are betwen 1.4 and 3.2 . They vary somewhat with color and occurrence. Doubly refractive gems (see the descriptions on page 40) have two refractive indices. (Also refer to the refractive indices chart on pages 38 and 39.)

Refractometer The light refraction is measured in practice with a refractometer. The values can be read directly from a scale. However, testing is only possible up to a value of 1.81 on a common instrument, and only stones with a flat face or facet are suitable. The determination of the refractive indices of other stones and of values over 1.81 requires the use of special devices.

The expert can find approximate values of cabachons using his own experience and knowledge.

Compact refractometer. In the small bottle next to it, the contact liquid.

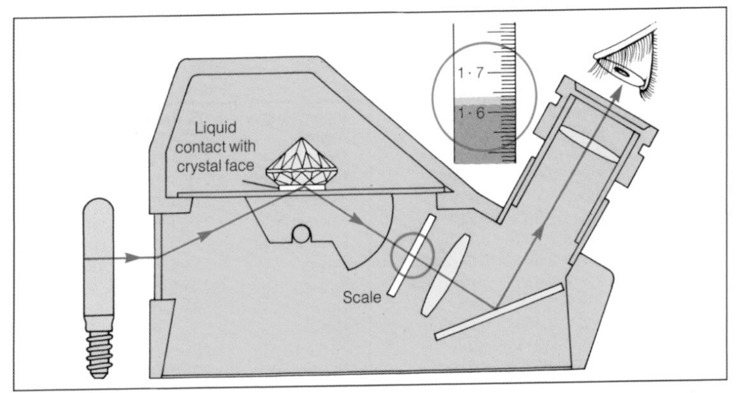

Diagram of a refractometer as commonly used in the trade, with inset of refractometer scale.

Immersion Method Without access to a costly apparatus, a rough measurement of the refractive index is fairly easy with the immersion method. The gem is viewed after immersing in a liquid with a known refractive index. On the basis of brightness, sharpness, and width of the contours/outlines as well as the facet edges, one can learn approximate values about the refractive index.

Immersion liquids with known refractive index

1.409	Amyl alcohol
1.45	Kerosene
1.473	Glycerine
1.498	Benzene
1.516	Ethyl iodide
1.525	Chlorobenzene
1.544	Clove oil
1.559	Dimethylaniline
1.569	Benzoesaurebenzylester
1.586	Aniline
1.618	Quinoline
1.633	Chloronaphthalene
1.663	o-Bromoiodobenzene
1.740	Diiodidemethane
1.778	Diiodidemethane saturated with sulfur
1.810	Tetraiodoethylene
1.843	Phenyldiiodoarsine
2.06	Diiodidemethane with sulfur and phosphorus

Gemstones in an immersion liquid.

1. White contour and dark facet edges: gem has lower refractive index.
2. Black contour and white facet edges: gem has higher refractive index.
3. Widened contour: refractive indices vary considerably.
4. Indistinct outline (tending to disappear): liquid and gem have same refractive index. (According to Dragstedt et al.)

Refractive Indices and Double Refraction for Selected Gemstones

	Refractive Index	Double Refraction
Hematite	2.940-3.220	0.287
Cinnabar	2.905-3.256	0.351
Proustite	2.881-3.084	0.203
Pyrargyrite	2.88-3.08	0.200
Cuprite	2.849	none
Rutile	2.616-2.903	0.287
Brookite	2.583-2.700	0.117
Anatase	2.488-2.564	0.046-0.067
Diamond	2.417-2.419	anomalous
Fabulite	2.409	none
Stibiotantalite	2.370-2.450	0.080
Sphalerite	2.368-2.371	none
Crocoite	2.29-2.66	0.270
Wulfenite	2.280-2.400	0.120
Tantalite	2.26-2.43	0.160
Linobate	2.21-2.30	0.090
Manganotantalite	2.19-2.34	0.150
Zirconia	2.150-2.180	none
Mimetite	2.120-2.135	0.015
Phosgenite	2.114-2.145	0.028
Senarmontite	2.087	none
Boleite	2.03-2.05	none
Zincite	2.013-2.029	0.016
Cassiterite	1.997-2.098	0.096-0.098
Simpsonite	1.994-2.040	0.058
GGG	1.970-2.020	none
Sulphur	1.958-2.245	0.291
Bayldonite	1.95-1.99	0.040
Scheelite	1.918-1.937	0.010-0.018
Andradite	1.88-1.94	none
Anglesite	1.878-1.895	0.017
Uvarovite	1.865	none
Purpurite	1.85-1.92	0.07
Titanite	1.843-2.110	0.100-0.192
YAG	1.833-1.835	none
Zircon	1.810-2.024	0.002-0.059
Cerussite	1.804-2.079	0.274
Gahnite	1.791-1.818	none
Spessartine	1.790-1.820	none
Painite	1.787-1.816	0.029
Monazite	1.774-1.849	0.049-0.055
Almandine	1.770-1.820	none
Gadolinite	1.77-1.82	0.01-0.04
Ruby	1.762-1.778	0.008
Sapphire	1.762-1.778	0.008
Benitoite	1.757-1.804	0.047
Shattuckite	1.752-1.815	0.063
Chrysoberyl	1.746-1.763	0.007-0.011
Periclase	1.74	none
Scorodite	1.738-1.816	0.027-0.030
Staurolite	1.736-1.762	0.010-0.015
Grossular	1.734-1.759	none
Chambersite	1.732-1.744	0.012
Hessonite	1.730-1.757	none
Epidote	1.729-1.768	0.015-0.049

	Refractive Index	Double Refraction
Pyroxmangite	1.726-1.764	0.016-0.020
Azurite	1.720-1.848	0.108-0.110
Pyrope	1.720-1.756	none
Hodgkinsonite	1.719-1.748	0.022-0.026
Taaffeite	1.719-1.730	0.004-0.009
Rhodonite	1.716-1.752	0.010-0.014
Gahnospinel	1.715-1.754	none
Spinel	1.712-1.762	none
Kyanite	1.710-1.734	0.015-0.033
Adamite	1.708-1.760	0.048-0.050
Diaspore	1.702-1.750	0.048
Serendibite	1.701-1.743	0.005
Sapphirine	1.701-1.734	0.004-0.007
Aegirine-augite	1.700-1.800	0.030-0.050
Vesuvianite	1.700-1.723	0.002-0.012
Tanzanite	1.691-1.700	0.009
Neptunite	1.690-1.736	0.029-0.045
Willemite	1.690-1.723	0.028-0.033
Rhodizite	1.690	none
Triphylite	1.689-1.702	0.006-0.008
Lithiophilite	1.68-1.70	0.01
Dumortierite	1.678-1.689	0.015-0.037
Legrandite	1.675-1.740	0.060
Hypersthene	1.673-1.731	0.010-0.016
Parisite	1.671-1.772	0.081-0.101
Clinozoisite	1.670-1.734	0.010
Sinhalite	1.665-1.712	0.036-0.042
Lawsonite	1.665-1.686	0.019-0.021
Diopside	1.664-1.730	0.024-0.031
Bustamite	1.662-1.707	0.014-0.015
Cornerupine	1.660-1.699	0.012-0.017
Hiddenite	1.660-1.681	0.014-0.016
Kunzite	1.660-1.681	0.014-0.016
Boracite	1.658-1.673	0.010-0.011
Axinite	1.656-1.704	0.010-0.012
Malachite	1.655-1.909	0.254
Sillimanite	1.655-1.684	0.014-0.021
Jadeite	1.652-1.688	0.020
Peridot	1.650-1.703	0.036-0.038
Ludlamite	1.650-1.697	0.038-0.044
Enstatite	1.650-1.680	0.009-0.012
Euclase	1.650-1.677	0.019-0.025
Phenakite	1.650-1.670	0.016
Dioptase	1.644-1.709	0.051-0.053
Jet	1.640-1.680	none
Eosphorite	1.638-1.671	0.028-0.035
Spurrite	1.637-1.681	0.039-0.040
Jeremejevite	1.637-1.653	0.007-0.013
Baryte	1.636-1.648	0.012
Siderite	1.633-1.875	0.242
Danburite	1.630-1.636	0.006-0.008
Clinohumite	1.629-1.674	0.028-0.041
Apatite	1.628-1.649	0.002-0.006
Andalusite	1.627-1.649	0.007-0.013
Friedelite	1.625-1.664	0.030
Smithsonite	1.621-1.849	0.228

	Refractive Index	Double Refraction
Datolite	1.621-1.675	0.040-0.050
Celestine	1.619-1.635	0.010-0.012
Tourmaline	1.614-1.666	0.014-0.032
Actinolite	1.614-1.653	0.020-0.025
Hemimorphite	1.614-1.636	0.022
Lazulite	1.612-1.646	0.031-0.036
Prehnite	1.611-1.669	0.021-0.039
Gaspéite	1.61-1.81	0.22
Turquoise	1.610-1.650	0.040
Topaz	1.609-1.643	0.008-0.016
Sugilite	1.607-1.611	0.001-0.004
Sogdianite	1.606-1.608	0.002
Brazilianite	1.602-1.623	0.019-0.021
Rhodochrosite	1.600-1.820	0.208-0.220
Odontolite	1.60-1.64	0.010
Nephrite	1.600-1.627	0.027
Pectolite	1.595-1.645	0.038
Montebrasite	1.594-1.633	0.22
Phosphophyllite	1.594-1.621	0.021-0.033
Meliphanite	1.593-1.612	0.019
Eudialyte	1.591-1.633	0.003-0.010
Chondrodite	1.592-1.646	0.028-0.034
Catapleiite	1.590-1.629	0.039
Wardite	1.590-1.599	0.009
Herderite	1.587-1.627	0.023-0.032
Colemanite	1.586-1.615	0.028-0.030
Howlite	1.586-1.605	0.019
Zektzerite	1.582-1.585	0.003
Amblygonite	1.578-1.646	0.024-0.030
Ekanite	1.572-1.573	0.001
Anhydrite	1.570-1.614	0.044
Augelite	1.570-1.590	0.014-0.020
Emerald	1.565-1.602	0.006
Aquamarine	1.564-1.596	0.004-0.005
Variscite	1.563-1.594	0.031
Precious beryl	1.562-1.602	0.004-0.010
Tremolite	1.560-1.643	0.017-0.027
Vivianite	1.560-1.640	0.050-0.075
Serpentine	1.560-1.571	0.008-0.014
Labradorite	1.559-1.570	0.008-0.010
Hambergite	1.553-1.628	0.072
Pyrophyllite	1.552-1.600	0.048
Muscovite	1.552-1.618	0.036-0.043
Beryllonite	1.552-1.561	0.009
Charoite	1.550-1.561	0.004-0.009
Amethyst	1.544-1.553	0.009
Aventurine	1.544-1.553	0.009
Rock crystal	1.544-1.553	0.009
Citrine	1.544-1.553	0.009
Prasiolite	1.544-1.553	0.009
Smoky quartz	1.544-1.553	0.009
Rose quartz	1.544-1.553	0.009
Andesine	1.543-1.551	0.008
Cordierite	1.542-1.578	0.008-0.012
Oligoclase	1.542-1.549	0.007
Talc	1.54-1.59	0.050
Scapolite	1.540-1.579	0.006-0.037
Amethyst	1.54-1.55	0009 Quartz
Petrified wood	1.54	none
Jasper	1.54	none
Amber	1.539-1.545	none
Ivory	1.535-1.570	none
Apophyllite	1.535-1.537	0.002
Tiger's-eye	1.534-1.540	none
Aragonite	1.530-1.685	0.155
Agate	1.530-1.540	0.004-0.009
Chalcedony	1.530-1.540	0.004-0.009
Chrysoprase	1.530-1.540	0.004-0.009
Moss agate	1.530-1.540	0.004-0.009
Sepiolite	1.53	none
Witherite	1.529-1.677	0.148
Milarite	1.529-1.551	0.003
Nepheline	1.526-1.546	0.0004
Sunstone	1.525-1.548	0.010
Amazonite	1.522-1.530	0.008
Pearl	1.52-1.69	0.156
Ammonite	1.52-1.68	0.155
Strontianite	1.52-1.67	0.150
Gypsum	1.520-1.529	0.009
Orthoclase	1.518-1.530	0.008
Sanidine	1.518-1.530	0.008
Moonstone	1.518-1.526	0.008
Pollucite	1.517-1.525	none
Carletonite	1.517-1.521	0.004
Stichtite	1.516-1.544	0.026
Thomsonite	1.515-1.542	0.006-0.025
Magnesite	1.509-1.717	0.208
Scolecite	1.509-1.525	0.007-0.012
Leucite	1.504-1.509	0.001
Mesolite	1.504-1.508	0.001
Dolomite	1.502-1.698	0.185
Petalite	1.502-1.519	0.012-0.017
Lapis lazuli	1.50	none
Haüyne	1.496-1.510	none
Tugtupite	1.496-1.502	0.006
Cancrinite	1.495-1.528	0.024-0.029
Celluloid	1.495-1.520	none
Ulexite	1.491-1.520	0.029
Yugawaralite	1.490-1.509	0.011-0.014
Whewellite	1.489-1.651	0.159-0.163
Kurnakovite	1.488-1.525	0.036
Inderite	1.486-1.507	0.017-0.020
Calcite	1.486-1.658	0.172
Coral	1.486-1.658	0.160-0.172
Moldavite	1.48-1.54	none
Natrolite	1.480-1.493	0.013
Sodalite	1.48	none
Analcime	1.479-1.489	none
Thaumasite	1.464-1.507	0.036
Creedite	1.461-1.485	0.024
Chrysocolla	1.460-1.570	0.023-0.040
Obsidian	1.45-1.55	none
Gaylussite	1.443-1.523	0.080
Glass	1.44-1.90	none
Fluorite	1.434	none
Sellaite	1.378-1.390	0.012
Opal	1.37-1.52	none

Double Refraction

In all gemstones, except opals, glasses, and those belonging to the cubic system, the ray of light is refracted when entering the crystal and at the same time divided into two rays. This phenomenon is called double refraction. It can be most clearly observed in the case of calcite. It is also easily seen in zircon, sphene, tourmaline, and peridot. When looking from above, one can see the doubling of the edges of the lower facets in transparent cut stones. Often, magnification is needed to see the effect. It is up to the lapidary to work the stone in such a way that the double refraction does not appear disturbing.

The double refraction can be useful in identifying gemstones. It is expressed as the difference between the highest and lowest refractive index. The expert also differentiates between positive and negative "optical character."

In the tabular information for the respective gemstones, this positive or negative property is indicated with either a + or a –. In the table "Refractive Indices and Double Refraction for Selected Gemstones," pp. 38–39, and likewise in the table of constants, pp. 296–311, we had to forego such a reference due to lack of space. Also, the declarations of value regarding the double refraction were listed, if applicable, in the table of constants simplified as an arithmetic mean. They must, therefore, possibly be read as a magnitude and not as a qualified number.

Anomalous Double Refraction As mentioned above, crystals listed in the cubic system do usually not reveal a double refraction. However, there are gemstones in the cubic system with a so-called anomalous double refraction. Among others, synthetic spinel (an important distinguishing characteristic!), YAG, garnet, and most diamonds belong to this group. A high double refraction is very typical for brownish diamonds. Even amorphous glass materials reveal an anomalous double refraction.

It appears that some kind of anomalies in the crystal lattice cause the anomalous double refractions, e.g. different growth, crystal lattice defects, foreign atoms, inclusions, and mechanical disruptions, such as crazes and fractures. Such anomalies can create tensions in the crystal (therefore, also called tension double refraction), which can be caused by improper handling, resulting in the stones shattering. This awareness is important for all those who work on gemstones.

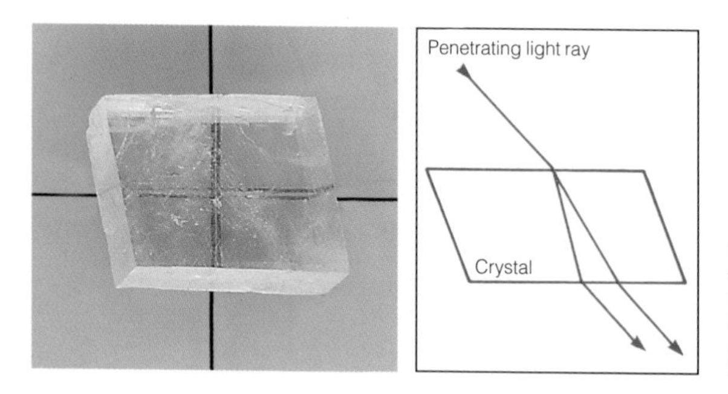

Calcite shows double refraction especially clearly (left), and diagram of double refraction (right).

Dispersion of the original white light when going through a prism; a breakdown into spectral colors.

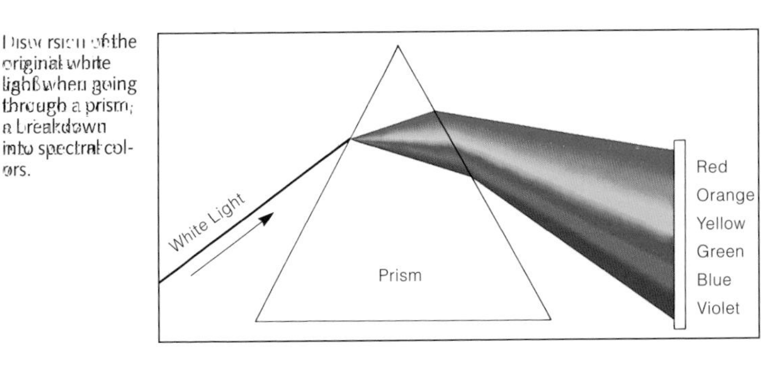

Dispersion

In colorless and cut gemstones, one can occasionally observe flashes of color, which come about through dispersion of the white light into the spectral colors. That is to say, the white light is not only refracted when penetrating a crystal, but it is also dispersed into its spectral colors, because each wavelength is refracted by a different amount. Violet is refracted more strongly than red.

The dispersion is different from one gemstone to another. A distinct dispersion usually occurs only in colorless or weakly tinted stones. Facets can enhance the dispersion. Colorful gemstones tend to mask the dispersion effect.

Color dispersion is especially high in diamonds, where it produces the so-called fire. Natural as well as synthetic gemstones with high dispersion (for instance, strontium titanate, synthetic rutile, sphalerite, sphene, and zircon) are used as substitutes for diamond; sometimes, though, they are supposititious (see page 91).

The measurement of the dispersion can be done with a refractometer and special devices. (However, this is not part of routine gem testing.) The dispersion of a stone is expressed in figures as the difference between the red and violet refractive indices. As the color usually comprises a wide spectrum, it is common to use certain lines (Fraunhofer lines) of the spectrum when doing the measurements. In the gemological literature, mainly the Fraunhofer lines B and G (BG dispersion) are used as a basis for the dispersion values, but occasionally the lines C and F (CF dispersion) are used.

Tables of dispersion values can be consulted for the determination of the gemstone. In the table on pages 42 and 43, the BG dispersion values are contrasted with those of the CF dispersion. In the description of the individual gemstones, the CF dispersion values are placed in brackets following the BG values.

It is important to remember that only transparent stones can show any dispersion at all. Additionally, in gemstones with dispersion lower than zircon's (0.039), the effect may not be visible except in large and colorless or very lightly tinted specimens.

Spectrum of the Fraunhofer lines. Dispersion refers either to the BG or the CF area.

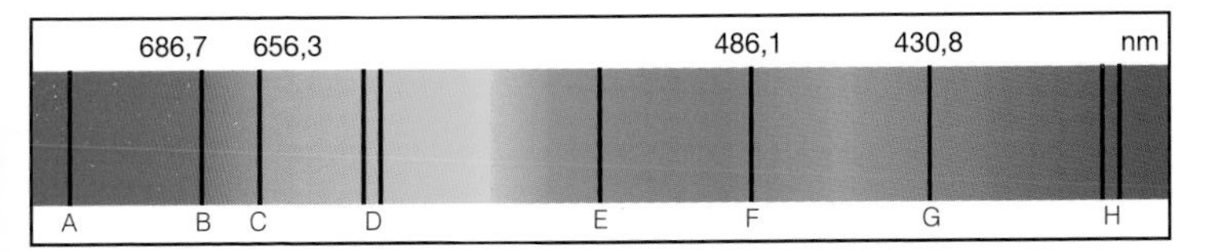

Dispersion of Selected Gemstones

	B-G	C-F
Cinnabar	0.40	
Synth. rutile	0.330	0.190
Rutile	0.280	0.120-0.180
Anatase	0.213-0.259	
Wulfenite	0.203	0.133
Vanadinite	0.202	
Fabulite	0.190	0.109
Sphalerite	0.156	0.088
Sulphur	0.155	
Stibiotantalite	0.146	
Goethite	0.14	
Brookite	0.131	0.12-1.80
Zincite	0.127	
Linobate	0.13	0.075
Synth. moissanite	0.104	
Cassiterite	0.071	0.035
Zirconia	0.060	0.035
Powellite	0.058	
Andradite	0.057	
Demantoid	0.057	0.034
Cerussite	0.055	0.033-0.050
Titanite	0.051	0.019-0.038
Benitoite	0.046	0.026
Anglesite	0.044	0.025
Diamond	0.044	0.025
Flint glass	0.041	
Hyacinth	0.039	
Jargoon	0.039	
Starlite	0.039	
Zircon	0.039	0.022
GGG	0.038	0.022
Scheelite	0.038	0.026
Dioptase	0.036	0.021
Whewellite	0.034	
Alabaster	0.033	
Gypsum	0.033	0.008
Epidote	0.030	0.012-0.027
Tanzanite	0.030	0.011
Thulite	0.03	0.011
Zoisite	0.03	
YAG	0.028	0.015
Almandine	0.027	0.013-0.016
Hessonite	0.027	0.013-0.015
Spessartine	0.027	0.015
Uvarovite	0.027	0.014-0.021
Willemite	0.027	
Pleonaste	0.026	
Rhodolite	0.026	
Boracite	0.024	0.012
Cryolite	0.024	
Staurolite	0.023	0.012-0.013
Pyrope	0.022	0.013-0.016
Diaspore	0.02	
Grossular	0.020	0.012
Hemimorphite	0.020	0.013
Kyanite	0.020	0.011
Peridote	0.020	0.012-0.013
Spinel	0.020	0.011
Synth. spinel	0.020	0.010
Vesuvianite	0.019-0.025	0.014
Clinozoisite	0.019	0.011-0.014
Labradorite	0.019	0.010
Axinite	0.018-0.020	0.011
Ekanite	0.018	0.012
Kornerupine	0.018	0.010
Corundum	0.018	0.011
Leucosapphire	0.018	
Rhodizite	0.018	
Ruby	0.018	0.011
Sapphire	0.018	0.011
Sinhalite	0.018	0.010
Sodalite	0.018	0.009
Synth. corundum	0.018	0.011
Diopside	0.017-0.020	0.012
Achroite	0.017	
Cordierite	0.017	0.009
Danburite	0.017	0.009
Dravite	0.017	
Elbaite	0.017	
Herderite	0.017	0.008-0.009
Hiddenite	0.017	0.010
Indicolite	0.017	
Liddicoatite	0.017	
Kunzite	0.017	0.010
Rubellite	0.017	
Schorl	0.017	
Scapolite	0.017	
Spodumene	0.017	0.010
Tourmaline	0.017	0.009-0.011
Verdelite	0.017	
Andalusite	0.016	0.009
Baryte	0.016	0.009
Euclase	0.016	0.009
Alexandrite	0.015	0.011
Chrysoberyl	0.015	0.011
Hambergite	0.015	0.009-0.010
Phenakite	0.015	0.009
Rhodochrosite	0.015	0.010-0.020
Sillimanite	0.015	0.009-0.012
Smithsonite	0.014-0.031	0.008-0.017
Amblygonite	0.014-0.015	0.008
Aquamarine	0.014	0.009-0.013
Beryl	0.014	0.009-0.013
Red beryl	0.014	
Brazilianite	0.014	0.008
Celestine	0.014	0.008
Precious beryl	0.014	0.009-0.013
Goshenite	0,014	
Heliodor	0.014	0.009-0.013
Morganite	0.014	0.009-0.013

	B-G	C-F
Emerald	0.014	0.009-0.013
Topaz	0.014	0.008
Amethyst	0.013	0.008
Amethystquartz	0.013	0.008
Anhydrite	0.013	
Apatite	0.013	0.010
Aventurine	0.013	0.008
Rock crystal	0.013	0.008
Citrine	0.013	0.008
Hawk's-eye	0.013	
Morion	0.013	
Prasiolite	0.013	0.008
Quartz	0.013	0.008
Smoky quartz	0.013	0.008
Rose quartz	0.013	0.008
Tiger's-eye	0.013	
Albite	0.012	
Bytownite	0.012	
Feldspar	0.012	0.008
Moonstone	0.012	0.008
Orthoclase	0.012	0.008
Pollucite	0.012	0.007
Sanidine	0.012	
Sunstone	0.012	
Beryllonite	0.010	0.007
Cancrinite	0.010	0.008-0.009
Leucite	0.010	0.008
Obsidian	0.010	
Strontianite	0.008-0.028	
Calcite	0.008-0.017	0.013-0.014
Fluorite	0.007	0.004
Hematite		0.500
Synth. cassiterite		0.041
Gahnite		0.019-0.021
Datolite		0.016
Pyroxmangite		0.015
Synth. scheelite		0.015
Dolomite		0.013

	B-G	C-F
Magnesite		0.012
Synth. emerald		0.012
Synth. alexandrite		0.011
Synth. sapphire		0.011
Phosphophyllite		0.010-0.011
Enstatite		0.010
Flat Glass		0.009-0.098
Anorthite		0.009-0.010
Actinolite		0.009
Jeremejevite		0.009
Nepheline		0.008-0.009
Apophyllite		0.008
Haüyne		0.008
Natrolite		0.008
Synth. quartz		0.008
Aragonite		0.007-0.012
Augelite		0.007
Tremolite		0.006-0.007
Adamite	strong	
Covelline	strong	
Dickinsonite	strong	
Crocoite	strong	
Lawsonite	strong	
Libethenite	strong	
Purpurite	strong	
Realgar	strong	
Creedite	moderate	
Pumpellyite	moderate	
Agalmatolite	weak	
Bronzite	weak	
Colemanite	weak	
Euchroite	weak	
Fuchsite	weak	
Lepidolite	weak	
Nambulite	weak	
Opal	weak	
Pyrophyllite	weak	
Sekaninaite	weak	
Vivianite	weak	

Absorption Spectra

The absorption spectrum of a gem can sometimes be one of the most important aids in identifying it. The absorption spectrum of a stone consists of the bands that appear in the spectral colors of light as they emerged from the gemstone (see page 45). As we can see, certain wavelengths (color bands) of the light are absorbed (see page 31) and the color of the gem is formed from the mixture of the remaining parts of the original white light. The human eye cannot recognize all minute color differences. Red tourmaline, for instance, can appear like red garnet, or even red-colored glass can appear deceivingly like the desirable red ruby. However, the absorption spectrum unmasks without any doubt the stone or glass used to imitate the ruby. Many gems have a very characteristic, even unique, absorption spectrum which is revealed (seen through a spectroscope) in black vertical lines or broad bands.

The great advantage of this testing method is the ease with which one can differentiate between gems of the same density and similar refractive index. One can also use this method for testing rough stones, cabochons, and even set stones. An important area where absorption spectroscopy is applied is in differentiating between natural stones, synthetic stones, and imitations.

Best results are obtained from strongly colored, transparent gemstones. The observation of the absorption spectrum of opaque stones is only possible when a very thin slice of the stone is prepared which can transmit light. Otherwise translucent edge must be presented, or light must be reflected from the surface.

The testing instrument is the spectroscope, with the help of which one can determine the wavelength of the absorbed light. The wavelength is measured in nanometers, symbol nm (1 nm = 10–9 m = 1 millionth millimeter). The former measurement Angstrom, symbol Å (1 Å = 10–10 m = 0.1 nm), is still widely used in the gemological literature. Because the absorption lines or bands are not always of equal strength, it is usual to note such measured differences. Strong absorption lines are in this book underlined, for instance, 653; medium strong absorption lines are in normal lettering, for instance, 594; weak lines are bracketed, for instance (432). (See the table of the absorption spectra of selected gemstones on pages 46 and 47.)

Hand spectroscope with wavelength scale in the small tube.

Absorption Spectra of Selected Gemstones

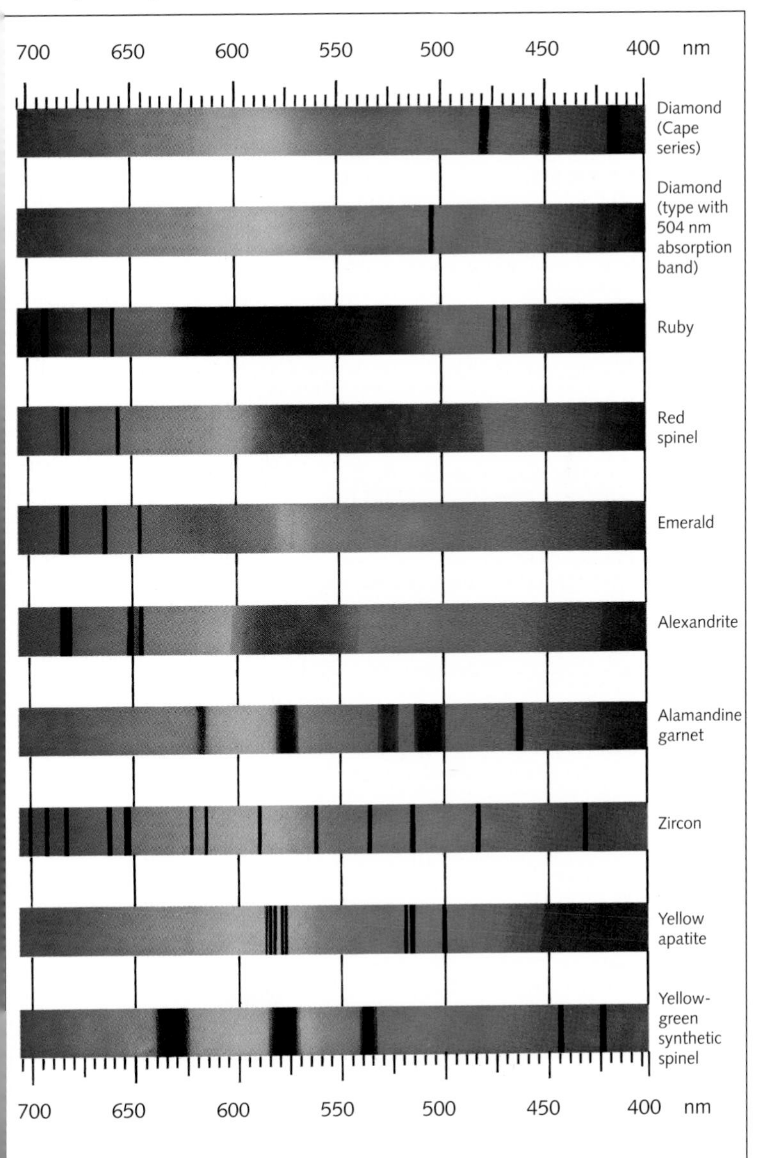

From *Gemmologists' Compendium* by R. Webster, NAG Press Ltd., London. Taken from *Edelsteinkundliches Handbuch* by Professor Dr. Chudoba and Dr. Gübelin, Stollfuss Verlag, Bonn.

Absorption Spectra of Selected Gemstones

All figures are in nanometers (nm).
Strong absorption lines are underlined; weak ones are in parentheses.

Actinolite: 503, 431
Agate, artificially colored green: 700, (665), (634)
Alexandrite, green direction: 680, 678, 665, 655, 649, 645, 640-555
Alexandrite, red direction: 680, 678, 655, 645, 605-540, (472)
Almandine: 617, 576, 526, 505,476, 462, 438, 428, 404, 393
Amethyst: (550-520)
Andalusite: 553, 550, 547, (525), (518), (495), 455, 447, 436
Andradite: 701, 693, 640, 622, 443
Apatite, yellow-green: 597, 585, 577, 533, 529, 527, 525, 521, 514, 469
Apatite, blue: 512, 507, 491, 464
Aquamarine: 537, 456, 427
Maxixe-Aquamarine: 695, 655, 628, 615, 581, 550
Aventurine, green: 682, 649
Axinite: 532, 512, 492, 466, 440, 415
Azurite: 500
Bowenite: 492, 464
Bronzite: 509, 506, 547, 502, 483, 459, 449
Calcit: 582
Chalcedony, artificially colored blue: 690-660, 627
Chalcedony, artificially colored green: 705, 670, 645
Chrysoberyl: 504, (495), 485, 445
Chrysoberyl-Cat's-Eye: 505, (495), 485, 460-450, 444
Chrysoprase, natural: 444
Chrysoprase, artificially colored with Nickel: 632, 444
Coral: 495
Cordierite: 645, 593, 585, 535, 492, 456, 436, 426
Corundum, synthetically Alexandrite colored: 687, (610-560), 570,475
Corundum, synthetically Peridote colored: 688, 678, 645
Crocoite: 555
Danburite: 590, 586, 585, 584, 583, 582, 580, 578, 573, 571, 568, 566, 564
Demantoide: 701, 693, 640, 622, 443
Diamond, naturally coloreless to yellow (>Cape<): 478, 465, 451,435, 423, 415, 401, 390
Diamond, naturally brown-greenish: (537), 504, (498)
Diamond, naturally yellowish brown: 576, 569, 564, 558, 550, 548, 523, 493, 480, 460
Diamond, artificially colored yellow: 594, 504, 498, (478), (415)
Diamond, artificially colored green: 741, 504, 498, 465, 451, 435, 423, 415
Diamond, artificially colored brown: (741), 594, 504, 498, 478, 465, 451, 435, 423, 415
Diaspore: 701, 471, 463, 454
Diopside: (505), (493), (446)
Chromdiopside: (690), (670), (655), (635), 508, 505, 490
Dioptase: 550, 465
Ekanite: 665, (637)
Emerald, natural: 683, 681, 662, 646, 637, (606), (594), 630-580, 477, 472
Trapiche-Emerald: 683, 680, 637, (625-580)
Emerald, synthetic: 683, 680, 662, 646, 637, 630-580, 606, 594, 477, 472, 430
Enstatite: 547, 509, 505, 502, 483, 459, 449
Chromenstatite: 688, 669, 506
Eosphorite, brownish-pink: 490, 410
Epidote: 475, 455, 435
Euclase: 706, 704, 650, 639, 468, 455
Fluorite, green: 634, 610, 582, 445, 427
Fluorite, yellow: 545, 515, 490, 470, 452
Friedelite: 556, (456)
Gahnite: 632, 592, 577, 552, 508, 480, 459, 443, 433
Grossular: 697, 630, 605, 505

Hematite: (700), (640), (595), (570), (480), (450), (425), (400)
Hessonite: 547, 490, 454, 435
Hiddenite: 690, 686, 669, 646, 620, 437, 433
Hypersthene: 551, 547, 505, 482, 448
Jadeite, artificially colored green: 665, 655, 645
Jadeite, natural green: 691, 655, 630, (495), 450, 437, 433
Kornerupine: 540, 503, 463, 446, 430
Kyanite: (706), (689), (671), (652), 446, 433
Nephrite: (689), 509, 490, 460
Precious Beryl, artificially colored blue: 705-685, 645, 625, 605, (587)
Obsidian, green: 680, 670, 660, 650, 635, 595, 555, 500
Opal, Fire Opal: 700-640, 590-400
Orthoclase: 448, 420
Peridot: 497, 495, 493, 473, 453
Petalite: (454)
Prehnite: 438
Pyrope: 687, 685, 671, 650, 620-520, 505
Quartz, synthetically blue: 645, 585, 540, 500-490
Rhodochrosite: 551, 449, 415
Rhodonite: 548, 503, 455, (412), (408)
Ruby: 694, 693, 668, 659, 610-500, 476, 465, 468
Sapphire, blue from Australia: 471, 460, 450
Sapphire, blue from Sri Lanka: (450)
Sapphire, yellow: 471, 460, 450
Sapphire, green: 471,460-450
Scapolite, pink: 663, 652
Scapolite, amethyst colored: (495), (488), (450)
Scheelite: 584
Sillimanite: 462, 441, 410
Sinhalite: 526, 492, 475, 463, 452
Sogdianite: (645-630), (493-488), 437, 419, 411
Spessartine: 495, 485, 462, 432, 424, 412
Sphalerite: 690, 667, 651
Spinel, naturally red: 685, 684, 675, 665, 656, 650, 642, 632, 595-490, 465, 455
Spinel, naturally blue: 632, 585, 555, 508, 478, 458, 443, 433
Spinel, synthetically blue: 634, 580, 544, 485, 449
Spinel, synthetically cobalt blue: 635, 580, 540, (478)
Spinel, synthetically green: 620, 580, 570, 550, 540
Spinel, synthetically yellow-green: 490, 445, 422
Stichtite: 665, 630
Sugilite: 570, 419, (411)
Taaffeite: 558, 553, 478
Tanzanite: 595, 528, 455
Titanite: 586, 582
Topaz, pink: 682
Tremolite: 684, 650, 628
Turquoise: (460), 432, (422)
Tourmaline, red: 555, 537, 525-461, 456, 451, 428
Tourmaline, green: 497, 461, 415
Variscite: 688, (650)
Verdite: 700, 699, 698, 455
Vesuvianite, green: (528), 461
Vesuvianite, brown: 591, 588, 584, 582, 577, 574
Vesuvianite, yellow-green: 465
Willemite: 583, 540, 490, 442, 431, 421
Zircon, High-Zircon: 691, 689, 662, 660, 653, 621, 615, 589, 562, 537, 516, 484, 460, 433
Deep-Zircon: 653, (520)
Zirconia, orange: 640, 630, (540), (536), (533), (530), 520, 517, 515, 512, 510, (503), 482, 480, 477, 475, (449), (447), (446)

Transparency

A factor in the evaluation of most gemstones is their transparency. Inclusions of foreign matter or fissures in the interior of the crystal affect the transparency or "clarity." The path of light through the crystal can also be impaired by strong absorption in the crystal. Grainy, stalky, or fibrous aggregates (such as chalcedony, lapis lazuli, and turquoise) are opaque because the rays of light are repeatedly refracted or reflected by the many tiny faces until finally they are completely reflected or absorbed. Where the light is only weakened by its passage through a stone, it is said to have translucency.

Luster

Many gemstones have a characteristic luster, cut or as an uncut natural stone. Thus, it can also serve to help classify the stone. The luster of a gem is caused by external reflection, i.e. the reflecting part of the incident light back from the surface. It is dependent on the refractive index of the stone and the nature of the surface. Generally, the higher the refraction, the higher the luster. Gems without luster are described as dull. However, there is no unified opinion regarding the influence of color on the luster.

While luster is not measurable, it is generally described with known items. We differentiate the following types of luster:

Metallic Luster The strongest luster of all. Just like polished metal or the luster of aluminum foil. Only existent with opaque gemstones, particularly with pure metal, sulfides and some oxides. Meets a refractive index of about 2.6 to more than 3.

Diamond Luster Sparkling luster as with cut diamonds or with lead crystal glass. Only existent with transparent or translucent gemstones. Meets a refractive index of about 1.9 to 2.6.

Greasy Luster Like the shine of grease spots on paper. Holds true for turbid stones with a low refractive index. Not widely common with gemstones, except on cleavage planes.

Pearly Luster Like the shine of mother-of-pearl, the innermost iridescent layer of some mussel, oyster, and snail shells. Especially found on cleavage planes of gemstones with perfect cleavage.

Silky Luster Intense ray of light, as can be observed with natural silk. Appears with parallel fibrous gemstones or corresponding aggregates.

Waxy Luster Dull shine of gemstones with coarser unevenness, such as flint.

Resinous Luster Resin-like, less intensive luster as with amber. Rare with gemstones.

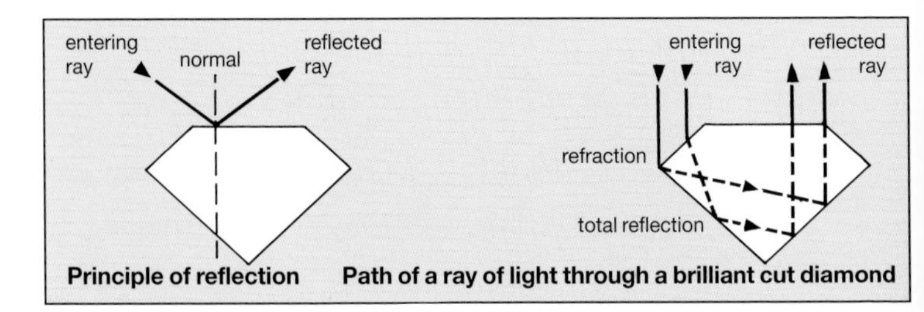

Principle of reflection **Path of a ray of light through a brilliant cut diamond**

Vitreous Luster Just like the shine of simple window glass. Meets a refractive index of about 1.3. With glass imitations, depending on the type of glass used, the refractive index reaches up to 1.9, that is, it almost reaches diamond luster.

Diamond luster is the most widely appreciated, while the most common is vitreous luster. Tarnish, annealing colors, and weathering can affect the luster. Therefore, the discerning eye will only judge the luster with an unaltered object and on fresh surfaces. The luster may vary on crystal, fractured, or cut areas, and the cut and polish of the stone can also increase the natural luster. In general usage, those light effects which are caused by the total internal reflection also count as luster. The lower facets of gemstones work as mirrors and reflect the entering light back to the surface, thus increasing the appearance of the luster. The expert calls this total light effect on the surface the gem's brilliance. With the diamond cut, the ideal total reflection and the highest brilliance has been reached.

Pleochroism

Some gems appear to have different colors or depth of color when viewed in different directions. This is caused by the differing absorption of light rays in doubly refractive crystals. Where two main colors can be observed (only in the tetragonal, hexagonal, and trigonal crystal systems), one speaks of dichroism; where three colors can be seen (only in the orthorhombic, monoclinic, and triclinic crystal systems) of trichroism or pleochroism. The latter term is also a collective description used for both kinds of the multi-coloredness.

Amorphous gems and those of the cubic crystal system show no pleochroism. Opaque stones and most that are translucent also do not show it. The appearances of the pleochroism can be weak, definite, or strong. It must be taken into consideration when cutting in order to avoid poor colors, or shades that are too dark or too light. The instrument of observation for pleochroism is the dichroscope.

The dichroscope is the instrument of choice for observing pleochroism. Two versions are common today: the hand-held spectroscope (see the illustration below) and the horizontal gemstone microscope equipped with a dichroscope eyepiece (the illustration on page 16). The hand-held spectroscope is sufficient for the amateur to get a general impression of the multihued coloration of gemstones. The expert, in need of more objective values, uses the gemstone microscope. In this way, he can exclude disturbing influences such as refraction, reflection, total reflection and false secondary light.

Hand-held spectroscope for the simple observation of pleochroism. Length of the instrument: 10cm.

Pleochroism of Selected Gemstones

Adamite	colorless, blue-green, yellow-green
Actinolite	yellow-green, light green, bluish green
Alexandrite	green alexandrite distinct in daylight: pigeon blood red-violet alexandrite distinct in lamplight: dark red, yellow-red, dark green
Amethyst	very weak; reddish violet, gray-violet
Anatase	distinct; yellowish, orange
Andalusite	very strong; yellow, olive, red-brown to dark red
Anhydrite	violet crystals: colorless to light yellow, dark violet to pink, violet
Apatite	yellow: weak; golden yellow, green-yellow; green: weak; yellow, green; blue: very strong; blue, yellow
Aquamarine	blue: distinct; almost colorless to light blue, blue to sky blue; green-blue: distinct; yellow-green to colorless, blue-green
Axinite	strong; olive green, red-brown, yellow-brown
Azurite	distinct; light blue, dark blue
Baryte	blue: weak
Benitoite	very strong; colorless, blue
Bronzite	pink, green
Cassiterite	weak to strong; green-yellow, brown, red-brown
Charoite	variable: colorless, pink
Chrysoberyl	very weak; red to yellow, yellow to light green, green
Citrine	natural: weak; yellow, light yellow
Clinohumite	golden yellow-red-yellow, light yellow-orange-yellow
Cordierite	very strong; blue: yellow, dark blue–violet, hazy blue; strong; hazy blue: almost colorless, dark blue, hazy blue
Corundum	synthetic: strong; blue-green, yellow-green
Danburite	weak; hazy light yellow, light yellow
Diaspore	strong; violet-blue, light green, pink to dark red
Diopside	weak; yellow green, dark green
Dioptase	weak; dark emerald green, light emerald green
Dumortierite	strong; black, red-brown, brown
Durangite	yellow, orange, colorless
Emerald	distinct; natural emerald: green, blue, blue-green to yellow-green synthetic emerald: yellow-green, blue-green
Enstatite	distinct; green, yellow-green
Euclase	very weak, green-white, yellow-green, blue-green
Epidote	strong; green-brown epidote: green, brown, yellow strong; green epidote: almost colorless, yellow-green, light brown
Hiddenite	distinct; blue-green, emerald green, yellow-green
Hodgkinsonite	variable: lavender colored, colorless
Hypersthene	strong; hyacinth-red, straw-yellow, sky blue
Jeremejevite	blue, colorless, light green
Kornerupine	strong; green, yellow, red-brown
Kunzite	distinct; amethyst colored, pale red, colorless
Kyanite	strong; colorless, light blue, dark blue
Lazulite	strong; colorless, deep blue
Linarite	
Malachite	very strong; almost colorless, yellow-green, deep green
Monazite	red-orange to yellow
Nephrite	weak; yellow to brown, green
Neptunite	strong; yellow, deep red
Precious beryl	golden beryl: weak; lemon yellow, yellow; green beryl: distinct; yellow-green, blue-green heliodor: weak; golden yellow, green-yellow; morganite: distinct; pink, blue-pink; violet beryl: distinct; violet, colorless
Orthoclase	yellow: weak

Painite	strong; ruby red, brown-orange
Peridote	very weak; colorless to pale green, vivid green, oily green
Phenakite	distinct; colorless, orange-yellow
Prasiolite	very weak; light green, pale green
Proustite	strong; red shadings
Purpurite	distinct; gray-brown, blood red
Rhodonite	distinct; red-yellow, pink-red
Rose quartz	weak; pink, pale pink
Ruby	strong; yellow-red, deep carmine red
Rutile	variable; red, brown, yellow, green
Sapphire	strong; orange sapphire: yellow-brown to orange, almost colorless; weak; yellow sapphire: yellow, pale yellow; weak; green sapphire: green-yellow, greenish yellow; distinct; blue sapphire: deep blue, greenish blue; distinct ; violet sapphire: violet, pale red synth. sapphire: dark blue, yellow to blue
Scapolite	pink scapolite: colorless, pink; distinct; yellow scapolite: colorless, yellow
Sheelite	weak; yellow to reddish brown
Sillimanite	strong; pale green, dark green, blue
Sinhalite	distinct; green, light brown, dark brown
Smoky quartz	dark smoky quartz: distinct; brown, reddish brown
Staurolite	strong; red-brown staurolite: yellowish, yellowish red, red
Stichtite	weak; light red, dark red
Tanzanite	very strong; purple, blue, brown, or yellow
Tantalite	strong; brown, red-brown
Thulite	strong; yellow, pink
Titanite	green titanite: colorless, green-yellow, reddish yellow; strong; yellow titanite: colorless, greenish yellow, reddish
Topaz	red topaz: strong; dark red, yellow, rose red; pink topaz: distinct; colorless, pale pink, pink; yellow topaz: distinct; lemon yellow, honey yellow, straw yellow; brown topaz: distinct; yellow-brown, dull yellow-brown; green topaz: distinct; pale green, light blue-green, greenish white; blue topaz: weak; light blue, pink, colorless; burnt topaz: distinct; pink, colorless
Tremolite	distinct; blue-red, pink, violet
Tugtupite	strong; bluish red, orange-red
Turquoise	weak; colorless, pale blue or pale green
Tourmaline	red tourmaline: distinct; dark red, light red; pink tourmaline: distinct; light red, reddish yellow; yellow tourmaline: distinct; dark yellow, light yellow; brown tourmaline: distinct; dark brown, light brown; green tourmaline: strong; dark green, yellow-green; blue tourmaline: strong; dark blue, light blue; violet tourmaline: strong; violet, light violet
Vesuvianite	green vesuvianite: weak; yellow-green, yellow-brown; yellow vesuvianite: weak; yellow, almost colorless; brown vesuvianite: weak; yellow-brown, light brown
Vivianite	strong; blue to indigo, pale yellow, green to blue-green, pale yellowish green
Willemite	variable
Zircon	red zircon: very weak; red, light brown; red-brown zircon: very weak; reddish brown, yellowish brown; yellow zircon: very weak; honey yellow, brown-yellow; brown zircon: very weak; red-brown, yellow-brown; brown-green zircon: very weak; rose yellow, lemon yellow; green zircon: very weak; green, brown-green; blue zircon: distinct; blue, yellow-gray to colorless

Light and Color Effects

Many gems show striated light effects or color effects which do not relate to their body-color and are not caused by impurities or their chemical composition. These effects are caused by reflection, interference, and refraction.

Adularescence Moonstone, being a variety of feldspar (see page 180), shows a blue-whitish opalescence (sometimes described as a "billowy" light) which glides over the surface when the stone is cut en cabochon. Interference phenomena of the layered structure are the cause of this effect, known as adularescence.

Asterism This is the effect of light rays forming a star (Latin *aster,* star); the rays meet in one point and enclose definite angles (depending on the symmetry of the stone). It is usually created through reflection of light by thin fibrous or needle-like inclusions that lie in various directions. (See photograph on page 53.) Ruby (page 98) and sapphire (page 102) cabochons can show effective six-rayed stars. There are also four-rayed stars and, rarely, twelve-rayed stars.

Spectrolite, a variety of the labradorite with especially effective labradorization; Finland.

Quartz Cat's-Eye. The finest rutile fibers cause the chatoyancy.

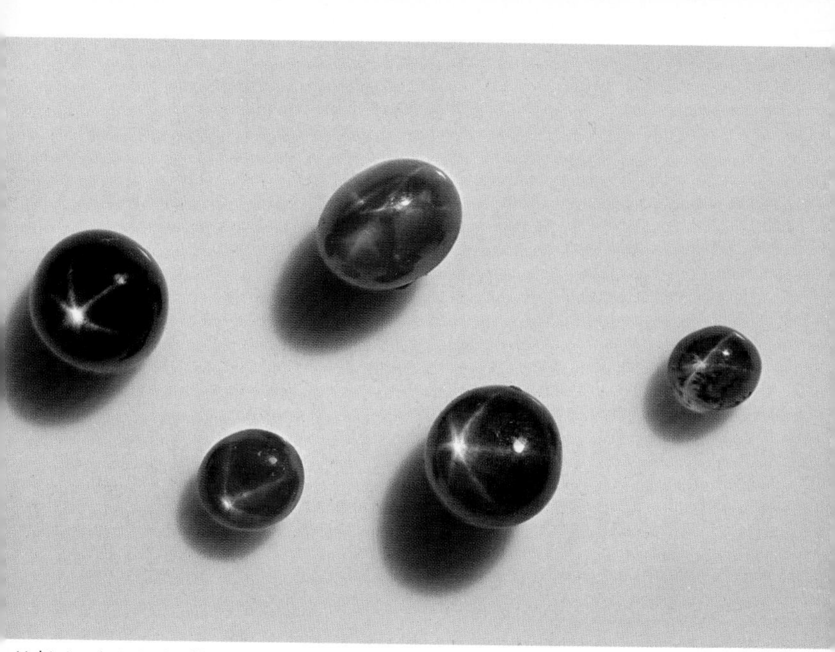

Light stars (asterism) in blue sapphires and in ruby.

If a piece of star rose quartz has been cut as a sphere, the rays move in circles over the whole surface; where included needles are partially destroyed, stunted stars, part circles, or light clusters are formed. Asterism also occurs in synthetic gems.

Aventurescence Colorful play of glittering reflections of small, plate-or leaf-like inclusions. The inclusions are hematite or goethite in the case of aventurine feldspar (page 182); fuchsite or hematite in aventurine quartz (page 138); and copper scrapings in aventurine glass.

Chatoyancy (cat's-eye effect) An effect which resembles the slit eye of a cat (French *chat* = cat, *oeil* = eye); this is caused by the reflection of light by parallel fibers, needles, or channels. This phenomenon is most effective when the stone is cut en cabochon in such a way that the base is parallel to the fibers. When the gem is rotated, the cat's-eye glides over the surface. The most precious cat's-eye is that of chrysoberyl (page 114). The effect can be found in many gemstones; especially well known are quartz cat's-eye, hawk's-eye, and tiger's-eye (page 140). If one talks simply of cat's-eye, one refers to a chrysoberyl cat's-eye. All other cat's-eye must have an additional designation.

Iridescence The rainbow-like hues (Latin *iris* = rainbow) seen in some gems, caused by cracks or structural layers breaking up light into spectral colors. Fire agate (page 150) is a natural gemstone that shows this phenomenon. Commercially, it is created by artificially producing cracks in rock crystal.

Labradorescence Iridescence in metallic hues, called schiller, found especially in labradorite (hence the name) and spectrolite (page 182). Blue and green effects are often found, but the whole spectrum can be observed. The cause of

Play of color in black precious opal; Australia.

the schillers is most probably due to lattice distortions accompanying alternating microscopic exsolution lamellae of high- and low-calcium plagioclase.

Opalescence Milky-blue or pearly appearance of common opal (page 168) caused by reflection of short wave, mainly blue, light. It should not be confused with play-of-color.

Orient Iridescence in pearls. It is created through diffraction and interference of the light by the shingle-like layers of aragonite platelets near the pearl surface (see page 256).

Play-of-Color Flashes of rainbow colors in opal (page 166) which change with the angle of observation. The electron-microscope shows the cause at a magnification of 20,000 X: small spheres of the mineral crystobalite included in a silica gel cause the diffraction interference phenomena. The diameter of these spheres is one to two ten-thousandths of a millimeter.

Silk Reflection of fibrous inclusions or canals causes a silk-like appearance. Especially desirable in faceted rubies and sapphires. Where the included needles are sufficiently numerous, the stone can display chatoyancy if cut en cabochon.

Luminescence

Luminescence (Latin *lumen*-light) is a collective term for the emission of visible light under the influence of certain rays, as well as by some physical or chemical reaction, but not including pure heat radiation.

The most important of these phenomena for the testing of gems is the luminescence under ultraviolet light, which is called fluorescence. The name fluorescence is derived from the mineral fluorite, which is the substance in which this light phenomenon was first observed. When the substance continues to give out light after irradiation has ceased, the effect is called phosphorescence (named after the well-known light property of phosphorus).

The causes of fluorescence are certain interference factors (impurities or flaws in the structure) in the crystal lattice. Many gemstones fluoresce to shortwave UV (254 nm). There are gemstones which exclusively react to shortwave, others only to longwave (366 nm), and again others which react to shortwave as well as longwave UV. Gemstones which contain iron do not show any fluorescence.

Fluorescence is not diagnostic because many specimans of a gemstone can fluoresce in completely different colors, while others of the same gemstone may not light up at all under UV. In the detection of synthetic gemstones, on the other hand, fluorescence can be determinative because synthesis under UV frequently react differently from natural gemstones. Also glued gemstones can sometimes be identified under UV when the glue fluoresces by itself or differently from the other parts. Diamonds (which fluoresce under X rays) can be separated from other stones (such as quartz or pyrope garnet). Occasionally, fluorescence can help establish a particular source locality, because sometimes typical phenotypes for a certain place are characteristic.

Luminescence caused by X rays can help to differentiate between natural and cultured pearls. The mother-of-pearl of saltwater pearl oysters does not luminesce while that of freshwater pearl mussels gives off a strong light, as the inserted nucleus of a cultured perl shows a luminescence which the natural ones do not have. (Refer to table of the fluorescing gemstones on pages 56–57.) X-ray testing of pearls is done only by trade laboratories.

Fluorescing minerals in white light (left) and under UV rays (right). Each from the left; top, aragonite, calcite; center, fluorite, halite; bottom, willemite.

Fluorescence of Selected Gemstones

Agate: differs within layers, partly strong; yellow, blue-white
Adamite: green, green-yellow
Aeschynite: green
Albite: violet-red, white to beige, pink, red, brown
Amazonite: weak; olive-green
Amber: bluish white to yellow-green, burmite: blue
Amblygonite: very weak; green
Amethyst: weak; greenish
Ammonite: mustard yellow
Analcime: creamy white
Andalusite: weak; green, yellow-green
Andesine: yellow, yellow-brown
Anglesite: weak; yellowish
Anhydrite: red
Ankerite: orange, weak dark red
Anorthite: red, beige, yellow
Anorthoklas:
Apatite, yellow A.: lilac to pink
Apophyllite: yellow, orange
Aragonite: weak; pink, yellow, yellow-brown, green, bluish (rare)
Aventurine, green A.: reddish
Axinite: red, orange
Baryte: white, blue-green, gray
Barytocalcite: light yellow, orange to red
Benitoite: strong, blue
Boracite: weak; greenish
Brucite: blue-white, light green
Bustamite: deep red, blue-violet
Calcite: red, pink, orange, white, yellow-white
Cancrinite: orange, dark violet, green-white, yellowish, white
Catapleiite: green
Cerussite: yellow, pink, green, bluish
Chabasite: white, gray, beige, green
Chalcedony: bluish white
Charoite: red
Chlorapatite: orange, yellow
Chondrodite: occasionally golden yellow, brown-orange
Chrysoberyl, green Chr.: weak; dark red;
Others: none
Cinnabar: brown
Colemanite: white, yellowish white
Coral: weak; violet
Creedite: beige, white, bluish
Crocoite: dark brown
Danburite: sky blue
Datolite: yellow, beige, white, dark blue
Diamond: varies considerably;
Colorless and yellow D.: usually blue;
brown and greenish D.: often green
synthetic D.: strong; yellow
Diaspore: light yellow, beige, bluish
Diopside: violet, orange, yellow, green
Dolomite: pink, orange-red
Dumortierite: weak; blue, blue-white, violet
Ekanite: yellow-green
Emerald: usually none;
synthetic E.: occasionally red
Ettringite: beige, gray, light blue, green
Eudialyte: red
Euclase: weak or none
Fluorapatite: orange, yellow
Fluorite: strong; blue to violet
Forsterite: yellow, white, brown, blue
Friedelite: reddish, also rarely green, yellow
GGG: weak; orange, yellow
Gaylussite: weak; creamy white
Greenockite: yellow, orange
Grossular: strong; red-orange
Gypsum: occasionally brownish, greenish
alabaster: green, yellow, pink, orange, violet
Hambergite: usually none, rarely orange
Haüyne: yellow-orange, red, pink
Hedenbergite: green
Helvite: red
Hemimorphite: weak
Herderite: weak; green, pale violet
Hiddenite: very weak; red-yellow
Hodgkinsonite: weak; red
Holtite: dark orange, luminous yellow
Howlite: brownish yellow
Humite: mostly none, rarely golden yellow, orange
Hübnerite: blue
Hyalophane: blue, light red
Hydroxylapatite: yellow, orange, brown
Hydroxylherderite: yellow to white
Ivory: various blues
Jadeite, green J.: very weak; whitish shimmer
Jeremejevite: blue-white, white
Kämmererite: dark green, dark orange
Kornerupine: usually none;
green K. from Kenya: yellow
Kurnakovite: yellow, light brown
Kunzite: strong; yellow-red, orange
Kyanite: weak; red
Labradorite: yellowish striations
Lengbeinite: weak; green-white
Lapis lazuli: strong; white, also orange, copper colored
Legrandite: green
Lepidolite: green, beige, yellow
Leucite: orange or none
Leucophanite: pink, lilac
Magnesite: blue, green, white
Mangan-Axinite: red
Marialite: red, pink, beige, white
Meionite: red, white, yellow, blue
Mellite: gray, light blue, yellow, green

Mesolite: white, yellow, orange, pink, beige, green
Microlite: green, yellow-green, dark green, pink
Moassanite: yellow, orange
Monazite: yellow, red-orange, white, brown, green
Moonstone: weak; bluish, orange
Montebrasite: pale blue, orange, green, light brown
Mordenite: white, beige, pink, blue
Morganite: weak; lilac
Moss agate: variable
Natrolite: beige, white, yellow, blue, violet, orange, green, pink
Nepheline: bright blue, weak orange
Norbergite: yellow, orange
Oligoclase: blue, violet-blue, dark red, yellow, brownish
Opal, white O.: white, bluish, brownish, greenish;
black opal: usually none;
fire opal: greenish to brown
Painite: powerfully red, weak red
Palygorskite: beige, gray-blue, white
Pargasite: greenish, blue, beige
Pectolite: greenish yellow to yellow
Periclase: weak; yellow
Pearl: ocean pearl: weak;
Naturally black pearl: red to reddish;
River pearl: strong; pale green
Petalite: weak; orange
Phenakite: pale greenish, blue
Phosgenite: yellow, orange-yellow
Phosphophyllite: violet
Pollucite: orange to pink
Powellite: yellowish, orange, brown
Prehnite: weak; orange
Prosopite: blue-white, orange-yellow
Pyrophyllite: yellow, orange
Rhodizite: yellow
Rhodochroisite: weak; red
Rose quartz: weak; dark violet
Ruby: strong; carmine
Sanidine: pink
Sapphire: blue S.: violet or none;
yellow S.: weak; orange;
colorless S.: orange-yellow, violet-blue
Scapolite: pink S.: orange, pink;
yellow S.: lilac, blue-red
Scheelite: strong; pale blue;
synthetic S.: pink, blue, orange-red
Scolecite: yellow, brown, bluish
Sellaite: yellowish, yellow
Senarmontite: brown
Shortite: yellow, orange, brown, pink, olive green
Siderite: olive green
Sillimanite: yellowish, brown-orange, brown
Simpsonite: blue, blue-white, pale yellow
Smithsonite: blue-white, pink, brown
Smoky quartz: generally none, seldom weak; brown-yellow
Sodalite: strong, orange
Sogdianite: weak; violet, dark red
Sphalerite: yellow to orange, red
Spinel: red Sp.: strong; red;
blue Sp.: weak, reddish, green
green Sp.: weak, reddish
Spodumene: orange, yellow, pink
Spurrite: white
Stolzite: green-white, yellow
Strontianite: white, olive green, blue-green
Sulphur: yellow-green, yellowish
Sunstone: dark red-brown
Taaffeite: variable green
Talc: beige, white, yellow, orange, greenish
Tephroite: green
Thaumasite: white
Thomsonite: white, beige, pale blue
Tiger's-eye: green, yellow
Topaz: pink T.: weak; brownish;
red T.: weak; brown-yellow;
yellow T.: weak; orange-yellow
Tremolite: white, blue, greenish, orange, red
Turquoise: weak; greenish yellow, pale blue
Tugtupite: luminous to weak orange
Tourmaline: Colorless T.: weak, green-blue;
pale yellow T.: weak, dark green;
red T.: weak; red-violet;
pink, brown, green, blue T.: none
Ulexite: green to yellow, blue
Vanadinite: deep green, blue
Variscite: pale green, green
Villiaumite: orange-yellow to dark red
Vlasovite: orange-yellow, brownish, orange
Wavellite: beige, blue, bluish white, orange, yellow-green
Weloganite: pale green, orange-red
Whewellite: blue-white
Willemite: green
Williamsite: weak; greenish
Wilkeite: yellowish, brownish
Witherite: blue, yellow-white, white
Wolframite: yellow
Wollastonite: blue-green
Wulfenite: red, orange
Wurtzite: yellow, orange
Xonotlite: blue, white, yellow, orange, pink
YAG: yellow
Zektzerite: light yellow
Zincite:
Zircon: blue Z.: very weak; pale orange;
red and brown Z.: weak, dark yellow
Zirconia: occasionally orange
Zoisite: red, gray
Zunyite: intense red, gray

Inclusions

There is hardly a gemstone which is completely "flawless." Most of the time, they contain foreign matter or some kind of dislocation or irregularity in the crystal lattice. Sometimes only under the microscope can these "flaws" be discovered. In the lingo of the trade, one does not speak of flaws, but rather of inclusions, because the term *flaw* has a derogatory meaning that is not justified.

Since inclusions are not accidental appearances but rather are subject to strict conformities with natural law, they can tell a lot about the origin of the gemstone deposits and, in addition to that, they help in their identification. Even though each gemstone has its individual inclusion history, there are nevertheless forms of inclusions which can be grouped together and can be associated with specific gemstone types and/or with gemstone imitations.

Inclusions of minerals are quite common, such as those of the same material (for instance, diamond in diamond) or of a foreign one (for instance, zircon in sapphire). Even if small, they provide a constructive picture of the formation of the surrounding crystal (the *host crystal*). Included minerals can be older than the host crystal, as they were just surrounded. They can also have been formed from a melt at the same time as the host crystal, which surrounded the smaller ones because its growth rate was greater. There are also mineral inclusions that are younger than the host crystal; these were formed out of liquids that entered into the crystal through cracks and fissures. In some cases (e.g. rutile needles in ruby or sapphire) inclusions form within the crystal as it cools.

Smoky quartz with star made of golden-yellow rutile needles; Minas Gerais/Brazil.

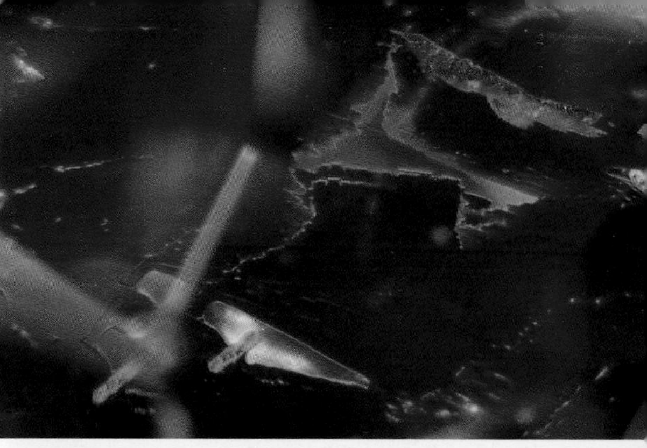

Natural emerald with oil residues that light up brightly in a natural crack (enlargement 40 X).

Synthetic emerald from a Russian production. Characteristic V-structure (enlargement 20 X).

Green glass emerald imitation with stars from small crystals, which came about through devitrification, as well as streaks, which are typical for glass (enlargement 35 X).

Amber with coal-like substance and insect inclusions; Samland, East Prussia/Russia.

Organic inclusions are only found in amber (see page 254 and the picture above). Parts of plants and insects are often preserved in it and bear witness to the life at the time, sometimes as much as 50 million years ago or more.

Irregularities in the crystal structure, marks of the crystallization phases, and color striations are all classed as inclusions. They can be formed by irregular growth from various crystallizing solutions.

Cavities also, when filled with liquid (water, liquid carbonic acid) or gases (carbon dioxide or monoxide), are also classed as inclusions. Where liquid and gas occur together, they are called two-phase inclusions. Liquid, gaseous, and small crystal inclusions are called three-phase inclusions. Completely empty cavities are not known. Air-filled bubbles are found frequently in obsidian, glass imitations, and synthetic gems, but not in mineral gems.

Even breaks and splits (called feathers), whether caused by internal stress or external pressure, are classed as inclusions. They can be observed in the interior of a stone, but they may also reach the surface. Air and solutions can enter a stone along such fissures and affect the color. If the stone is "heated," the foreign substance can be extruded, but scars show the old crack.

The trade and layman consider most inclusions as devaluing a stone because, under certain circumstances, color, optical properties, and mechanical resistance can be affected. However, some inclusions cause light phenomena that can produce the most valuable properties of some gems; for instance, cat's-eye effect, asterism, silk (page 52), and dendrites (page 146). The golden inclusions of rutile in rock crystal or smoky quartz are most effective, especially when they form a star (page 58).

Only for diamonds do generally accepted standards for grading clarity exist (page 93). For all other gemstones, the individual speciman is judged for the quality effect of inclusions.

Deposits and Production of Gemstones

Gemstones are found in many parts of the world, singly or grouped together. Groups large enough to be worked are called *deposits*. Places with a single find are simply called the location of discovery, place of discovery, or point of discovery. The work *occurrence* refers to all four terms.

The Finsch diamond mine, located in Northern Cape Province/South Africa, exploits a kimberlite pipe. Discovered in 1960, first put into production in 1965, now over 20 prospect levels deep.

Fluorite mine near Chiang Mai/Thailand. Without using machines, the strongly weathered rocks are being loosened, carried off, and sorted.

Types of Deposits

According to the way the gemstones have been formed, we differentiate between igneous or magmatic (formed out of the magma), sedimentary (formed by sedimentation), and metamorphic (re-formed from other rocks) deposits.

It is often more useful to speak of primary and of secondary deposits. That way, one differentiates between occurrences in which gemstones are still at the "first" location, i.e., at the place of the original formation, and those places to which they were transported to a "second" location.

In primary deposits the stones still have their original relationship with their host rock. The crystals are usually well preserved, but often the yield of the deposit is not very large. Many tons of *deaf* (nongem-bearing) rock have to be removed in the search for the gemstones.

In the case of secondary deposits, the gems have been transported from the place of their formation and deposited somewhere else. During this process, harder crystals become rounded, and softer ones are made smaller or even destroyed. According to the way the stones have been transported or according to the place of deposit, they are differentiated as *fluvial* (river), *marine* (in the sea), *littoral* (coastal), or *aeolian* (by the wind) deposits. Rivers can transport gem-bearing rocks many hundreds of miles. When the current diminishes and with it the transporting energy, the denser gems (for instance, diamond, zircon, garnet, sapphire, chrysoberyl, topaz, peridot, and tourmaline) are deposited before the lighter quartz sand. The gems left behind thus concentrate in certain places. This usually makes mining the deposit easier and more productive than working primary deposits.

Gemstone deposits which were deposited by river are called placers or alluvial deposits. Similarly, alluvial deposits can be found in the surf-pounded strands of the sea. In southwest Africa (Namibia), such diamond deposits are being worked very successfully (see photo, page 66). Small gem crystals can even be transported by the wind and enrich a particular place by sorting.

Seen genetically, between the primary and secondary deposits there are places where decomposition or weathering takes place. The gems are often found at the foot of steep cliffs or high mountains, and have collected in such places as part of the weathered debris; then, because the specifically lighter host rock was gradually carried away by rain and wind, the gems were left. These are called eluvial deposits.

Distribution of gems over the earth is irregular. Some regions are more favored than others: South Africa, south and southeast Asia, the Urals, Australia, Brazil, and the mountainous zones of the United States.

A gemstone mine in Sri Lanka, run by the government. The only machine run by an engine is the pump that sucks up the water entering into the mine.

Mining Methods

Most gemstone deposits were discovered by accident. Even today systematic prospecting is primarily limited to diamond occurrences. The chief reason for this is that the diamond production is run by international companies and the price is controlled on a worldwide scale. This makes a large capital investment possible for prospecting.

Prospecting for nondiamondiferous deposits is usually accomplished by simple means, without modern techniques or scientific basis. The success of local prospectors in finding new deposits is surprising. A deposit being worked is called a mine.

With the exception of diamonds, mining methods in most countries are very primitive. In some regions, they have not changed in the last 2000 years (see picture on page 62). The increasing demand for gemstones has led some countries to modernize their prospecting. In the emerald mining of South Africa (see picture on page 108) and in the opal mines of Australia (see picture on page 65), for instance, there have been limited advances in ore processing. In Brazil at some places hydraulic mining is used for the sorting of the slope rubble.

The easiest way to collect gemstones is to discover stones at the surface, for instance, in a dry riverbed, in rock crevices, or in caves. But more often, the winning of gemstones takes a tremendous effort and a lot of hard work.

Gemstone mine in Ratnapura in Sri Lanka. With a simple windlass hoist, the earth containing gemstones, which is dug up deep down, is brought up for further processing.

Extraction machines bring the mined rocks containing opal to the surface; Cooper Pedy/South Australia.

Opal mining with quarryman; Lightning Ridge, New South Wales, Australia.

Crystals grown into the mother-rock are loosened with hand tools, compressed-air (pneumatic) tools, or explosives. Underground mining has been done, to some extent, for centuries with both vertical shafts and transverse tunnels.

The mining of gemstones from secondary placer deposits is comparatively simple. With pickax and shovel the material containing gemstones is loosened, dug up, and carried off in baskets, to be worked at another location and examined for gemstones (see picture on page 63). In many countries where there are workable gemstone deposits, it is generally most cost-effective to employ unskilled labor rather than use expensive machinery. This is especially true in many developing nations.

If the secondary deposits containing gemstones are underneath a surface layer, the clay and sand layer is removed or shafts are built downwards. Such shafts, with minimal bracing, can be up to 30 ft (10 m) deep (see picture on page 64). The only modern piece of equipment in such gemstone mines are sump pumps to remove water which enters into the pit by rain or seepage.

Another way to exploit placer deposits is through prospecting in riverbeds. Various water-flow conditions are created through sluiceways and small dams for the purpose of selecting minerals. The less dense clay and sands are swept away, while the heavier gems remain. Workers agitate the material with long sticks to accelerate the sorting (see picture on page 101).

The actual separation of the gemstones from the gem-bearing material is done through another sorting out of the lighter minerals with the help of the moving water in step-like sluiceways and collection pits. Here, the gem-bearing material is further concentrated in baskets, which are washed, allowing lighter material to pan off over the edge of the baskets (see picture on page 105).

The final selection of a gemstone-quality material is always done by hand. The yield of gemstones is usually very low. Often only two or three little stones of gemstone quality are contained in the remaining concentration.

Widely differing ideas prevail in the respective countries regarding prospecting, worker payment, and, if applicable, profit sharing. Generally, however, it can be said that the work of gem prospecting falls to the disadvantaged classes.

A special problem of gem production is the taking of gemstone-quality stones by the workers themselves, which can imperil every mine by undercutting economically viable prices. There is endless ingenuity employed to smuggle gems out of the mines, but, at the same time, methods for prevention are becoming more stringent. Diamond mines are the best protected. (See page 88 for more about the deposits and mining of diamonds.)

Diamond mine of Oranjemund/Namibia. Huge retaining barriers hold back the water of the Atlantic Ocean, so that the placers containing diamonds in the strip along the coast can be mined once a 100-ft. (30-m)-tall sand cover has been removed.

After the use of large-scale equipment, an excavation of the former ocean floor is carried out by hand.

Offshore mining of diamond containing rock. Special ships from the De Beers fleet pick up diamond-bearing conglomerates with suction dredges from the ocean floor off the coast of Namibia and South Africa. In 1998, 6 ships were employed; afterward, usually 4 ships.

Cutting and Polishing of Gems

The oldest way of decorating a gemstone is the scratching of figures, symbols, and letters on it. From this, the art of stone engraving developed. The origins of gem cutting can be found in India. Up to about 1400, only the natural crystal faces or cleavage planes of transparent stones were polished in order to give them a higher luster and improved transparency. But even before then, opaque stones—mainly agates—were cut and polished with hard sandstone either flat or slightly arched (cabochon).

Stone cutting culminates in the faceted stone. There were reports of a faceted diamond in Venice as early as 800 A.D., but according to other opinions, the facet cut was only developed around the 15th century. For a long time the technique of faceting was kept a strict secret. (See also diamond cutting on page 96.)

Since the beginning of modern times, Amsterdam and Antwerp have developed as diamond cutting centers. Idar-Oberstein/Rhineland-Palatinate became the center of agate and colored stone manufacturing in the 16th century. Nowadays numerous cutting centers are being developed around the world. In order to encourage these centers, many countries have prohibited the export of rough stones.

In the manufacture of gemstones, one differentiates between engraving (page 69), working of agates (page 70), colored stone (page 71), ball or cylinder cutting (pages 74 and 75), working of diamond (page 77), and the piercing of gemstones (page 76). Commercially, there is no strict division between these fields of work.

Rolling off in clay of a seal made from limestone. The figures depict fighters, animals, and other creatures. The roller-seal stems from the early third century A.D.; discovered in Syria.

Cutting of agate with diamond sawblade.

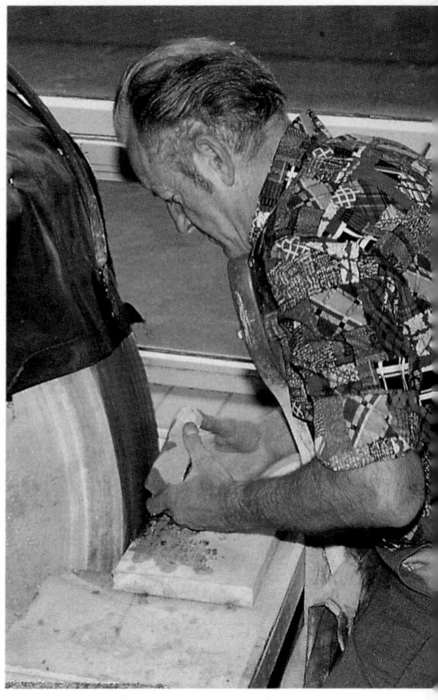

Fine grinding of agate on the sandstone wheel.

Engraving on Stones

The art of stone engraving or carving (also called glyptography) refers to the cutting of cameos and of intaglios, as well as to the creation of small objets d'art and other ornamental pieces.

The oldest stone engravings were cylinders engraved with symbols and figures which were used as seals or amulets (see picture on page 68). They were made in the ancient kingdoms of Sumeria, Babylon, and Assyria. The oldest figures are the scarabs of the Egyptians.

Stone engraving was practiced in ancient Greece and reached a high standard in ancient Rome. Even though the glyptography received renewed activity through heraldry in the Middle Ages, in general, development stagnated. Only with the Renaissance was the art of stone-cutting revived in Italy. Today, glyptography is appreciated as an art, especially due to its creative use of modern tools. In antiquity agate, amethyst, jasper, carnelian, and onyx were first used for stone engraving; gradually other gemstones were also used. Today, virtually all gemstones are engraved, including diamond. (For modern engraving, see page 158; for the technique, see page 160.)

Polishing of agate on slowly rotating wheels without or with only a little cooling fluid.

Cutting and Polishing Agate

Large and heavy stone blocks used to be split with hammer and chisel along fissures or other lines marked in the stone; nowadays they are almost without exception sawed with a diamond-charged circular saw. In modern factories, the cutting disc is cooled with special cutting fluid that emulsifies in water. This is advantageous since gemstones and equipment can be easily cleaned with water. This coolant replaces the use of petroleum that harms the environment, gives off a strong smell, and poses fire hazards.

Agate is first roughly shaped on a carborundum wheel. To keep the stone steady, the cutter holds it between his knees. The wheel is cooled with water. The final shaping is obtained on a sandstone wheel. The cutter sits on a chair with a support for his chest (see picture on page 69). Grooves in the wheel also enable the stone to be cut en cabochon.

The final process is polishing, giving the stone luster and showing up the fine structural lines. Polishing is done on a slowly rotating cylinder or wheel of beech wood, lead, felt, leather, or tin with the help of a polishing powder such as chromium oxide, tripoli, or other pastes. No coolant is used in this process, so special care must be taken; otherwise the stone could be damaged by the resultant heat.

Machines have now been developed which cut flat stones automatically. Cabochons can also be cut this way with the aid of a model. (For coloring of agates, see page 152. For history of agate polishing, see page 154.)

Cutting and Polishing Colored Stones

In the manufacturing process, the term "colored stones" refers to all gems with the exception of diamonds. (In Germany, they exclude agates as well.) The cutting of the colored stones is called lapidary work; the cutter is known as a lapidary. Most lapidaries specialize in a certain gem or stone group, so that consideration can be best given to the characteristics of the stone, such as depth of color, pleochroism, or hardness.

Circular saws with edges containing diamond powder instead of teeth are used first to cut the stone roughly to the required size. The final shape is given to the rough stone on a vertical, roughly grained carborundum wheel, cooled with water. In modern factories, special environmentally friendly cutting fluids serve as cooling fluid; in earlier times, soapy water, oil, or petroleum was also used. Opaque stones, or those with inclusions, are cut on grooved carborundum wheels.

Cabochon Cutting

Opaque gemstones and those with value-decreasing inclusions receive a rounded cut en cabochon. On grooved carborundum wheels, the stones are given their shape with dome-like arched top and an even or slightly convex bottom (see the illustration on page 81) with the help of putty sticks or "dops" (see the illustration on page 72). The final result is called "en cabochon" according to the French expression.

Its use is versatile, as a ring stone, for brooches, colliers, or as pendant. It is commonly used as costume jewelry with fantastic and imaginative shapes.

Prepolishing of colored stone at the vertically running, slightly moistened wheel.

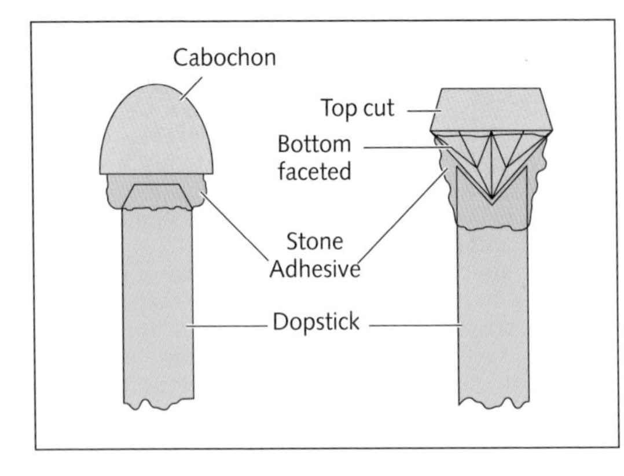

Principle for cementing for cabochon cutting and faceting. Here, great precision is demanded. If a stone is attached in a crooked manner or has a faulty cemented position, it will be inevitably cut poorly.

Faceting

Transparent stones, once roughly shaped, are faceted on horizontal grinding wheels. For this purpose, they are cemented with a special cement of shellac on so-called putty sticks.

These are 80-120 mm long round sticks, usually made from wood and only rarely made from metal. The diameter of the sticks depends on the size of the gemstone. It should be about ⅓ smaller than the diameter of the stone. These putty sticks are guided at an angle with the help of a board with numerous holes that is placed off the side of the grinding wheel (see below). Depending on the desired angle of the facets being cut, the opposite end of the holder is inserted in the appropriate hole.

In order to be able to cut various sides of a stone, the stone must repeatedly be cemented anew. Special fasteners, such as cements and adhesives, various solvents for releasing the stones, and other tools facilitate this process.

Instead of the board with holes, these are used for mass production and also

Faceting on a horizontally rotating polishing wheel with the aid of a board that has holes for the gem holder.

Faceting machine with faceting head in the white circle. Upper right, receptacle with cooling fluid.

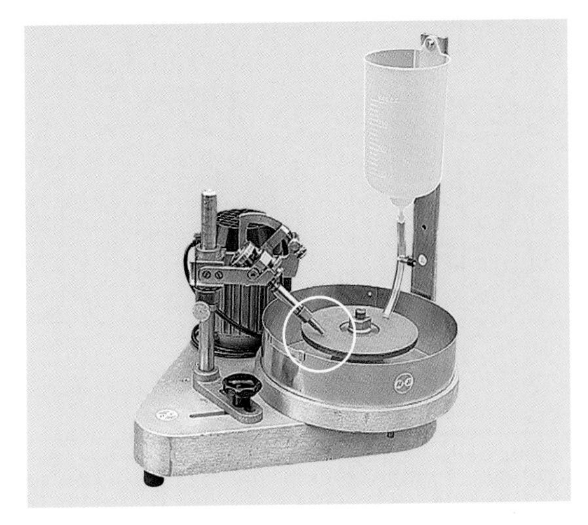

for the amateur cutting machines with a holder for a faceting head. With this machine, angles can be adjusted to precision for faceting (see illustration above).

The material for the polishing wheel (lead, bronze, copper, tin, etc.), the type of polishing powder (carborundum, aluminum oxide, silicon carbide, boron carbide, diamond), and the speed of the wheel all vary with the stone to be worked.

The last process is polishing on a horizontally rotating wheel, or wooden cylinder, or on leather straps to remove the final traces of scratches on the surface and to gain pure luster. Melting processes at the stone surface and the formation of a very thin layer (the so-called Beilby layer) enhance the fine polishing effect.

Gemstone cutting with vertically running wheel and fiddle-bow drive; Sri Lanka.

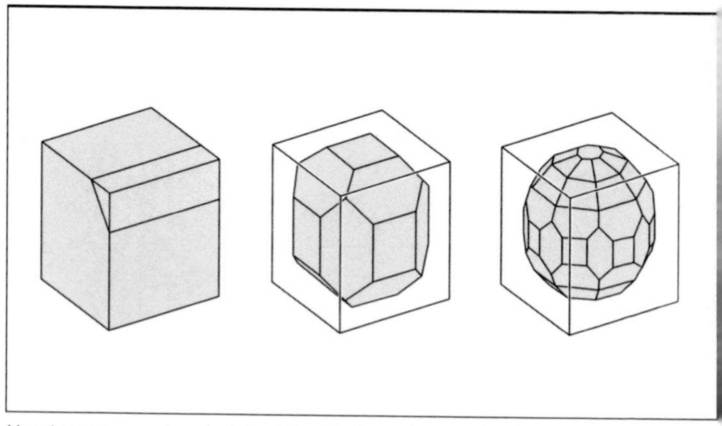

Mapping a raw gemstone for ball cutting. Starting with an equal-sided cube, a rounded, multi-edged body is created through continual beveling. (According to H. Bieler)

Ball Cutting

Equal-sided cubes are the raw pieces used as the basic form for ball cutting (see ill. above). The edges are cut over and over again and are later polished until a rounded, yet multi-edged body results. A cutting machine, supported by corresponding polishing agents, brings the gem into its final ball-like shape (see ill. below). Larger stone balls can only be worked individually by hand. They have to be in constant rotating motion with both rough and fine polishing. There are automatic cutting machines, so-called ball mills, for smaller balls up to about 10 mm in diameter. Interestingly, these balls are called "pearls" on the market, but they have, apart from their small size, nothing in common with organic pearls (page 256).

Finished ball cuts. Upper row from left: Snow Flake Obsidian, Labradorite, Blue Colored Agate, Black Colored Agate.
Bottom row from left: Jasper, Smoky Quartz with Rutile, Rose Quartz.

System for cylinder cutting, left with rotating steel barrels (rotating barrels) around a horizontal axis, right with vibratory tumblers and vertical axis.

Cylinder Cutting

A modern way of gemstone processing is by tumbling. With the help of rotating barrels (ill. top left) or vibratory containers (ill. top right), irregularly shaped, rounded-off stones (ill. bottom) are produced that are perfectly suited for costume jewelry. In the trade, these are known as baroque stones. With the horizontally hung tumblers made from wood, stone, or steel, the stones polish each other through tumbling movements, which is further supported by polishing substances of various grits. With machines running nonstop, this cutting procedure can last between two and five weeks. The cutting goes ⅔ faster with vibratory tumblers. In this case, however, only smaller stones can be processed. Optimal end products can be achieved when the pieces being polished in the tumblers consist of raw pieces of the best quality of about the same size and the same hardness.

Baroque stones cut and polished, after being processed in rotating barrels. Variety of pectolite, larimar.

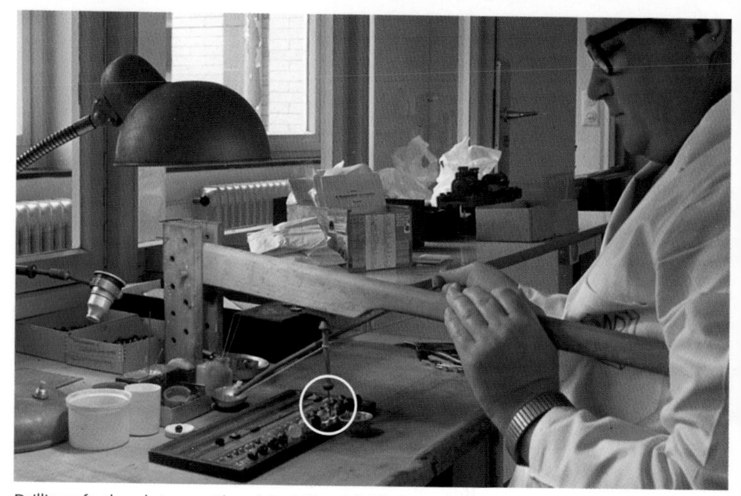

Drilling of colored stone with pointy drill and fiddle-bow drive.

Drilling Gemstones

Occasionally, gemstones have to be pierced in order to make jewelry. Formerly, this was done individually by hand with a fiddle-bow-driven drill. Today, mainly high-speed electrical drills are used. They have to be continuously provided with a coolant. Diamond powder and polishing paste aid the drilling procedure. One usually drills from both sides, so that the stone does not splinter at the original hole. In specialized companies, ultrasound drilling machines which work with vibration are used. The special advantage of these robotic devices is not only that the drilling time is shortened but that it is also possible to drill differently formed holes. (For more about the drilling of pearls, see page 263.)

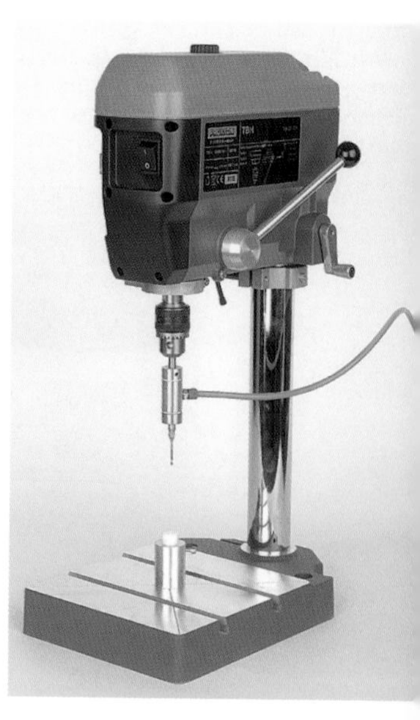

Electronically operated drilling machine. Cooling fluid is added through the hollow drill.

Cutting and Polishing Diamond

The following processes can be distinguished in the working of diamond: pre-examination, cleaving or sawing, bruting, cutting, and polishing.

Pre-Examination Before the actual working of a diamond is started, the cutter first must study the "inner life" of a rough piece; he must get to know the structure of the crystal and judge the inclusions. That is done with a magnifying glass (loupe), and it can take several weeks in the case of more valuable pieces. Diamonds with a mat surface first must be pre-cut for this purpose; they receive, so to speak, a window through which one looks into the diamond. Once it is decided whether a stone, which is quite large, will remain at that size or whether it is better to divide it into several fragments, the intended cleaving or sawing direction is marked on the crystal with ink.

At one time larger diamonds were divided by cleaving. To do this, an expert prepares the diamond, then carefully positions a metal blade and strikes it with a mallet. The cleavage surfaces are always octahedral planes. The Cullinan diamond, the largest diamond ever found—as large as a fist—was cleaved in 1908 in Amsterdam by the firm of Asscher, in the first stage into three parts, then again, so that finally nine large and 96 smaller brilliants could be cut from it.

Clearing, Sawing and Lasering Although the technique of cleaving is common, it often happened that a rough stone was broken because inner tensions and hidden cracks had not been recognized. Because of this, sawing diamonds has gradually become common practice since the turn of the century. A special

Cleaving of diamond with wedge and with a blow by hand.

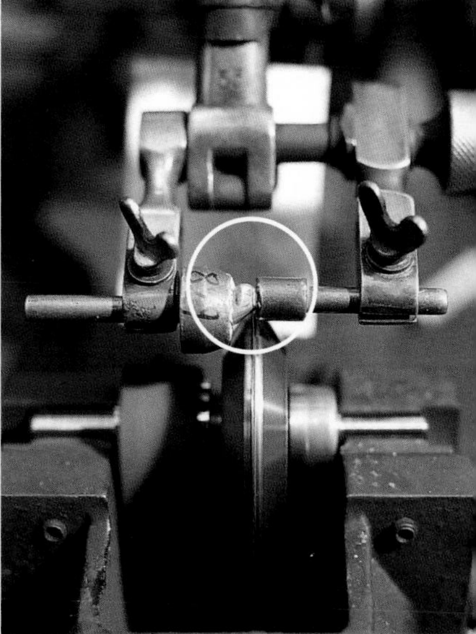

Sawing of diamond on a circular saw covered with diamond powder.

Two diamonds receive their raw shape for the brilliant cut by rubbing them against each other.

Faceting of diamond on horizontally rotating cutting wheel with the help of a dop.

advantage obtained from sawing diamonds is that a higher yield can be obtained from the rough stone. Well-formed octahedrons are sawn, for instance, through the central plane, or just above it, thus yielding a more favorable rough form for cutting brilliant. The sawing plane becomes the future table of the diamond-brilliant.

The disk of a diamond saw (50–70 mm diameter) is made of copper, bronze, or another alloy, is paper-thin, and is impregnated with diamond dust, which has to be continually renewed. It rotates at about 4500 to 6000 revolutions per minute (r.p.m.), but can reach up to 10,000. The rough stone is held in a vise-like arrangement. The sawing process is time-consuming: for a one-carat stone of ¼– to ¾–in (6–7 mm) diameter, about five to eight hours are needed. Lately, diamonds are only divided with laser beams. The advantage of this procedure is that one does not have to adhere to a structural direction of the crystal, unlike when cleaving or sawing.

Girdling The next process is girdling, which gives the diamond its rough shape with girdle, crown, and pavillion. Two diamonds are used in bruting: one fixed in a small lathe, the other cemented onto a dopstick and held in the operator's hand. These are ground against each other so that the stone edges become rounded, corresponding to the double cone of the brilliant shape. Those diamonds which are not supposed to get the brilliant cut are ground on a disk that is covered with diamond dust. Polishing or sawing of diamond is only possible with other diamond, because diamond varies in hardness on different crystal faces and in different directions (see the sketch on page 21).

It is necessary to examine the diamond thoroughly in order to utilize the differences in hardness when cutting it. According to statistical probability, diamond powder contains grains lying in all directions, so that there are directions at all cutting angles. The powder can thus be used to cut the less hard directions of the diamond crystal, since those will be scratched/cut much more than the harder directions.

Grinding The technique of polishing requires much experience. The facets of the diamond, held in a dop, are polished on a horizontally running wheel, about 12 in (30 cm) in size, primed with diamond powder and oil, and rotating at about 2000 to 3000 r.p.m.

Polishing Traditionally, and still in much of the industry, the corners of all facets and placement of angles are judged by eye by experienced cutters without elaborate instruments, only using a loupe to gauge the cut. Since the 1970s automatic polishing machines have been increasingly used for stones up to about a half carat. Larger rough and fine stones continue to be cut by the old methods.

The loss of material during cutting of diamonds is very large, as high as 50 to 60 percent. In the case of the Cullinan, it was about 65 percent. The cut-off diamond powder is always captured and used for other cutting procedures. The brilliant is finally polished on the same scaife, but on another ring with finer diamond powder. (For the history and development of diamond cutting, see page 96.)

Types and Forms of Cut

There is no general rule that can be applied to the various cuts of gemstones. Therefore, one finds in the literature conflicting attempts to classify them. Nevertheless, a distinction according to types of cut and forms of cut is quite possible.

Types of Cut

Based on the optical impression of the cut gemstones, three main groups, or types of cut, can be named: the faceted cut, the plain cut, and the mixed cut. (Compare with the drawings on page 81.)

Each diamond polisher usually operates several dops on the polishing wheel at one time.

Faceted Cut The faceted cut receives its character from a variety of small planes, the facets. It is applied mostly for transparent gems. Most faceted cuts can be divided into three basic types, brilliant cuts with mostly rhomboid (lozenge) and triangular facets in a radial pattern, step cuts with trapezoid or rectangular facets in concentric rows, and mixed facet cuts combining both brilliant-and step-type facets. (For more about faceted cuts, see pages 72, 78, and 96.)

Plain Cut The plain cut can be executed plain (level tablet), or arched as a cabochon or as a sphere. No facets interrupt the even stone surface. The plain cut serves mainly for the cutting of agate and other opaque gemstones. Often a plain cut suffices where the otherwise insignificant surface of an opaque stone should be made visible, vivified, or embellished.

A cabochon, however, can already be of great appeal through its shape alone. Choosing the cut from among a number of possible basic shapes is almost arbitrary, even though raw material, color, and the pattern of the gemstone will influence the respective form. Balls are made from both transparent and opaque stones. For more about the practice of employing the plain cut, see pages 70, 71, and 74.

Mixed Cut The mixed cut is a combination of the faceted and plain cut. On the upper (or lower) part, the gemstone carries facets; on the other part, it is smooth, roundish, or plain. Occasionally there are cutting forms followed in which the cut is "mixed" on the same side of the stone.

Forms of Cut

An abundance of forms is derived from the basic types of the gemstone cuts. They can be round, oval, cone-shaped, square (carré), rectangular (baguette), triangular, and multi-cornered.

In addition, there are many forms that imitate familiar shapes, such as olive, pear (drop or pendeloque), little ships (navette or marquise), heart, trapeze, or barrel. A number of purely fantasy cuts can hardly be overlooked, and new forms are increasingly created by designers. (A selection of the well-known forms of cuts is pictured on the front inside cover.)

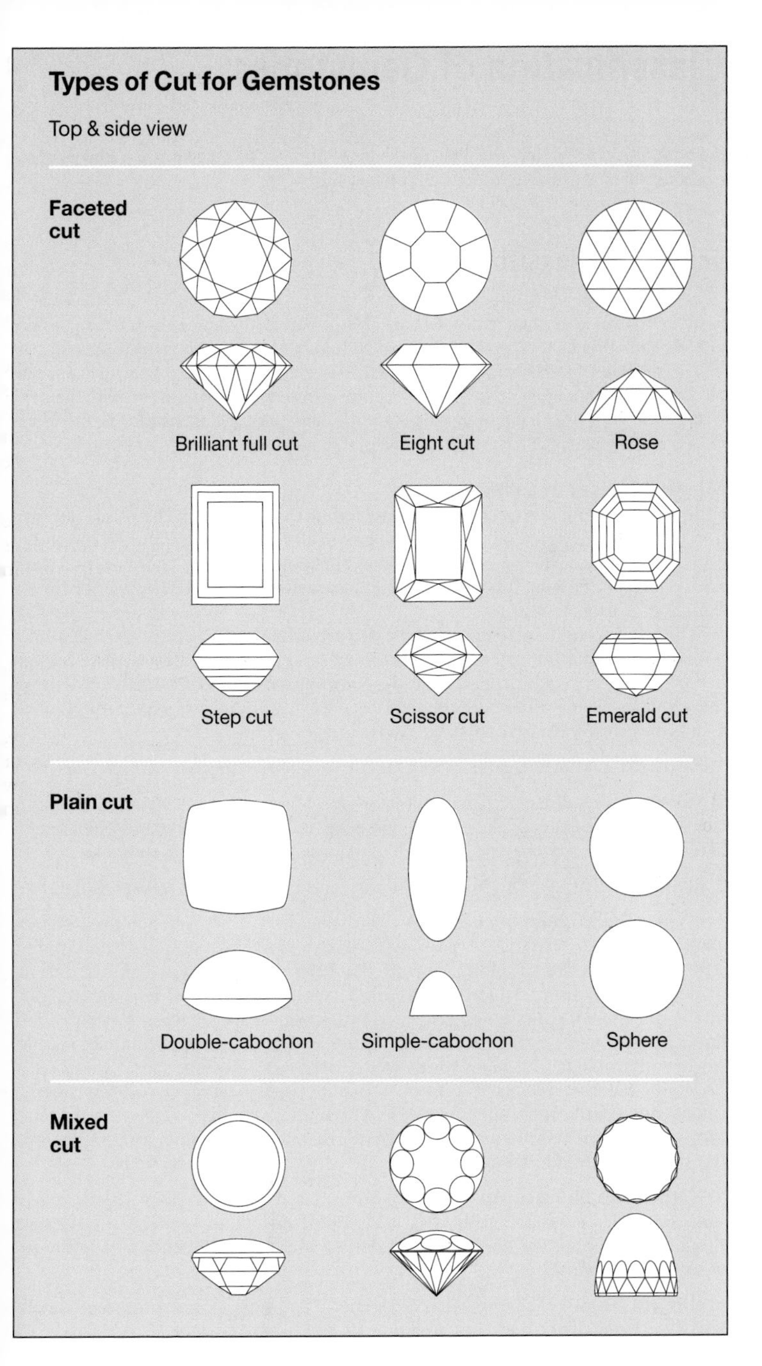
Types of Cut for Gemstones
Top & side view
Faceted cut
Brilliant full cut
Eight cut
Rose
Step cut
Scissor cut
Emerald cut
Plain cut
Double-cabochon
Simple-cabochon
Sphere
Mixed cut

Classification of Gemstones

In order to have a better overview of the large number of gemstones, it makes sense to group them according to their characteristics.

Scientific Classification

In the science of gemology, gemstones are classified according to their structure, origin, history, and other factors, or what is called petrology. Some organic gemstones and artificial gemstone materials are specially classified outside of the typical structuring principle. Difficulties can often arise when allocating gemstones to a specific system, since the term *gemstone* is not an unambiguous scientific definition. A classification is, therefore, not necessarily perforce. Overlaps and subjective assignments are thus inevitable.

Mineral Classes of Gemstones

Until recently, minerals were classified according to their crystal chemical composition into 9 classes based on the Mineralogical Tables by Huge Strunz. At the end of 2001, Strunz published an improved classification system in collaboration with E.H. Nickel. Following this system, minerals are divided into 10 classes. We follow this classification system.

While the individual classes are subdivided according to their crystal chemical compounds in the scientific system, the gemstones belonging to each class are listed alphabetically. As a concession to gemstone lovers without specialized scientific knowledge, the following list mentions both the names of the gemstone species (e.g., corundum) as well as the gemstone types of these groups (e.g., ruby).

1. **Elements** Diamond, gold, silver, sulfur

2. **Sulfides and sulfosalts** Algodonite, bornite, breithauptite, chalcosine, cobaltite, covelline, greenockite, chalcopyrite, marcasite, millerite, melonite, nickeline, pentlandite, proustite, pyrargyrite, pyrite, realgar, sphalerite, wurtzite, cinnabar

3. **Halides** Boleite, chiolite, creedite, fluorite, cryolite, prosopite, sellaite, villiaumite

4. **Oxides, hydroxides** Agate, aschynite, alexandrite, amethyst, amethyst quartz, ametrine, anatase, aventurine, petrified wood, mountain crystal, bismutotantalite, bixbyite, blue quartz, brookite, brucite, chalcedone, chromite, chrysoberyl, chrysoprase, citrine, cuprite, davidite, diaspore, euxenite, hawk's-eye, fergusonite, gahnite, gahnospinel, galaxite, goethite, hematite, heliotrope, hercynite, bloodstone, hubnerite, ilmenite, jasper, carneol, cassiterite, cat's-eye quartz, corundum, magnesiochromite, magnetite, manganotantalite, microlite, morion, moss agate, opal, periclase, picotite, pleonaste, prase, prasiolite, psilomelane, pyrolusite, quartz, smoky quartz, rose quartz, ruby, rutile, samarskite sapphire, sard, senarmontite, simpsonite, spinel, stibiotantalite, taaffeite, tantalite, thorianite, tiger's-eye, wolframit, yttrotantalite, zincite

5. **Carbonates, nitrates** Ankerite, aragonite, azurite, barytocalcite, calcite, cerussite, dolomite, gaspeite, gaylussite, magnesite, malachite, parisite, phosgenite, rhodochrosite, shomiokite, shortite, siderite, smithsonite, stichtite, strontianite, weloganite, witherite

6. **Borates** Boracite, chambersite, colemanite, hambergite, inderite, jeremejevite, kurnakovite, painite, rhodozite, sinhalite, ulexite

7. **Sulfates, chromates, molybdates, wolframates** Anglesite, anhydritspar, barite, celestite, crocoite, ettringite, gypsum, langbeinite, linarite, powellite, scheelite, stolzite, sturmanite, wulfenite

8. **Phosphates, arsenates, vanadates** Adamite, amblygonite, apatite, augelite, bayldonite, beryllonite, brazilianite, childrenite, chlorapatite, cerulite, descloizite, dickinsonite, durangite, eosphorite, euchroite, fluorapatite, herderite, heterosite, hureaulite, hurlbutite, hydroxylherderite, cacoxenite, lazulite, legrandite, libethenite, lithiophilite, ludlamite, manganapatite, mimetite, monazite montebrasite, natromontebrasite, phosphophyllite, purpurite, schlossmacherite, scorzalite, skorodite, triphylite, turquoise, vayrynenite, vanadinite, variscite, vivianite, wardite, wavellite, wilkeite

9. **Silicates** Achroite, aegirine, aegirine-augite, actinolite, albite, allanite, almandite, amazonite, analcim, andalusite, andesine, andrachite, anortite, anorthoclase, anthophyllite, apophyllite, aquamarine, axinite, benitoite, beryl, bikitaite, bloodstone, buergerite, bustamite, bytownite, canasite, cancrinite, carletonite, catapleiite, chabasite, charoite, chloromelanite, chondrodite, chromdravite, chrysocol, clinochrysotile, clinoenstatite, clinohumite, clinozoisite, cordierite, danburite, datolite, demantoid, diopside, dioptase, dravite, dumortiertite, precious beryl, ekanite, elbaite, emerald, enstatite, epidote, eudialite, euclase, feldspar, ferroaxinite, ferrosilite, forsterite, friedelite, fuchsite, gadolinite, gold beryl, goshenite, garnet, grandidierite, grossularite, hackmanite, hancockite, hauyne, hedenbergite, helidore, helvine, hemimorphite, hessonite, hiddenite, hodgkinsonite, holtite, hornblende, humite, hyalophane, hyacinth, hypersthene, indicolite, jadeite, jeffersonite, kämmererite, katoite, kornerupine, kunzite, kyanite, labradorite, lapis lazuli, lasurite, lawnsonite, lepidolite, leucite, leuco garnet, leucophanite, liddicoatite, lizardite, magnesio-axinite, manganaxinite, marialite, meionite, melanite, melinophane, mesolite, microcline, milarite, moonstone, mordenite, morganite, muskovite, nambulite, natrolite, nepheline, nephrite, neptunite, norbergite, oligoclase, orthoclase, palygorskite, papagoite, pargasite, pectolite, peridot, peristerite, petalite, pezzottaite, phenakite, piemontite, pollucite, poudretteite, povondraite, prehnite, pumpellyite, pyrope, pyrophyllite, pyroxmangite, rhodolite, rhodonite, richterite, rinkite, rubellite, salite, sapphirine, sarcolite, schefferite, sanidine, schorl, schorlomite, sekaninaite, sepiolite, serandite, serendibit, serpentine, shattuckite, siberite, sillimanite, scapolite, scolecite, smaragdite, sodalite, sogdianite, spessartite, spurrite, staurolite, sugilite, sunstone, talc, tanzanite, tawmawite, tephroite, thaumasite, thomsonite, thorite, thulite, tinzenite, titanite, topaz, topazolite, tourmaline, tremolite, tsavorite, tsilaisite, tugtupite, ussingite, uvarovite, uvite, verdelite, vesuvianite, vlasovite, willemite, wollastonite, xonolite, yugavaralite, zektzerite, zircon, zoisite, zunyite

10. **Organic Substances** Ambroid, amber, whewellite

Rock Families of Gemstones

Most gemstones are minerals. Only very few gems belong to the rock family.

Igneous Graphic granite, obsidian (rock glass), orbicular diorite (diorite), syenite, unakite (granite)

Sedimentary Alabaster (gypsum), cannel coal (pit coal), gagate (brown coal), goodletite (rock debris), landscape marble (limestone), onyx marble (limestone), pietersite (breccia), sea foam (concretion), schungite (pit coal), sepiolite (breccia), tufa (limestone)

Metamorphic Agalmatolite (dike rock), anyolite (zoisite amphibolite), connemara (serpentinite), corundum-fuchsite, eclogite, gneiss, nuummite (anthophyllite rock), ricolite (serpentinite), verdite (serpentinite)

Meteorites and Tektites Moldavite (tektite)

Commercial Classification

For practical reasons, the descriptions of gemstones are arranged into five main groups according to their significance in the trade.

Best-Known Gemstones

These include all the stones that have been traditionally traded and otherwise generally known. These stones are set and worn as pieces of jewelry or worked into an objet d'art.

The order within this main group is determined by Mohs' hardness, which is particularly significant for the different trade value of these gemstones.

The representatives of this group are described on pp. 86-193.

Lesser-Known Gemstones

Most of the representatives of this second group had been rare in the trade until a few years ago, and are becoming more popular today. Formerly appreciated mostly by collectors, they are today made into jewelry and worn just like any of the other better-known gemstones.

The sequence within this main group is also determined by Mohs' hardness, because it gives an indication of their practical suitability as gemstones.

The representatives of this group are described on pp. 194-219.

Gemstones for Collectors

There are many minerals which are faceted by collectors, amateurs, and professional lapidaries; however, they have no practical use in jewelry since they are too soft, too brittle, otherwise endangered, or even too rare. Gem lovers, though, collect such faceted or en cabochon stones as rarities or as small pieces of art.

A selection of these gemstones is presented on pp. 220-243.

Rocks as Gemstones

This group (pp. 244-249) is seen as the fringe zone of gems. As costume jewelry, rocks (mineral aggregates in large combinations) are becoming increasingly important. Striking structures and beautiful colors of such rocks can be seen in the gemstone trade.

Organic Gemstones

This group includes materials which, even though they are of organic origin, have preserved or acquired a certain stone character. They are an important part of the trade, especially with respect to amber or pearl.

Presentation of organic gemstones can be found on pp. 250-265.

Value of Gemstones

There are no firm prices regarding gemstones. They are subject to fluctuations just as with other products. In the end, supply and demand regulate the gemstone market. In addition, economic purchasing power and fashion trends can heavily influence the demand for certain gemstones. There are great differences in pricing and purchase between diamonds and the other so-called colored gemstones.

Diamond Prices With diamonds, the consumer can compare prices easily. Value-determining elements, such as color, purity, and cut, are internationally defined and can be controlled. Of course, there are profit margins and pursuits of profit. One can, however, avoid outright fraud by dealing with diamond wholesalers and established jewelers. Here a code of honor prevails like nowhere else in world trade. On the diamond stock exchanges, the wholesale diamond markets, business has been sealed from even the earliest times with a firm handshake and a person's word. See the report about the diamond trade on p. 90.

Colored Stone Prices With colored stones, price often fluctuates greatly, even in the serious trade, since gem extraction sometimes depends on chance happenings that cannot be predicted. As always, beauty, rarity, and gemstone size determine prices. However, in contrast with the diamond trade, there are no controlling organizations, no possible exertion of influence on honest trade. Furthermore, many small businesses are active with extraction and production. They often seek direct contact with the consumer, particularly tourists who are in the buying mood, and deceitfully exploit their ignorance.

Price Lists Numerous institutions such as diamond stock exchanges, gemological facilities, and professional journals publish current market prices and long-term price tendencies. This pricing information is intended for trade and traders and not for the consumer who incurs substantial surcharges set by the traders.

Prices of stones are usually expressed "per carat" (see p. 30). The carat price often progressively increases with the size of the stone. For example, a one-carat piece might cost $750, but a two-carat piece is not necessarily worth twice the amount, or $1500, but perhaps $3000 or more.

Less valuable stones such as agate are sold by grams and kilograms, not carats.

Description of Gemstones

When describing gemstones, it is important to give as much information as possible.

The significance of abbreviations, such as "lat." (Latin), "gr." (Greek), and shortened forms, such as "crystals" (instead of crystal system or crystal form), "chemism" (instead of chemical composition), and the waiving of additions of measurements are clearly recognizable in the respective context. The shading applied to sketches of crystals is purely schematic.

Best-Known Gemstones

This group includes all those stones that have been traditionally represented in the trade or which are otherwise generally known. (Compare also to page 83.)

Diamond

Color: Colorless, yellow, brown, rarely green, blue, reddish, orange black
Color of streak: White
Mohs' hardness: 10
Density: 3.50–3.53
Cleavage: Perfect
Fracture: Conchoidal to splintery
Crystals: (Cubic), mainly octahedrons, also cubes, rhombic dodecahedrons, twins, plates
Chemical composition: C, crystallized carbon

Transparency: Transparent to opaque
Refractive index: 2.417–2.419
Double refraction: None
Dispersion: 0.044(0.025)
Pleochroism: None
Absorption: Colorless and yellow D.: 478 465, 451, 435, 423, 415, 401, 390
Brown and greenish D.: (537), 504 (498)
Fluorescence: Very variable;
Colorless and yellow D.: mostly blue
Brown and greenish D.: often green

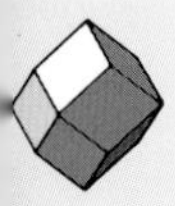

The name diamond refers to its hardness (Greek—*adamas,* the unconquerable). There is nothing comparable to it in hardness. Its cutting resistance is 140 times greater than that of ruby and sapphire, the gemstones next in hardness after diamond. However, the hardness of a diamond is different in the individual crystal directions. This allows one to cut diamond with diamond and/or with diamond powder. Because of the perfect cleavage, care must be taken not to accidentally bang against an edge of a diamond, and also when setting it. Its very strong luster sometimes enables the experienced eye to differentiate between a diamond and its imitations. (See p. 266.)

Diamond is generally insensitive to chemical reactions. High temperatures, on the other hand, can induce etchings on the facets. Therefore, special care must be taken during soldering!

In the last fifty years, it has been recognized that there are various types of diamonds with different characteristics. Science differentiates between type la, Ib, Ila, and IIb. This is of little importance to the trade, but does assist the cutter. Due to the optical effects, the high hardness, and its rarity, the diamond is considered the king of gemstones. It has been used for adornment since ancient times.

See the following for further details:

Diamond cutting, page 77
Diamond deposits, page 88
Diamond trade, page 90
Quality evaluation, page 92
Famous diamonds, page 94
Development of the brilliant cut, page 96
New diamond imitations, page 270
Testing for genuineness, page 274

1 Diamond, brilliant 0.49ct
2 Diamond, marquise 0.68ct
3 Diamond, brilliant 2.22ct
4 Diamond, 3 baguettes, total 0.59ct
5 Diamond, brilliant 0.21ct
6 Diamond, 2 old cut 0.97ct
7 Diamond, 2 brilliants 0.57ct
8 Diamond, brilliant 2.17ct
9 Diamond, 3 roses, total 0.67ct
10 Diamond, 2 marquises, total 0.69ct
11 Diamond, twice sawn, total 1.43ct
12 Diamond, 10 brilliants
13 Diamond, 9 brilliants
14 Diamond, white rough, total 6.37ct
15 Diamond, colored rough, total 10.22ct
16 Diamond aggregate, 8.26ct
17 Diamond crystal on kimberlite
18 Diamond crystal, 8.14ct

Figs. 1–15 twice the original size; Figs. 16–18 reduced by one-third in size.

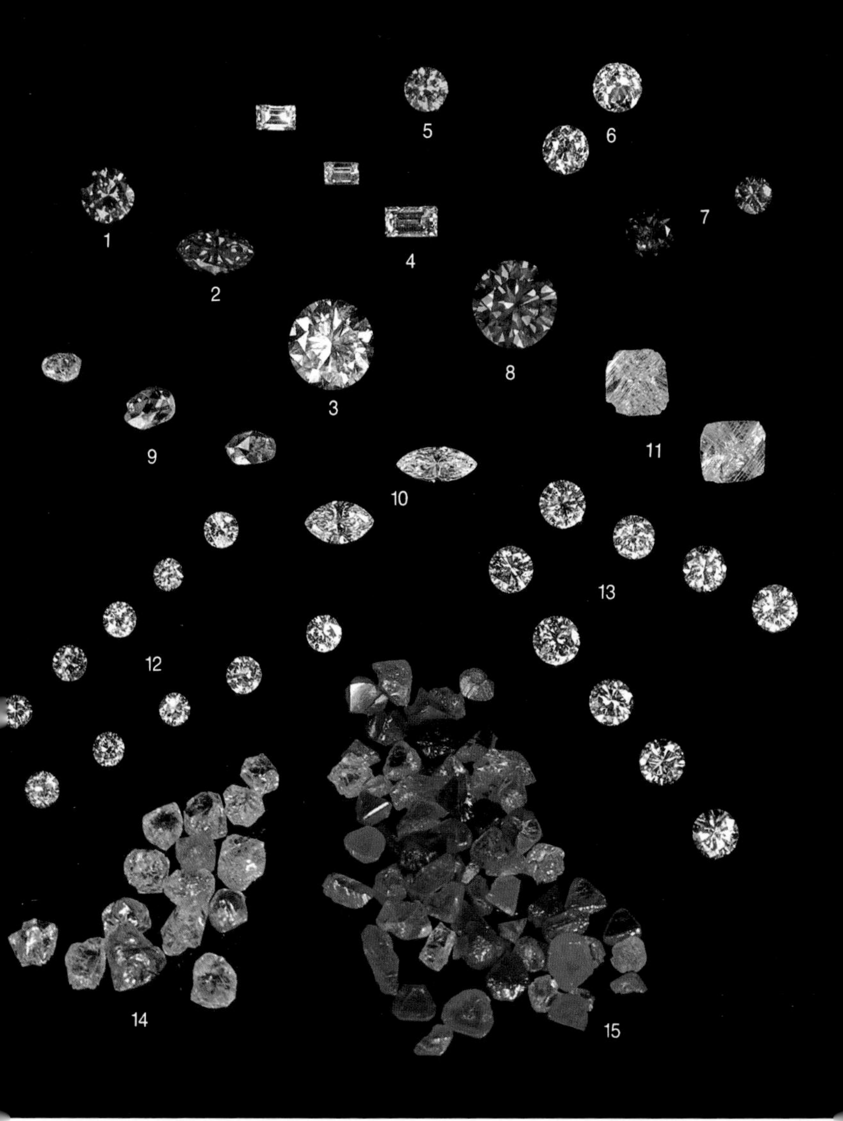
1
2
3
4
5
6
7
8
9
10
11
12
13
14
15

16
17
18

Diamond Occurrences

Diamonds are found in primary and secondary deposits. Until 1871 diamonds were only washed out of diamondiferous placers. By chance a primary deposit was discovered in South Africa: volcanic pipes filled with diamond-bearing rock, kimberlite (more rarely lamproite, a greenish-gray speckled mafic rock).

Diamond is formed at great depth (as much as 150–300 km) at very high temperatures and high pressures. By a particular kind of volcanic eruption, they came to the surface of the earth or close to it with rising magma. With gradual erosion, the original volcanic cone was removed, exposing the kimberlite pipes (see illustration below).

Up to the 18th century, some diamonds came from Borneo, but most from India. In 1725 the first diamonds were found in the South American continent in Minas Gerais in Brazil. In 1843 a brown-black carbonado was discovered in Bahia; this is a microcrystalline aggregate that is very tough and therefore in demand for industry. Today, diamonds can also be faceted (see page 282).

Brazil was superseded by South Africa, which until about 1970 had an unsurpassed position as to production and trade. The first diamond find in 1866 was in the region near the source of the Orange River. At first only the alluvial deposits were worked. In the meantime many kimberlite pipes (see picture, page 61) have been discovered. These contain on top a claylike rock, which is called yellow ground because of its color. It is the weathered product of the blue ground, kimberlite, below. The kimberlite is an igneous rock, rich in olivine in a porphyritic texture.

The most famous and noteworthy pipe of South Africa, the Kimberley mine, was worked from 1871 to 1908 without any machinery. This produced what was once the largest man-made hole, the so-called Big Hole: the surface hole has a diameter of 1510 ft (460 m), and the depth is 3510 ft (1070 m). Today it is half-filled with water. Altogether 14.5 million carats (2900 kg or about three tons) of diamonds were found there. After having been worked for a short time underground, the mine was discontinued in 1914, because it proved not economically feasible (see photo on opposite page). Other South African pipe mines changed from open pit to underground mining early on. A shaft is sunk next to the kimberlite pipe, and then the diamond-bearing rock is mined along transverse levels.

There are important alluvial deposits in Namibia at the western edge of the desert Namib. In 1908 the first diamonds were found in Lüderitz. With extensive use of machinery, a 100-ft (30-m)-thick overburden was removed, exposing the diamond-bearing sands below (see the picture on page 66). The Namibian diamonds were eroded from the pipe sources in South Africa. They were carried along the rivers to the coast, deposited in the surf zone of the Atlantic Ocean, and partially covered again by sand dunes. Specialized boats are also used here to prospect for

Development and mining of a kimberlite pipe. 1. Volcanic eruption, building a cone; 2. Erosion of the volcanic cone; 3. Levelling of the cone (frequently covered with a sedimentary layer); 4. Exploitation of the pipe matrix above ground and along tunnels.

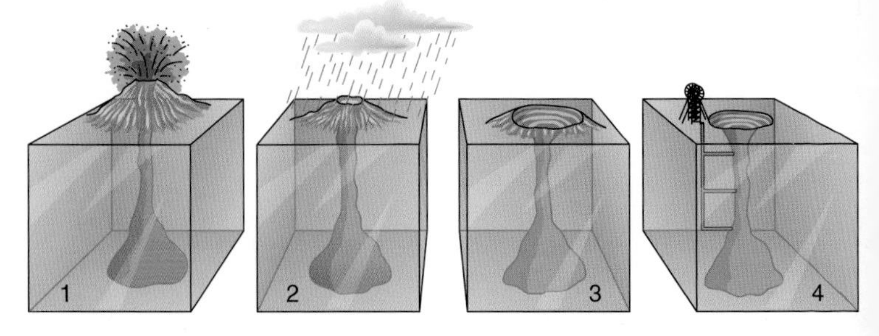

diamonds beyond the surf zone in the ocean. The share of diamonds with gemstone quality is unusually high in Namibia, being around 90 to 95 percent.

Meanwhile, the circle of diamond suppliers has become quite large. Today, more than twenty countries belong to it. Important diamond producers include Botswana, South Africa, Angola, the Democratic Republic of the Congo, Namibia, Sierra Leone, Central African Republic, Guinea, Russia, China, the United States, Brazil, Venezuela, Canada, and the Argyle and Merlin diamond mines in Australia.

Years ago Russia became one of the big diamond producers. The first diamonds had been found in the Ural Mountains in 1829, but these deposits proved to be of no commercial value. In 1949 a new era began when important alluvial deposits were found in Siberia (Yakutsk), and a few years later also diamond-bearing pipes. In 2005, Russia was first in the list of world diamond suppliers.

China now belongs to the important diamond traders as well. Alluvial deposits were first exploited in 1940, but the big breakthrough came only when several kimberlite pipes were discovered. Diamond production in Australia has also reached spectacular levels. Since 1986 the Argyle mine in the northwest of the country has been in full production. In 1998 alone 41 million carats were produced, putting Australia in first place with one-third of world production. However, the value proportion is only one-tenth of world value, because only five percent of Australia's diamond production is of gemstone quality, 45 percent is of inferior jewelry quality, and fully 50 percent is good enough only for industrial use. In February, 1999, the Merlin Mine in the Northern Territory of Australia began its production. The quality of diamonds is better than those from the Argyle mine.

Based on what is known today, many experts believe there are great diamond-bearing pipes to be found in the near future in Canada. The production in the Ekati mine in the Northwest Territories started in October 1998 amidst great expectations.

Diamond Production

The winning of diamonds from the host rock is done today with a great deal of machinery. The least expensive method is to loosen the diamonds from placers and weathered kimberlite, the yellow ground. Because of the loose texture, the diamonds, due to their high density, can be separated and accumulated in pans. The blue ground, the pipe matrix, must first be cut into small pieces with stone crushers, before it can be washed in the same way. The separation of the diamonds from the concentrate, originally exclusively done by hand, is done today almost fully automatically. The tendency of diamonds to adhere to grease is also utilized. Diamond is, in contrast to other materials, hard to moisten, i.e., even in water it actually does not get wet. Therefore, from the concentrate, which glides over grease vibrating tables, the diamonds get stuck to the grease, while other minerals continue to glide along.

The former Kimberley mine is called the Big Hole. Diameter 1510 ft (460 m), depth 3510 ft (1070 m). Today it is half-filled with water. Overall, 14.5 million carats (2900 kg or about three tons) of diamonds were found here.

Other diamond sorting methods include electrostatic separation, with optical selection by photo-cells, or by using X rays or ultraviolet rays, utilizing the fluorescence.

Despite the usage of such modern techniques, the final selection must be done by hand, because other minerals end up in the final concentrate besides diamonds.

The diamond content of the host rocks varies, depending on the occurrences. It is usually higher in placers than in kimberlite pipes. In some placer deposits, a yield of ⅕ carat per cubic meter of rock is still economic. However, diamond content is misleading. Australia's Argyle mine (the world's largest diamond mine) is hosted by a lamproite and provides 7 carats per rock. Since diamonds are individually valued, the degree of size and quality of stones can dramatically affect the value such that yields defined in number of carats per cubic meter or penton are not very informative.

Diamond Trade

At times, up to 80 percent of the world diamond production and rough supply is managed, i.e., controlled, by an enterprise, which is known by the short name "De Beers." Numerous splinter associations of a densely intertwined mammoth enterprise, trade associations as well as institutions and companies outside of the diamond trade, all hide behind the term "De Beers," for instance, "De Beers Consolidated Alines Limited," "DeBeers Centenary AG," "Diamond Trading Company" [until 1999, called the "Central Selling Organization" (CSO)].

Derogatorily, one sometimes speaks of a diamond syndicate. It is factually correct, however, to call De Beers a monopoly, if one wants to indicate the concentration of power within the diamond trade.

All sorted and valued diamonds go to London to the DTC. Here, they are put together into lots (parcels) for the purpose of selling and are offered for a fixed price. Partial purchase does not exist, only by lot.

Only a few diamond cutters and diamond traders, the "sightholders," are entitled to buy. Currently there are 120 such individuals who are accredited at the DTC. They purchase the merchandise, which was put together by the DTC, at ten annual selling events, the "sights," in London, Lucerne, and Johannesburg, according to the registered wishes of the customers. The DTC delivers only rough stones.

The resale and the dividing of the lots is done by the "direct buyers," preferably on diamond exchanges (sometimes called diamond clubs or bourses). Such exchanges exist in Antwerp, Amsterdam, New York, and Ramat Gan in Israel, in addition to those in Johannesburg, London, Milan, Paris, and Vienna, and, since 1974, also in Idar-Oberstein/Rhineland-Palatinate. By far the most important are the four diamond exchanges in Antwerp, the largest diamond-trading location in the world. Diamond exchanges are not speculation exchanges in the usual sense of the word, but rather are large diamond markets.

De Beers also controls and influences through its selling system the prices of the brokers. The goal is to keep the value of the diamond high and to call a halt to any dubious maneuvers in the diamond trade.

Such collaboration among the leading production countries under the organized leadership of De Beers can be said to be in the interest of everyone, not just the diamond producers, but even, perhaps, the customer, the buyer of the diamond jewelry, because through this cartel prices are kept high and extreme variations in price are avoided.

Certain problems for the controlled market situation arise because some countries (currently Angola, Congo, and Russia) do not have their diamond production or sale restricted through the cartel. Thus any diamonds which are freely offered at whatever rate a more or less free market will support can influence the prices

Weekly production of a diamond mine in Namibia, about 30,000 carats.

throughout the world beyond the control of the organization. So-called conflict or, blood, diamonds, which are produced in civil war areas, are excluded from the world trade following an international agreement. Nevertheless, diamonds have survived as a dependable investment through all political and economic ups and downs of recent decades. Through this, not only capital was secured, but also millions of jobs, directly or indirectly connected with the diamond trade, were preserved. After all, the market value of diamond production is over 90 percent of the entire gemstone trade.

Diamond Imitations The fact that the diamond can be confused, in appearance, with many gemstones can lead to fraud, although not in the legitimate retail trade. A colorless diamond looks similar to rock crystal (page 132), precious beryl (page 112), cerussite (page 216), sapphire (page 102), scheelite (page 212), sphalerite (page 216), topaz (page 118), and zircon (page 124). Also many yellowish stones can look like diamonds to the eye of the layman.

Apart from these, there are various synthetic stones (compare to page 270) that are used to imitate diamonds. Especially strontium titanate (fabulite), YAG, GGG (galliant), linobate, and cubic zirconia (CZ, djevalite) must be mentioned. A well-known diamond imitation made from glass is the so-called strass (page 266).

In 1970 the first gem-quality diamonds were synthesized, but these still do not compete with natural diamonds. At first, they served only scientific purposes since they are too small to be worn as jewelry. In 2003, however, synthetic diamonds up to 2.5 ct appeared on the market. For further details on this change, see page 274. In the trade one also finds natural diamonds, colored artificially by various irradiation treatments.

Diamond-doublets are made with: upper part-diamond; lower part-synthetic colorless sapphire, rock crystal, or glass. Other doublets have synthetic spinel as upper part, and strontium titanate underneath.

Valuation of Diamonds

All diamonds that De Beers brings on the market are classified beforehand into one of the 16,000 different standards of quality, according to form, quality, color, and size. Besides London, there are such sorting centers in Lucerne in Switzerland, Gaborone in Botswana, Windhoek in Namibia, and Kimberley in South Africa.

Formerly, 20 percent of all diamonds were considered suitable for jewelry, having been of "gemstone quality." The rest were sold to the industry as so-called industrial diamonds to be used for drilling crowns, milling machines, cutting wheels, etc. Since 1983, another 20 percent have been classified as "almost gemstone quality" and are cut mainly in India. The smallest diamonds and diamonds of lesser quality can be worked and offered to buyers other than the strictly controlled diamond market.

In the valuation of faceted diamonds, color, clarity, cut, and carat are taken into consideration. These four c's decide the value of a diamond.

Grading for Color Diamonds are found in all colors. Mostly they are yellowish. In the grading, these are evaluated together with the purely colorless diamonds. The rarer strong colors (green, red, blue, purple, and yellow), the so-called fancy colors, are valued individually and fetch collector's prices. Brown and black diamonds also occur.

Formerly, terms and definitions in grading for color were not uniform and often confusing, until an international agreement was reached to cover the so-called "yellow series." This was published in 1970 as RAL 560 A5E. Since then, various institutions have come up with improved guidelines, especially the Gemological Institute of America (GIA), the International Diamond Council (IDC), and the Confédération Internationale de la Bijouterie, Joaillerie, Orfevrerie des Diamants, Perles, et Pierres (CIBJO). Today, the IDC regulations, written in English, are accepted worldwide. In the United States, the GIA system is most commonly used.

The old grading terms, the "Old Terms," should not be used anymore. In fact, however, they are still in usage in the gemstone trade.

Experts use standard sample collections for consistent, comparison color grading.

Grading for Clarity In Germany only the inner perfection is understood as "clarity," while in the United States and in Scandinavia aspects of the quality of the outer finish are taken into consideration. Enclosed minerals, cleavages, and growth lines affect clarity; they are collectively called inclusions, but formerly were called "flaws"

Color Grading of Faceted Diamonds

CIBJO	IDC	GIA	Old Terms	RAL 560 A5E
very fine white+	exceptional+ white	D	River	blue-white
very fine white	exceptional white	E		
fine white+	rare white+	F	Top Wesselton	fine white
fine white	rare white	G		
white	white	H	Wesselton	white
		I	Top Crystal/	
slightly tinted white	slightly tinted white	J	Crystal	weakly tinted white
		K		
tinted white	tinted white	L	Top Cape	tinted white
		M		
tinted 1	tinted color 1	N	Cape	weakly yellow
		O		
tinted 2	tinted color 2	P	Cape	yellowish
		Q		
tinted 3	tinted color 3	R	Light Yellow	weakly yellow
tinted 4	tinted color 4	S-Z	Yellow	yellow

or "carbon spots." Polished diamonds without any inclusions under a 10 × loupe are considered "flawless." Inclusions visible with larger magnification are not taken into account for grading.

According to CIBJO, it is permitted to subdivide the clarity grades VVS, VS, and SI into two subgroups each, for stone sizes over 0.47 ct. Definitions assume a trained grader working under favorable conditions.

Grading for Clarity of Faceted Diamonds

According to CIBJO, it is permitted to subdivide the clarity grades WS, VS, and SI into two subgroups each, for stone sizes over 0.47 ct. Definitions assume a trained grader working under favorable conditions.

CIBJO	Definition	GIA
Lr flawless	Free of inclusions under 10 X magnification and absolutely transp.	If internally flawless
VVS very very small inclusions	Very few, very small inclusions, under 10 X magnification very difficult to see	VVS 1 very very VVS 2 slightly included
VS very small inclusions	Very small inclusions, under 10 X magnification difficult to find	VS 1 very slightly incl. VS 2 slightly incl.
SI small inclusions	Small inclusions, easily recognized under 10 X magnification	SI 1 slightly included SI 2 slightly included
PI distinct inclusions	Inclusions, immediately recognized under 10 X magnification, but not diminishing brilliance	I1 included I
PII larger inclusions	Larger and/or numerous inclusions slightly diminishing the brilliance recognizable with the naked eye	I2 included II
PIII large	Large and/or numerous inclusions, diminishing the brilliance considerably	I3 included III

Grading for Cut To grade for cut, the type and shape of cut, proportions, and symmetry as well as outer marks are taken into consideration. In Germany the normal cut is the "fine brilliant cut" (page 96), in the rest of Europe the "Scandinavian-standard brilliant." In the United States, the only widely used cut grading system is that of the American Gem Society, based on the "AGS Ideal Cut." The following table shows the terms and definitions for grading in a simplified form, according to the RAL 560 A5E, published in 1971.

Grading of Cut of the Diamond Brilliants

RAL 560 A5E	Definition
Very good	Exceptional brillance. Few and only minor outer marks. Very good proportions.
Good	Good brilliance. Some outer marks. Proportions with some deviations.
Medium	Slightly less brilliance. Several larger outer marks. Proportions with considerable deviations.
Poor	Brilliance considerably less. Larger and/or numerous outer marks. Proportions with very distinct deviations.

Famous Diamonds

A number of diamonds are well known and famous because of their size, beauty, or their adventurous past. (Glass replicas shown opposite.)

1 **Dresden** Green, 41 carats. Probably from India; early history not known. Supposedly bought in 1742 by Friedrich August II, Duke of Saxony, for 400,000 taler. Name derived from its place of safe-keeping, the Green Vaults in Dresden, Germany.

2 **Hope** 45.52 carats. Appeared 1830 in the trade and was bought by the banker H. Ph. Hope (hence the name). Probably recut from a stolen stone. Since 1958 in the Smithsonian Institution, Washington, D.C.

3 **Cullinan I, or Star of Africa** 530.20 carats. Cut from the largest rough gem diamond ever found of 3106 carats, the Cullinan (named after Sir Thomas Cullinan, chairman of the mining company) together with 104 other stones, by the firm Asscher in Amsterdam in 1908. Adorns the sceptre of the English king's insignia. Kept in the Tower of London; largest cut diamond of fine quality.

4 **Sancy** 55 carats. Said to have been worn by Charles the Bold around 1470. Bought in 1570 by Seigneur de Sancy (hence the name) from the French ambassador to Turkey. Now on display at the Louvre in Paris.

5 **Tiffany** 128.51 carats. Found in Kimberley mine, South Africa, in 1878; rough weight 287.42 carats. Bought by the jewelers Tiffany in New York and cut in Paris with 90 facets.

6 **Koh-i-Noor** 108.92 carats. Originally a round stone of 186 carats belonging to the Indian Raj. Plundered in 1739 by the Shah of Persia, who called it "Mountain of Light" (*Koh-i-Noor*). Came into possession of the English East India Company, which presented it to Queen Victoria in 1850. Recut, it was set in the crown of Queen Mary, wife of George V, and then in a crown made for the mother of Queen Elizabeth II; now in the Tower of London.

7 **Cullinan IV** 11.50 carats. One of the 105 cut stones from the largest gem diamond ever found, the Cullinan (see 3 above). Also in the crown of Queen Mary; can be removed from this and worn as a brooch. Kept in the Tower of London.

8 **Nassak** 43.38 carats. Originally over 90 carats and in a Temple of Shiva near Nassak (hence the name) in India. Looted in 1818 by the English; recut 1927 in New York. Acquired by the King of Saudi Arabia in 1977.

9 **Shah** 88.70 carats. Came from India, shows cleavage planes, partially polished. Has three inscriptions of monarchs' names (amongst them the Shah of Persia's—hence the name). Given in 1829 to Tsar Nicholas I. Kept in the Kremlin, Moscow.

10 **Florentine** 137.27 carats. Early history steeped in legend. In 1657 in the possession of the Medici family in Florence (hence the name). During the 18th century in the Habsburg crown, then used as brooch. Whereabouts after 1919 are unknown.

Other important cut diamonds (in carat): Cullinan II. (317.40, picture on page 9), Centenary (273.85), De Beers (234.50), Great Mogul (280.00), Jonker (125.35), Jubilee or Reitz (245.35), Nizam (277.00), Orloff (189.60), Regent or Pitt (140.50), Victoria (184.50), Blane Wittels-bache (35.56)

The greatest rough stones found suitable for gem purposes (in carats): Cullinan (3106), Excelsior (995.2), Star of Sierra Leone (968.9), Incomparable (890), Great Mogul (787.5), Woyie River (770), President Vargas (726.6), Jonker (726), Jubilee (or Reitz 650.8), Dutoitspan (616), Baumgold (609), Lesotho (601.25), Centenary (599.0), Nizam (440.0), De Beers (428.5).

1
2
3
4
5
6
7
8
9
10

Development of the Brilliant Cut

Although diamond as a gemstone has been known for over 2,000 years, the first cut to improve the optical effect was developed in the 13th century. Previously, only rough stones were used, perhaps with the rough edges smoothed away. It is also possible that cleaving was sometimes used to improve the shape of rough crystals. Then, flat surfaces at slight angles to the faces of the natural octahedron crystal were polished. This "point cut," so called after the shape of the crystal, is the first real diamond cut. It represents the beginning of a series, the final product being the modern brilliant cut.

About 1400 the "table cut" was developed; this was an octahedral crystal with a large flat surface on top, the table, and with a smaller facet underneath called the culet. At about the same time the "thin cut" made its appearance through the cutting off of the two octahedral points.

Since the end of the 15th century, a cutting wheel was used to technically improve the facet cut. New facets were cut in addition to the natural crystal facets to enhance the optical effect of the stone. By about the middle of the 16th century, the table cut had been developed into a faceted stone with a multi-cornered table. The four lateral edges of the upper and lower part are cut off to give a faceted surface. This stone, now called the "old single cut," had 18 facets, including the pointy surface at the bottom. By adding more facets above the lateral edges, the "old double cut" was created, with 34 facets overall and a rounded outline (rondiste). This cut was supposedly based on the inspiration of the French Cardinal Mazarin around 1650 (therefore also called the "Mazarin cut").

Toward the end of the 17th century, a diamond with 58 facets was developed and was attributed to a Venetian cutter named Vicenzio Peruzzi. Even though the outline (rondiste) is not round and the facets are irregular, this cut, the "triple cut" or "Peruzzi cut," had already come very close to the modern brilliant cut, and not only because the number of facets is the same.

Brilliant Cut Perfection of the diamond as a gem is only realized by the modern brilliant cut, developed around 1910 from the so-called old cuts of the last century (see below). Its characteristics are as follows: round girdle, at least 32 facets on the

Historic Cut of the Diamond.

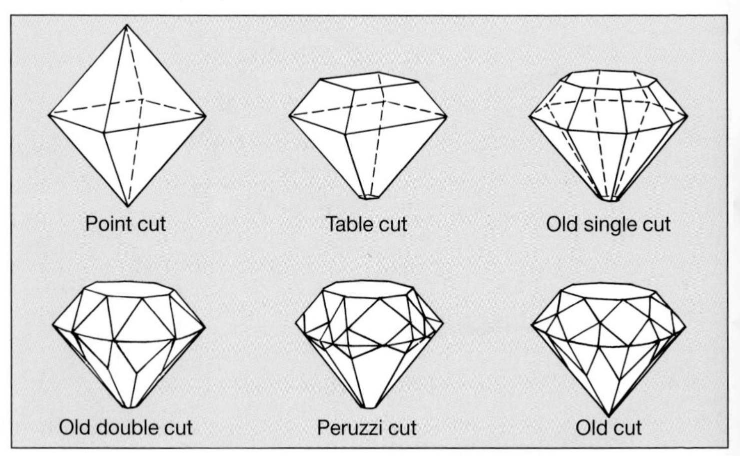

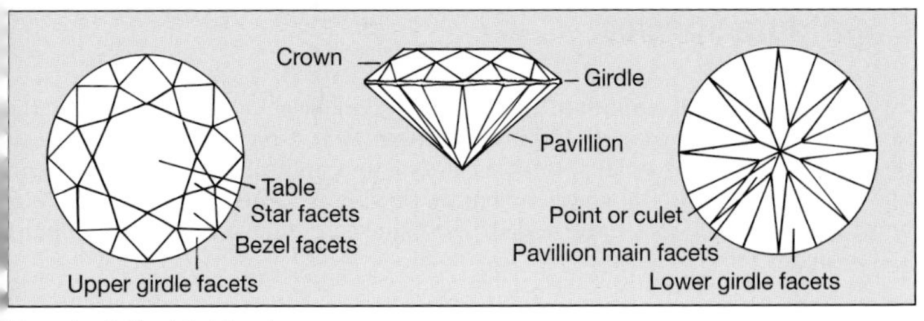

Terms for Brilliant Cut Facets

crown and table, at least 24 facets and sometimes a culet (i.e., a very small facet at the point) on the pavilion.

The term *brilliant* without any additional information can only refer to the round diamond with brilliant cut. All other cut types must be rightly marked as such (CIBJO, 1991).

By calculation and experience, several variations of the modern brilliant cut have been developed. The following ones are the best known:

Tolkowsky brilliant (1919, Tolkowsky) Very good light reflection. Best brilliancy. Basis of the American Gem Society "AGS Ideal Cut."

Ideal brilliant (1926, Johnson and Rösch) By no means as advantageous as could be assumed by the name. Not very high brilliancy, appears not uniform.

Fine-cut brilliant (also referred to as fine cut of the practice or practical fine cut, 1949, Eppler) Proportions were calculated from cut stones with highest brilliance, also developed by experience. Basis of cut grading in Germany.

Parker brilliant (1951, Parker) Good reflection, but due to flat crown, has small dispersion and thus little play of colors.

Scandinavian standard brilliant (1968) Basis of polished diamond grading in Scandinavia. Values established from cut diamonds.

Brilliants having more facets than normally common King cut (1941) with 86 facets, Magna cut (1949) with 102 facets, Highlight cut (1963) with 74 facets, Princess 144-cut (1965) with 146 facets, Radiant (1980) with 70 facets.

Fine-Cut Brilliant in practice

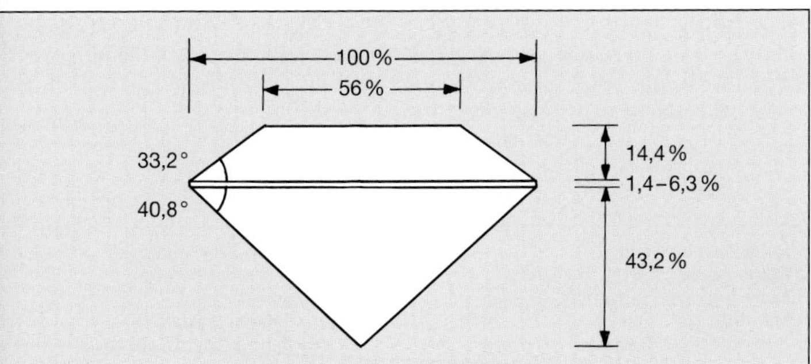

Corundum Species

Two color varieties of corundum are used for making jewelry, the red ruby and the sapphire which comprises all other colors (page 103). Common corundums, i.e., those not of gemstone quality, serve as cutting and polishing material. The well-known polishing material emery is mainly fine-grain corundum, to which magnetite, hematite, and quartz are added. The name corundum has its origin in India and probably referred to ruby.

Ruby

Corundum Species

Color: Varying red
Color of streak: white
Mohs' hardness: 9
Density: 3.97–4.05
Cleavage: None
Fracture: Small conchoidal, splintery, brittle
Crystal system: (Trigonal) hexagonal prisms or tables, rhombohedrons
Chemical composition: Al_2O_3, aluminum oxide

Transparency: Transparent to opaque
Refractive index: 1.762–1.778
Double refraction: —0.008
Dispersion: 0.018 (0.011)
Pleochronism: Strong; yellow-red, deep carmine red
Absorption: 694, 693, 668,659, 610-500, 476, 475, 468
Fluorescence: Strong: carmine red

Ruby is thus named because of its red color (Latin—*ruber*). It was not until about 1800 that ruby, as well as sapphire, was recognized as belonging to the corundum species. Before that date, red spinel and the red garnet were also designated as ruby.

The red color varies within each individual deposit, so it is not possible to determine the source area from the color. The designations "Burma ruby" and "Siam ruby" are therefore strictly erroneous, and refer more to quality than origin. The most desirable color is the so-called "pigeon's blood," pure red with a hint of blue. The distribution of color is often uneven, in stripes or spots. The substance that provides the color is chromium, and in the case of brownish tones, iron is present as well. As a rough stone, ruby appears dull and greasy, but, when cut, the luster can approach that of diamond. Heat treatment is commonly used to improve the color.

Ruby is the hardest mineral after diamond. However, the hardness varies in different directions. Ruby has no cleavage, but has certain preferred directions of parting. Because of brittleness, care must be taken when cutting and setting.

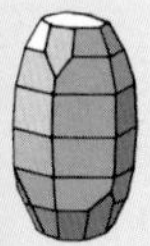

Inclusions are common. They are not always indicative of lower quality, but show the difference between a natural and a synthetic stone. The type of inclusion (minerals, growth structures, canals, or other cavities) often indicates the source area.

Included rutile needles bring about either soft sheen (called silk) or, when cut en cabochon, the rare cat's-eye effect (see page 99, no. 5), or more often the very desirable asterism—a six-rayed star (see page 99, no. 4), which shimmers over the surface of the stone when it is moved. Nowadays there are also Trapiche Rubies on the market. Their appearance is equal to the Trapiche Emeralds (page 108).

1 Ruby, 5 faceted stones
2 Ruby, 2 drops, 2.51ct, Thailand
3 Ruby, engraved cabochon, 30.97ct
4 Star ruby,
5 Ruby cat's-eye, 6.64ct
6 Ruby, 4 tabular crystals
7 Ruby, 3 prismatic crystals
8 Ruby, rolled crystal
9 Ruby, tabular crystal
10 Ruby in host rock, Sri Lanka

1
2
3
4
5
6
7
8
9
10

Deposits The host rocks of ruby are metamorphic dolomite marbles, gneiss, and amphibolite. The yield of rubies from such primary deposits is not economically profitable. Rather, secondary alluvial deposits are worked. Because of its high density, ruby is normally separated through the washing of river gravels, sands, and soil, then concentrated, and finally picked out by hand.

Production methods are still as primitive as they were a hundred years ago in many locations. In state-owned mines, on the other hand, the usage of machinery is not exactly the rule, but much more frequent than in private companies. Some state-regulated companies (e.g., Mogok in Myanmar) lately even work with highly mechanized machinery both above- and underground.

Some of the most important deposits are in Myanmar, Thailand, Sri Lanka, and Tanzania. For centuries, the most important have been in upper Myanmar near Mogok. The ruby-bearing layer runs several yards under the surface. Apparently only about one percent of production is of gem quality. Some of the rubies are of pigeon's blood color. They are considered to be the most valuable rubies of all. Large stones are rare. Minerals found together with ruby, often also of gemstone quality, are precious beryl, chrysoberyl, garnet, moonstone, sapphire, spinel, topaz, tourmaline, and zircon. In the early 1990s large new deposits were discovered at Mong Hsu in Myanmar. Rubies from Thailand often have a brown or violet tint to them. They are found southeast of Bangkok in the district of Chantaburi in clayey gravels. Shafts are sunk to a depth of 26 ft (8 m). However, in recent years Thai ruby production has been declining.

In Sri Lanka deposits are situated in the southwest of the island in the district of Ratnapura. Rubies from these deposits (called *illam* by the local population) are usually light red to raspberry-red. Some of the rubies are recovered from the river sands and gravels.

Since the 1950s Tanzania has produced a decorative green rock, a zoisite (anyolite), with quite large, mostly opaque rubies (page 176, numbers 12 and 14). Only a few crystals are cuttable, most being used as decorative stones. On the upper Umba River (northwest Tanzania), on the other hand, rubies with gemstone quality have been found that are violet to brown-red.

Other mining deposits are in Afghanistan, Australia (Queensland, New South Wales), Brazil, India, Cambodia, Kenya, Madagascar, Malawi, Nepal, Pakistan, Zimbabwe, Tajikistan, the United States (Montana, North Carolina), and Vietnam.

Small ruby deposits can also be found in Switzerland (Tessin), in Norway, and on the Southwest Coast of Greenland.

Famous Rubies Ruby is one of the most expensive gems, large rubies being rarer than comparable diamonds. The largest cuttable ruby weighed 400ct; it was found in Burma and divided into three parts. Famous stones of exceptional beauty are the Edwardes ruby (167ct) in the British Museum of Natural History in London, the Rosser Reeves star ruby (138.7ct) in the Smithsonian Institution in Washington, D.C., the De Long star ruby (100ct) in the American Museum of Natural History in New York, and the Peace ruby (43 ct), thus called because it was found in 1919 at the end of World War I.

Many rubies comprise important parts of royal insignia and other famous jewelry. The Bohemian St. Wenzel's crown (Prague), for instance, holds a nonfaceted ruby of about 250ct. But some gems, thought to be rubies, have been revealed as spinels, such as the "Black Prince's ruby" in the English State crown (page 9) and the "Timur Ruby" in a necklace among the English crown jewels. The drop-shaped spinels in the crown of the Wittelsbachs dating from 1830 were also originally thought to be rubies.

Loosening of gemstone-bearing sediment in a river; Sri Lanka

Working Today rubies are often cut in the countries where they were found. Because the cutters usually aim for maximum weight, the proportions are not always satisfactory, so that many stones have to be recut by dealers in other countries. Transparent qualities are cut in step and brilliant cut; less transparent stones, en cabochon or they are formed to carvings. Only synthetic rubies are used for watches and bearings, formerly the most important technical application for natural stones.

Possibilities for Confusion With almadite (page 120), pyrope (page 120), spinel (page 116), topaz (page 118), tourmaline (page 126), and zircon (page 124).

Since the beginning of the 20th century, there have been synthetic rubies with gemstone quality (page 267); these resemble natural ones especially in their chemical, physical, and optical properties. But most of them can be recognized by their inclusions as well as by the fact that they, in contrast to natural rubies, transmit shortwave ultraviolet light.

Numerous imitations currently are on the market, especially glass imitations and doublets. These have a garnet crown and glass underneath, or the upper part is natural sapphire with synthetic ruby underneath. There are many false names in the trade such as Balas ruby (= spinel), Cape ruby (= pyrope), and Siberian ruby (= tourmaline).

Sapphire

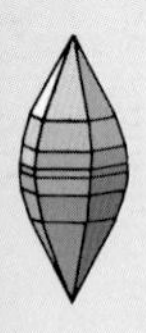

Color: Blue in various tones, colorless, pink, orange, yellow, green, purple, black
Color of streak: White
Mohs' hardness: 9
Density: 3.95–4.03
Cleavage: None
Crystal system: (Trigonal), doubly pointy, barrel-shaped, hexagonal pyramids, tabloid-shaped
Chemical composition: Al_2O_3 aluminum oxide
Transparency: transparent to opaque
Refractive index: 1.762–1.788
Double refraction: —0.008
Dispersion: 0.018 (0.011)
Pleochroism: Blue: definite; dark blue, green-blue
Yellow: weak; yellow, light yellow
Green: weak; green-yellow, yellow
Purple: definite; purple, light red
Absorption spectrum: Blue, Sri Lanka, 471, 460, 455, 450, 379 Yellow, 471, 460, 450 Brown, 471, 460–450
Fluorescence: Blue S.: none; Colorless S.: orange-yellow, violet

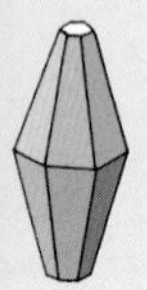

The name sapphire (Greek—blue) used to be applied to various stones. In antiquity and as late as the Middle Ages, the name sapphire was understood to mean what is today described as lapis lazuli. Around 1800 it was recognized that sapphire and ruby are gem varieties of corundum. At first only the blue variety was called sapphire, and corundums of other colors (with the exception of red) were given special, misleading names, such as "Oriental peridot" for the green variety and "Oriental topaz" for the yellow type.

Today corundums of gemstone quality of all colors except red are called sapphire. Red varieties are called rubies (page 98). The various colors of sapphire are qualified by description, e.g., green sapphire or yellow sapphire. Colorless sapphire is called leuko-sapphire (Greek—white), pinkish orange sapphire Padparadscha (Sinhalese for "Lotus Flower").

There is no definite demarcation between ruby and sapphire. Light red, pink, or violet corundums are usually called sapphires, as in this way they have individual values in comparison with other colors. If they were grouped as rubies, they would be stones of inferior quality. The coloring agents in blue sapphire are iron and titanium; and in violet stones, vanadium. A small iron content results in yellow and green tones; chromium produces pink, iron, and vanadium orange tones. The most desired color is a pure cornflower-blue. In artificial incandescent light, some sapphires can appear to be ink-colored or black-blue.

Through heat treatment at temperatures of about 3100–3300 degrees F (1700–1800 degrees C), some cloudy sapphires, nondistinct in color, can change to a bright blue permanent color.

Hardness is the same as ruby and also differs clearly in different directions (an important factor in cutting). There is no fluorescence characteristic for all sapphires.

Inclusions of rutile needles result in a silky shine; oriented, i.e., aligned, needles cause a six-rayed star sapphire (page 103, no. 2).

1 Sapphire, oval, 5.73ct, Thailand
2 Star sapphire, 9.46ct, Burma
3 Sapphire, brilliant cut, 2.81ct
4 Sapphire, 6 faceted stones, together 2.34ct, Tanzania
5 Sapphire, oval, 1.62ct, Sri Lanka
6 Sapphire, drop, 6.09ct
7 Sapphire, yellow, 11.32ct, Sri Lanka
8 Sapphire, antique cut, 5.18ct
9 Sapphire, antique cut, 3.74ct
10 Sapphire, 5 crystal shapes

The illustrations are 30 percent larger than the originals.

1
2
3
4
5
6
7
8
9
10

Deposits Host rocks of sapphire are dolomotized limestones, marble, basalt, or pegmatite. It is mined mainly from alluvial deposits or deposits formed by weathering, rarely from the primary rock. Production methods are usually very simple. The underground gem-bearing layer is worked from hand-dug holes and trenches. The separation of clay, sand, and gravel is done by washing out the gemstones due to their higher density. Final selection is made by hand. Sapphire is much more common than ruby, as the substances which lend color to sapphire are more common than those of ruby. Today economically important sapphire deposits are in Australia, Myanmar, Sri Lanka, and Thailand.

Australian deposits have been known since 1870. The host rock is basalt; the sapphires are washed out of the weathered debris. Quality is modest. Under artificial light, the deep blue stones appear inky, blue-green, nearly black; lighter qualities have a green tint. In recent decades black star sapphires have been found in Queensland. Accompanying minerals are pyrope, quartz, topaz, tourmaline, and zircon. Since 1918 good blue qualities have been found in New South Wales.

The alluvial deposits in upper Myanmar near Mogok are partially worked with modern methods and yield rubies and spinels as well as sapphires. The host rock is pegmatite. In 1966 the largest star sapphire was found here, a crystal of 63,000ct (281b/12.6kg).

Sapphires have been found in Sri Lanka since antiquity. The deposits are in the southwest of the island in the region of Ratnapura. The mother rock is a dolomotized limestone, which is enclosed in granite gneiss. There are also 10–20-in (30–60cm) thick river gravel placers (called *illam* locally) that are exploited from a depth of 3–33 ft (1–10 m). Sapphires are usually light blue, with a tinge of violet. There are also yellow and orange (Padparadscha) varieties as well as green, pink, brown, and nearly colorless stones, also star sapphires. Accompanying minerals are apatite, epidote, garnet, quartz, ruby, spinel, topaz, tourmaline, and zircon.

There are two sapphire deposits in Thailand: one in the region of Chantaburi, southeast of Bangkok, the other one near Kanchanaburi, northwest of Bangkok. The host rock is marble and/or basalt; placers and deposits formed by weathering are mined. The stones are of good quality in various colors, including star sapphire. Blue sapphires have a deep color, but tend to have a tinge of blue-green.

The most desired sapphires used to come from Kashmir (India), where the deposits were situated at a height of 16,500 ft (5000 m) in the Zaskar mountains. Production varied since 1880, and the deposits have apparently been worked out. The host rock is a kaolin-rich pegmatite in crystalline schist. The decomposition product yields sapphires of deep cornflower-blue color, often with a silky sheen. Most stones sold today as Kashmir sapphire come from Myanmar.

In the late 1800s sapphire deposits were discovered in Montana (United States). The host rock is andesite dikes. Mining is carried out on the dike rock, also from weathered material. Color of sapphire varies and is often pale blue or steel-blue. Mining has been interrupted repeatedly since the end of the 1920s, but has been steadier in recent years.

There are also significant sapphire deposits in Brazil, Cambodia, China, Kenya, Madagascar, Malawi, Nigeria, Pakistan, Rwanda, Tanzania, Vietnam, and Zimbabwe. Isolated star sapphires have been found in Finland.

Gemstone-washers washing out gem-bearing earth in a basin, Sri Lanka.

Famous Sapphires Large sapphires are rare. They are sometimes named in the same way as famous diamonds. The American Museum of Natural History (New York) owns the "Star of India," perhaps the largest cut star sapphire (536ct); also the "Midnight Star," a black star sapphire (116ct). The "Star of Asia," a star sapphire weighing 330ct, is owned by the Smithsonian Institution in Washington, D.C. Two famous sapphires (St. Edward's and the Stuart sapphire) are part of the English Crown Jewels. In the United States, the heads of presidents Washington, Lincoln, and Eisenhower have been carved out of three large sapphires, each weighing roughly 2000ct.

Possibilities for Confusion Sapphires can be confused with various stones. The blue sapphire looks similar to benitoite (page 200), indicolite (page 126), iolite (page 196), kyanite (page 212), spinel (page 116), tanzanite (page 176), topaz (page 118), and zircon (page 124), as well as blue glass.

Some imitations are made from doublets-blue cobalt glass with a crown of garnet or a crown of green sapphire and a pavilion of synthetic blue sapphire. Lately doublets have appeared using two small natural sapphires. Star sapphire is imitated by using star rose quartz with blue enamel on a flat back; alternatively the star is engraved on the flat back of a synthetic cabochon or on glass.

Synthetic sapphire was produced with properties identical to the natural stone at the beginning of this century. Since 1947 synthetic star sapphires of gem quality have been sold.

Beryl Species

Several color varieties of beryl are used as gemstones. Deep green beryls are called emerald, blue aquamarine (page 110). All beryls of other colors of gemstone quality are called precious beryl (page 112). The name beryl comes from India and has always been associated with the gemstone.

Emerald

Beryl Species

Color: Emerald green, green, slightly yellowish-green
Color of streak: White
Mohs' hardness: 7½–8
Density: 2.67–2.78
Cleavage: Indistinct
Fracture: Small concoidal, uneven, brittle
Crystal system: (Hexagonal), hexagonal prisms
Chemical composition: $Al_2Be_3[Si_6O_{18}]$ aluminum beryllium silicate

Transparency: Transparent to opaque
Refractive index: 1.565–1.602
Double refraction: –0.006
Dispersion: 0.014 (0.009–0.013)
Pleochroism: Definite; green, blue-green to yellow-green
Absorption spectrum: 683, 681, 662, 646, 637, (606), (594), 630–580, 477, 472
Fluorescence: Usually none

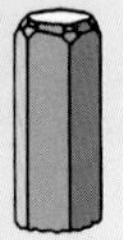

The name emerald derives from Greek *smaragdos*. It means "green stone" and, in ancient times, referred not only to emeralds but also probably to most green stones.

Emerald is the most precious stone in the beryl group. Its green is incomparable, and is therefore called "emerald green." The coloring agent for the "real emerald" is chrome. Beryls that are colored by vanadium ought to be called "green beryl" and not emerald. The color is very stable against light and heat, and only alters at 1292–1472 degrees F (700–800 degrees C). The color distribution is often irregular; a dark slightly bluish green is most desired.

Only the finest specimens are transparent. Often the emerald is clouded by inclusions (compare to page 58). These are not necessarily classified as faults, but are evidence as to the genuineness of the stone as compared with synthetic and other imitations. The expert refers to these inclusions as *jardin* (French—garden).

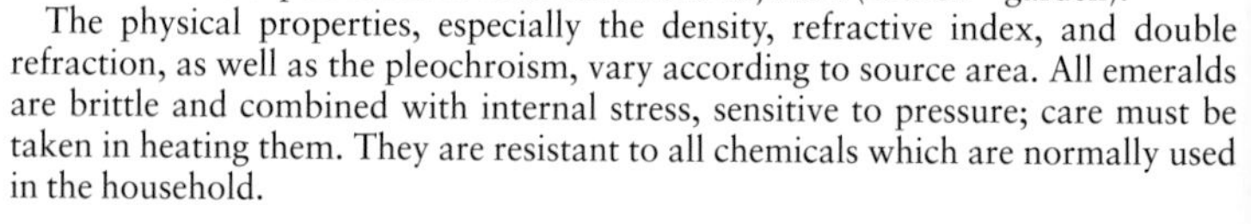

The physical properties, especially the density, refractive index, and double refraction, as well as the pleochroism, vary according to source area. All emeralds are brittle and combined with internal stress, sensitive to pressure; care must be taken in heating them. They are resistant to all chemicals which are normally used in the household.

Deposits Emeralds are formed by hydrothermal processes associated with magma and also by metamorphism. Deposits are found in biotite schists, clay shales, in lime-stones, with pigmatites. Mining is nearly exclusively from host rock, where the emerald has grown into small veins or on walls of cavities. Alluvial placers are very unlikely to come about as the density of emerald is near that of quartz. Therefore, rare secondary deposits are mostly formed by weathering.

Significant deposits are in Colombia, especially the Muzo mine northwest of Bogotá. First mined by native tribes, the Muzo deposit was abandoned and rediscovered in the 17th century. The mine yields fine-quality stones of a deep green color. Mining, apart from shafts, is mainly by step-form terraces. The emerald-bearing, soft broken rock is loosened with sticks, lately also through blasting or

1 Emerald in host rock
2 Emerald, oval, 0.91ct, Colombia
3 Emerald, 2 pear shapes, 1.59ct
4 Emerald, 2 emerald cuts
5 Emerald, antique cut, 4.14 ct. South Africa
6 Emerald, oval, 1.27ct
7 Emerald, cabochon, 5.24ct
8 Emerald, cabochon, 4.26ct
9 Emerald, cabochon, 3.11ct
10 Emerald, crystal, Brazil

1
2
4
3
5
6
7
8
9
10

Work done in a sorting plant of an emerald mine; Cobra mine/South Africa

with bulldozers, and the emeralds picked out by hand. The host rock is a dark carbonaceous limestone. Accompanying minerals are albite, apatite, aragonite, barite, calcite, dolomite, fluorite, and pyrite.

Another important Colombian deposit, the Chivor Mine, is northeast of Bogotá. It was also mined by Native Americans. The host rock is gray-black shale and gray limestone. It is mined in terraces, and also from shafts.

During recent decades further emerald deposits, which promise to be successful, have been found in Colombia. There is always a high demand for the rare, so-called Trapiche emeralds found exclusively in Colombia, a wheel-like growth of several prismatic crystals. Only a third of the Colombian emeralds are worth cutting. Stones larger than nut size are usually low quality or broken.

In Brazil there are various deposits in Bahia, Goias, and Minas Gerais. Stones are lighter than Colombian ones, mostly yellow-green, but they are often free of inclusions. Through deposits newly discovered since the beginning of the 1980s, Brazil has become one of the most important suppliers of emeralds.

Since the second half of the 1950s, emerald deposits have been exploited in Zimbabwe. Most important is the Sandawana mine in the south. Crystals are small, but of very good quality.

In the northern Transvaal (South Africa), emeralds are mined by modern methods using machinery (Cobra and Somerset Mines). Only five percent of the production is of good quality. Most stones are light or turbid and only suitable for cabochons.

Emerald deposits were discovered in 1830 in Russia in the Urals north of Yekaterinburg (Sverdlovsk), but the yield has fluctuated widely over the years. Good qualities are rare; most crystals are opaque, turbid, and slightly yellow-green. The host rock is a biotite mica shale, interfoliated with talc and chlorite.

Further emerald deposits are in Afghanistan, Australia (New South Wales, Western Australia), Ghana, India, Madagascar, Malawi, Mozambique, Namibia, Nigeria, Pakistan, Zambia, Tanzania, and the United States (North Carolina). The emerald mines of Cleopatra (perhaps worked as early as 2000 B.C.), east of Aswan in upper Egypt, are of historical interest only.

The Austrian deposits in the Habachtal Valley near Salzburg are well known.

The host rock is biotite hornblende shale. The stones are of interest only to mineral collectors, as cuttable material is rare. The stones are mostly turbid; color quality is good. Some individual emeralds have been found in Norway, north of Oslo.

Famous Emeralds There are many well-known large emeralds, as valuable and famous as diamonds and rubies. Some beautiful specimens of several hundred carats are kept by the British Museum of Natural History in London, by the American Museum of Natural History in New York, in the treasury of Russia, in the state treasury of Iran, and in the Treasury room in Topkapi Palace, Istanbul, Turkey. In the Viennese treasury is a vase, 4¾ in (12 cm) high and weighing 2205ct, cut from a single emerald crystal.

Working Because emerald is so sensitive to knocks, a step cut was developed, the four corners being truncated by facets (so-called emerald cut). Clear, transparent qualities are sometimes brilliant cut. Turbid stones are used only for cabochons or as beads for necklaces. Occasionally emeralds are worn in their natural form, and sometimes engraved.

For some years, a number of cutters in Israel have been specializing in cutting emerald, after diamond cutters lost much of their production to India.

Possibilities for Confusion Due to very similar colors, possibilities of confusion exist with aventurine (page 138), demantoid (page 122), diopside (page 206), dioptase (page 210), fluorite (page 214), grossularite (page 122), hiddenite (page 130), peridot (page 174), uvarovite (page 122), and verdelite (page 126).

Numerous doublets are on the market, mostly two genuine pale stones (rock crystal, aquamarine, beryl, or pale emerald) cemented together with emerald-green paste. The pavilion or both top and bottom may also be glass or synthetic spinel. The upper parts of natural stones are determined by inclusions and hardness, which are features of genuineness. When set, these doublets can be difficult to detect.

The first emerald synthesis was made in 1848. Since the beginning of the 20th century, various methods have been developed, and, since the 1950s, commercial products of excellent quality have appeared on the market. An important aid to differentiation is ultraviolet light. Synthetic emerald transmits ultraviolet light more than natural emerald.

In order to hide very fine hairline fractures and other faults, emeralds are often dipped into special oils or prepared with artificial resin in a vacuum; typically this is done at the country of origin. In the United States, FTC guidelines require disclosure of this treatment.

An emerald substitute through synthetically colored quartz can be found on page 280.

Trapiche emerald with a wheel-like growth of several crystals.

Aquamarine

Beryl Species

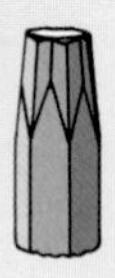

Color: Light blue to dark blue, blue-green
Color of streak: White
Mohs' hardness: 7½–8
Density: 2.68–2.74
Cleavage: Indistinct
Fracture: Conchoidal, uneven, brittle
Crystal system: (Hexagonal), hexagonal prisms
Chemical composition: $Al_2Be_3[Si_6O_{18}]$ aluminum beryllium silicate

Transparency: Transparent to opaque
Refractive index: 1.564–1.596
Double refraction: –0.004 to –0.005
Dispersion: 0.014 (0.009–0.013)
Pleochroism: Definite; nearly colorless, light blue, blue, light green
Absorption: 537, 456, 427
Maxixe-A: 695, 655, 628, 615, 581, 550
Fluorescence: None

Aquamarine (Latin—water of the sea) is so named because of its seawater color. A dark blue is the most desired color. The coloring agent is iron. Lower qualities are heated to 725–850 degrees F (400–450 degrees C) to change them to the desired, permanent aquamarine blue. Higher heat will lead to discoloration. Care must be taken when making jewelry! Colors can also be improved with neutron and gamma irradiation, but these changes do not last. Aquamarine is brittle and sensitive to pressure. Inclusions of fine, oriented hollow rods or aligned foreign minerals rarely cause a cat's-eye effect or asterism with six-rayed stars with a vivid sheen.

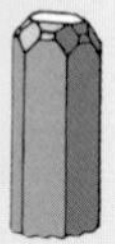

Santa Maria Trade term (i.e., not an established variety name) for especially fine aquamarine. Named after the mine with the same name in Ceara (Brazil).

Santa-Maria-Africana Trade term for fine aquamarine from Mozambique, on the market since 1991. Name derived from the "Santa Maria" quality of Ceara (Brazil).

Maxixe Deep blue beryl; color fades in daylight. Originally (since 1917) it came only from the Maxixe Mine in Minas Gerais (Brazil). Since the 1970s it has been more widely available, but obviously made more beautiful through irradiation, but the color does not last.

Deposits The most important deposits are in Brazil, spread throughout the country. The well-known deposits in Russia (the Urals) seem to be worked out. Other deposits of some commercial significance are in Australia (Queensland), Burma (Myanmar), China, India, Kenya, Madagascar, Mozambique, Namibia, Nigeria, Zambia, Zimbabwe, and the United States. The host rock is pegmatite and coarse-grained granite as well as their weathered material.

The largest aquamarine of gemstone quality was found in 1910 in Marambaya, Minas Gerais (Brazil). It weighed 243 lb (110.5 kg), was 18 in (48.5 cm) long and 15½ in (42 cm) in diameter, and was cut into many stones with a total weight of over 100,000 ct. There have been finds weighing a few tons, but these aquamarines are opaque and gray, not suitable for cutting.

The preferred cuts are step (emerald) and brilliant-cut with rectangular or long oval shapes. Turbid stones are cut en cabochon or are used for necklace beads.

Possibilities for Confusion With eudase (page 194), kyanite (page 212), topaz (page 118), tourmaline (page 126), zircon (page 124), and glass imitations. Synthetic aquamarine can be produced but is uneconomical. The "synthetic aquamarine" sold in the trade is really aquamarine-colored synthetic spinel.

1 Aquamarine, emerald cut, 72.46ct
2 Aquamarine, emerald cut, 17.41ct
3 Aquamarine, antique cut, 45.38ct
4 Aquamarine, marquise, 25.58ct
5 Aquamarine, antique cut, 18.98ct
6 Aquamarine, briolette, 6.65ct
7 Aquamarine, cyrstal, 68.5mm, 45g
8 Aquamarine, 3 crystals, together 77g

1
2
3
4
5
6
7
8

Precious Beryl

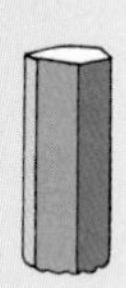

Color: Gold-yellow, yellow-green, yellow, pink, colorless
Color of streak: White
Mohs' hardness: 7½–8
Density: 2.66–2.87
Cleavage: Indistinct
Fracture: Conchoidal, brittle
Crystal system: (Hexagonal), hexagonal prisms
Chemical composition: $Al_2Be_3[Si_6O_{18}]$ aluminum berylium silicate
Transparency: Transparent to opaque
Refractive index: 1.562–1.602
Double refraction: –0.004 to –0.010
Dispersion: 0.014 (0.009–0.013)
Pleochroism: Golden: weak; lemon-yellow, yellow
Heliodor: weak; golden-yellow, green-yellow
Morganite: definite; pale pink, bluish-pink
Green: definite; yellow-green, blue-green
Absorption spectrum: Not diagnostic
Fluorescence: Morganite: weak; violet

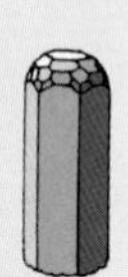

Precious beryl refers to all color varieties of the beryl group that are not called emerald (page 106) and aquamarine (page 110).

Precious beryls are brittle and therefore sensitive to pressure and resistant to chemicals used in the household, and they have a vitreous luster, occasionally displaying the cat's-eye effect and asterism. They are typically found with aquamarine (page 110). Often used with a step cut. Color varieties have either special names in the trade, or the respective color precedes the word beryl (e.g., yellow beryl, green beryl).

Bixbite [3] Raspberry-red. Origin of the name is unknown. Many scientists reject this as separate variety. Occurrences in Utah (United States).

Golden Beryl [1] Color varies between lemon-yellow and golden-yellow. Inclusions are rare. Decolorization at 482 degrees F (250 degrees C). Deposits in Brazil, Madagascar, Namibia, Nigeria, Zimbabwe, Sri Lanka.

Goshenite [5] Colorless beryl, named after locality in Goshen, Massachusetts (United States). Used as imitation for diamond and emerald by applying silver or green metal foil to the cut stone. Occurrences in Brazil, China, Canada, Mexico, Russia, and the United States.

Heliodor [2,6] Light yellow-green (Greek—present of the sun). Discovered in 1910 as apparent new variety in Namibia, but beryls of the same color were previously known in Brazil and Madagascar. Since there is no clear distinction possible in the yellow and green-yellow tones in comparison to golden beryl, heliodors are generally rejected as an independent precious beryl variety and rather are counted among the weak-colored golden beryls.

Morganite [4] (Also called pink beryl) Soft pink to violet, also salmon-colored. Inclusions are rare. Named after the American banker and collector J. P. Morgan. Density between 2.71 and 2.90. Inferior qualities can be improved by heating above 752 degrees F (400 degrees C). Deposits in Afghanistan, Brazil, China, Madagascar, Mozambique, Namibia, Zimbabwe, and the United States (Utah, California). Synthetic morganite is known.

Possibilities for Confusion With many gemstones because of the richness of colors of the precious beryls. Greenish precious beryls are transferred into blue aquamarine stones through heating at 725–930 degrees F (444–500 degrees C).

The demarcation between emerald (page 106) and green beryl depends on the depth and intensity of color as well as hue. To be considered emerald, the stone's color must be reasonably strong and green, bluish green, or perhaps slightly yellowish green. If the color is too pale or too yellowish, the stone is classified as green beryl.

1 Golden beryl, antique cut, 28.36ct
2 Heliodor, antique cut, 45.24ct
3 Bixbite, antique cut, 49.73ct
4 Morganite, antique cut, 23.94ct
5 Goshenite, marquise, 25.58ct
6 Heliodor, oval, 29.79ct
7 Beryl, 2 crystals, together 32.5g
8 Morganite, rough, 24.5g

1
2
3
4
5
6
7
8

Chrysoberyl

Color: Golden-yellow, green-yellow, green, brownish, red
Color of streak: White
Mohs' hardness: 8½
Density: 3.70–3.78
Cleavage: Good
Fracture: Weak conchoidal, uneven
Crystal system: (Orthorhombic), thick-tabled, intergrown triplets
Chemical composition: $BeAl_2O_4$ beryllium aluminum oxide
Transparency: Transparent to opaque
Refractive Index: 1.746–1.763
Double refraction: +0.007 to +0.011
Dispersion: 0.015 (0.011)
Pleochroism: Very weak: red lo yellow, yellow to light green, green
Absorption: 504, 495, 485, 445
Fluorescence: Usually none
Green: weak; dark red

Chrysoberyl (Greek—gold) has been known since antiquity; the varieties alexandrite and chrysoberyl cat's-eye are especially valued. The host rock is granite pegmatite, mica schist, and placers. Deposits of the actual chrysoberyl (numbers 3, 9, and 10 below) are in Brazil (Minas Gerais, Espirito Santo) and Sri Lanka, as well as Burma (Myanmar), Madagascar, Russia, Zimbabwe, and the United States.

Stones are fashioned mainly in step, Ceylon, and brilliant cuts. The famous Hope chrysoberyl (London), a light green, faceted stone of 45ct, is completely clean. A blue-green variety can be found on page 280.

Possibilities for Confusion With andalusite (page 194), braziliante (page 206), golden beryl (page 112), hiddenite (page 130), peridot (page 174), sapphire (page 102), sinhalite (page 202), scapolite (page 204), spinel (page 116), topaz (page 118), tourmaline (page 126), and zircon (page 124).

Chrysoberyl Cat's-Eye [2, 4] (Also called cymophane; Greek—waving light) Fine, parallel inclusions produce a silver-white line, which appears as a moving light ray in a cabochon cut stone. The name chrysoberyl cat's-eye is derived from this effect, which reminds one of the pupil of a cat. The short term "cat's-eye" always refers to chrysoberyl; all other cat's-eye must be designated by an additional name. There are deposits in Sri Lanka and Brazil, as well as in China, India, and Zimbabwe.

Possibilities for Confusion With quartz cat's-eye (page 140), prehnite cat's-eye (page 204). Synthetic chrysoberyl cat's-eye and doublets are known.

Alexandrite [5–8] (Named after Czar Alexander II) Was discovered only as recently as 1830 in the Urals. It is green in daylight, and light red in artificial incandescent light. This changing of color is best seen in thick stones. Alexandrite displaying the cat's-eye effect is a great rarity. Care must be taken when working with it, as it is sensitive to knocks and color changes are possible with exposure to great heat. High-quality alexandrite is one of the most expensive of all gemstones.

The deposits in the Urals are worked out. Today it is mined in Sri Lanka and Zimbabwe, and since the end of the 1980s especially in Brazil (Minas Gerais). Deposits are also found in Burma, Madagascar, and Tanzania. The largest stone, 1876ct, was found in Sri Lanka. The largest cut alexandrite weighs 66ct; it is in the Smithsonian Institution in Washington, D.C.

Possibilities for Confusion With synthetic corundum, synthetic spinel (page 116), andalusite (page 194), pyrope (page 120). Synthetic alexandrite is available on the market.

1 Chrysoberyl in host rock
2 Chrysoberyl cat's-eye, 24.09ct
3 Chrysoberyl, 3.36 and 2.23ct
4 Chrysoberyl cat's-eye, 4.33ct
5 Alexandrite in daylight and artificial light
6 Alexandrite, oval, 0.80ct
7 Alexandrite, intergrown triplet crystal
8 Alexandrite cat's-eye, 2.48ct
9 Chrysoberyl, oval, 9.24ct
10 Chrysoberyl, antique cut, 2.1ct
11 Chrysoberyl crystal
12 Chrysoberyl in host rock

1
2
4
3
5
6
10
7
8
9
11
12

Spinel

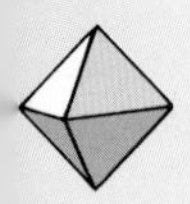

Color: Red, orange, yellow, brown, blue, violet, purple, green, black
Color of streak: White
Mohs' hardness: 8
Density: 3.54–3.63
Cleavage: Indistinct
Fracture: Conchoidal, uneven
Crystal system: (Cubic), octahedron, twins, rhombic dodecahedron
Chemical composition: $MgAl_2O_4$ magnesium aluminium oxide
Transparency: transparent to opaque
Refractive index: 1.712–1.762
Double refraction: None
Dispersion: 0.020 (0.011)
Pleochroism: Absent
Absorption: red Sp.: 685, 684, 675, 665, 656, 650, 642, 632, 595–490, 465, 455
Fluorescence: Red Sp. strong: red; Blue Sp. weak: reddish, green

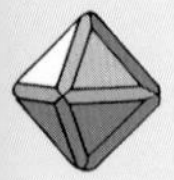

In mineralogy, spinel classifies a whole group of related minerals; only a few are o gemstone quality. The origin of the name is uncertain, perhaps Greek "spark" o Latin "thorn." The species gemologists designate as spinel occurs in all colors, th favorite being a ruby-like red. The coloring agents are iron, chromium, vanadium and cobalt. Large stones are rare and star spinels are very rare.

Flame Spinel Trade term for bright orange to orange-red spinel. In the past, some times called rubicelle.

Balas Spinel [4] Falsely called Balas ruby (after a region of Afghanistan). Pale re variety.

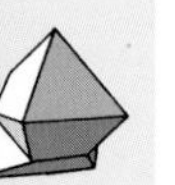

Pleonaste [1] Also called Ceylonite. Dark green to blackish, opaque spinel whic contains iron $(Mg,Fe)\ A1_2O_4$; density 3.63–3.90.

Hercynite Dark green to black spinel-group species, containing iron. $Fe_2{+}Al_2O_4$ density 3.95.

Gahnite Also known as Zinc spinel. Blue, violet, or dark green to blackish spinel group mineral. $ZnAl_2O_4$; density 4.00–4.62. (See page 220, no. 1).

Gahnospinel [3,7] Blue to dark blue or green species between spinel and gahnite i composition, colored by iron. $(MgZn)Al_2O_4$; density 3.58–4.06.

Picotite Also called Chrome spinel. Brownish, dark green, or blackish spinel-grou member. $Fe\ (Al,Cr_2)O_4$; density 4.42.

Spinel was recognized as an individual mineral only 150 years ago. Before then it wa classed as ruby. Some well-known "rubies" are really spinels, such as the "Black Prince's Ruby" in the English Crown (see photo, page 9) and the 361ct "Timur Ruby" in a necklace in the English Crown Jewels. The drop-shaped spinels in the Wittelsbach crown of 1830 were also originally thought to be rubies.

It occurs with ruby and sapphire in placer deposits, mainly in Burma (Myanmar near Mogok), Cambodia, and Sri Lanka (near Ratnapura). Other deposits are found in Afghanistan, Australia, Brazil, Madagascar, Nepal, Nigeria, Tadzhikistan, Tanza nia, Thailand, and the United States (New Jersey).

The two largest spinels (formed as roundish octahedrons) weigh 520ct each and are in the British Museum of London. The Diamond Fund in Moscow owns a spinel that weighs over 400ct.

Possibilities for Confusion With amethyst (page 134), chrysoberyl (page 114) pyrope (page 120), sapphire (page 102), topaz (page 118), tourmaline (page 126) and zircon (page 124). Synthetic spinels have been on the market since the 1920s (compare to page 267). They imitate not only natural spinel but also many other gemstones, especially ruby.

1 Pleonaste crystals in host rock
2 Spinel, 28.47 and 4.16ct
3 Spinel, 3 faceted stones
4 Spinel, so-called balas ruby 17.13ct
5 Spinel, antique cut, 5.05ct
6 Spinel, two ovals, 7.96 and 5.32ct
7 Spinel, blue, 15.08 and 30.11ct
8 Spinel, 12 different reds
9 Spinel, yellow, 3.14 and 5.07ct
10 Spinel, crystals and other rough stones

1
2
3
4
5
6
7
8
9
10

Topaz

Color: Colorless, yellow, orange, red-brown, light to dark blue, pink-red, red, violet, light green
Color of streak: White
Mohs' hardness: 8
Density: 3.49–3.57
Cleavage: Perfect
Fracture: Conchoidal, uneven
Crystal system: Orthorhombic, prisms with multi-faceted ends, often 8-sided in cross-section striations along length
Chemical composition: $Al_2[(F,OH)_2|SiO_4]$ fluor containing aluminium silicate

Transparency: Transparent, translucent
Refractive index: 1.609–1.643
Double refraction: +0.008 to +0.016
Dispersion: 0.014 (0.008)
Pleochroism: Yellow: definite; lemon-honey, straw-yellow
Blue: weak; light and dark blue,
Red: strong; dark red, yellow, pink-red
Absorption spectrum: Pink: 682
Fluorescence: Pink: weak; brown
Red: weak; yellow-brown
Yellow: weak; orange-yellow

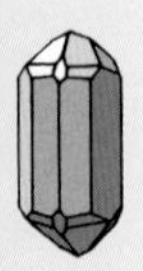

Formerly, the name topaz was not applied consistently or specifically; one called all yellow and golden-brown, and sometimes also green, gemstones topaz. The name topaz is most probably derived from an island in the Red Sea, now Zabargad but formerly Topazos, the ancient source of peridot (see page 174).

Colors of the gemstone that is today called topaz are rarely vivid. The most common color is yellow with a red tint; the most valuable is pink to reddish-orange. The coloring agents are iron and chromium. Some yellowish-brown varieties of certain deposits gradually fade in the sunlight.

Care must be taken during polishing and setting because of the danger of cleavage. They are also not resistant to hot sulphuric acid. The luster is vitreous.

Deposits are associated with pegmatites or secondary placers. During the 18th century, the most famous topaz mine was at Schneckenstein in the southern Voigtland in Saxony. Today, Brazil (Minas Gerais) is the most important supplier. Other deposits are in Afghanistan, Australia, Burma (Myanmar), China, Japan, Madagascar, Mexico, Namibia, Nigeria, Pakistan, Russia (the Urals, Transbaikalia), Zimbabwe, Sri Lanka, and the United States. Light blue topazes are found also in Northern Ireland, Scotland, and Cornwall, England.

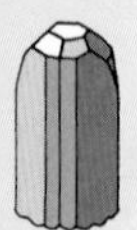

Topazes weighing several pounds are known. In 1964 some blue topazes were found in the Ukraine, each weighing about 220 lb (100 kg). The Smithsonian Institution in Washington, D.C., owns cut topazes of several thousand carats each.

Colored stones are usually step-(emerald) or scissor-cut, and colorless ones or weakly colored ones are brilliant-cut. Topazes with disordered inclusions are cut en cabochon.

Possibilities for Confusion With apatite (page 210), aquamarine (page 110), brazilianite (page 206), chrysoberyl (page 114), citrine (page 136), danburite (page 198), diamond (page 86), precious beryl (page 112), fluorite (page 214), kunzite (page 130), orthoclase (page 180), phenakite (page 196), ruby (page 98), sapphire (page 102), spinel (page 116), tourmaline (page 126), and zircon (page 124).

Blue synthetic topazes have been known since 1976; however, they are too expensive for the gemstone trade. Other synthetic stones show good topaz colors. Almost all blue topaz sold today is produced by first irradiating and then heating natural colorless topaz. Since the quartz variety citrine (page 136) is in the trade often falsely called "gold topaz" or "Madeira topaz," real topaz is sometimes called precious topaz, in order to clearly distinguish them.

---

1 Topaz, rectangular step cut, 46.61ct, Brazil
2 Topaz, rough, 225.50ct, Brazil
3 Topaz, crystal cleavage, 18.00ct, Brazil
4 Topaz, oval, 93.05ct, Afghanistan
5 Topaz, emerald cut, 88.30ct, Brazil
6 Topaz, fancy cut, 32.44ct
7 Topaz, 2 faceted ovals, 53.75ct, Russia
8 Topaz, cabochon, 17.37ct, Brazil
9 Topaz, crystal, 65.00ct, Brazil
10 Topaz, crystal in host rock

1
2
3
4
5
6
7
8
9
10

Garnet Group

This is a group of differently colored minerals with similar crystal structure and related chemical composition. The main representatives are pyrope, almandite and spessartite (pyralspite series), grossularite, andradite, and uvarovite (ugrandite series). Within the series are also mixed members. The name derives from the Latin for grain because of the rounded crystals and similarity to the red kernals of the pomegranate. Garnet, in the popular sense, is usually understood only as the red "carbuncle stones" pyrope and almandite (see page 270).

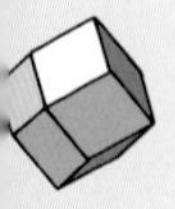

Data common to all garnets:
Color or streak: White
Mohs' hardness: 61/2–71/2
Cleavage: Indistinct
Crystal system: (Cubic) rhombic dodecahedron, icositetrahedron
Fracture: Conchoidal, splintery, brittle
Transparency: Transparent to opaque
Double refraction: Normally none
Pleochroism: Absent
Fluroescence: Mostly none
Luster: Vitreous

Pyrope [4, 5] Garnet Group

Color: Red, frequently with brown tint
Density: 3.62–3.87
Chemical composition: $Mg_3Al_2[SiO_4]_3$ magnesium aluminium silicate
Refractive index: 1.720–1.756
Dispersion: 0.022 (0.013–0.016)
Absorption: 687, 685, 671, 650, 620–520, 505

Pyrope (Greek—fiery) was the fashion stone of the 18th and 19th centuries, especially the "Bohemian Garnet." Deposits are found in Burma (Myanmar), China, Madagascar, Sri Lanka, South Africa, Tanzania, and the United States. Can be confused with almandite (see below), ruby (page 98), spinel (page 116), and tourmaline (page 126). Imitations are made with red glass.

Rhodolite [6, 7, 8] Purplish red or rose-color garnet between pyrope and almandite in composition.

Almandite [9, 10] Garnet Group

Color: Red with violet tint
Density: 3.93–4.30
Chemical composition: $Fe_3^{2+}Al_2[SiO_4]_3$ iron aluminium silicate
Refractive index: 1.770–1.820
Dispersion: 0.027 (0.013–0.016)
Absorption: 617, 576, 526, 505, 476, 462, 438, 428, 404, 393

Its name is derived from the town in Asia Minor. Deposits are found in Brazil, India, Madagascar, Sri Lanka, and the United States, as well as the Czech Republic and Austria. Can be confused with pyrope (see above), ruby (page 98), spinel (page 116), and tourmaline (page 126).

Spessartite [2, 3] Garnet Group

Color: Orange to red-brown
Density: 4.12–4.18
Chemical composition: $Mn_3^{2+}Al_2[SiO_4]_3$ manganese aluminum silicate
Transparency: Transparent, translucent
Refractive index: 1.790–1.820
Dispersion: 0.027 (0.015)
Absorption: 495, 485, 462, 432, 424, 412

Its name is derived from occurrence in the Spessart (= forest), Germany. Deposits are found in Burma (Myanmar), Brazil, China, Kenya, Madagascar, Sri Lanka, Tanzania, and the United States.

The best specimens come from Namibia ("Mandarin Spessartite" on page 279). Can be confused with andalusite (page 194), chrysoberyl (page 114), fire opal (page 168), hessonite (page 122), sphene (page 210), and topaz (page 118).

1 Range of garnet colors, green-yellow-brown-red
2 Spessartite crystal in host-rock
3 Spessartite, 3 cabochons
4 Pyrope crystal, icosatetrahedron
5 Pyrope, 3 faceted stones
6 Rhodolite, brilliant cut, 4.02ct
7 Rhodolite, marquise, 2 ovals
8 Rhodolite crystal, rolled
9 Almandite in mica
10 Almandite, trapezoid, oval, emerald cut

1
2
3
4
5
6
7
8
9
10

Grossularite (Grossular) [4, 5]

Garnet Group

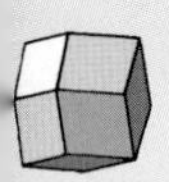

Color: Colorless, green, yellow, brown
Density: 3.57–3.73
Chemical composition: $Ca_3Al_2[SiO_4]_3$ calcium aluminum silicate

Refractive index: 1.734–1.759
Dispersion: 0.020 (0.012)
Absorption spectrum: 697, 630, 605, 505
Fluorescence: Dense G. Strong: red-orange

Grossularite (Latin—gooseberry) deposits are found in Canada, Kenya, Mali, Pakistan, Russia (Siberia), Sri Lanka, South Africa, Tanzania, and Vermont. Can be confused with demantoid (page 122), emerald (page 106), and tourmaline (page 126).

Hessonite [1, 2, 3] (also called cinnamon stone and Kaneel stone) Brown-red variety. Deposits are found in Sri Lanka, as well as Brazil, India, Canada, Madagascar, Tanzania, and the United States. Can be confused with chrysoberyl (page 114), cassiterite (page 200), spessartite (page 120), and zircon (page 124).

Leuco garnet [6] Colorless variety. Deposits are found in Canada, Mexico, and Tanzania.

Hydrogrossular Originally a variety of grossularite; in 1984, identified as a proper mineral and named katoite (see page 280).

Tsavorite (Tsavolite) Green to emerald green variety from Kenya and Tanzania; discovered in the early 1970s. Ill. on page 279.

Andradite

Garnet Group

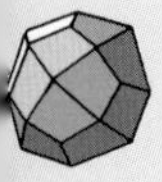

Color: Black, brown, yellow-brown
Density: 3.7–4.1
Chemical composition: $Ca_3Fe_2^{3+}[SiO_4]_3$ calcium iron silicate

Refractive index: 1.88–1.94
Dispersion: 0.057
Absorption: 701, 693, 640, 622, 443

Andradite (named after a Portuguese mineralogist) varieties include the following:

Demantoid [7] the most valuable garnet (name means "diamond-like luster"), green to emerald green. Deposits found in China, Korea, Russia, the United States, and Zaire. Can be confused with grossularite (see above), peridot (page 174), emerald (page 106), spinel (page 116), tourmaline (page 126), and uvarovite (see below).

Melanite [9] Opaque black variety (Greek—black). Deposits are in Germany (Kaiserstuhl/Baden-Wurttemberg), France, Italy, and Colorado. Used for mourning jewelry. Can be confused with black glass.

Topazolite [10] Yellow to lemon yellow, topaz-like (therefore the name) variety. Deposits found in Switzerland, the Italian Alps, and California.

Uvarovite [8]

Garnet Group

Color: Emerald-green
Density: 3.77
Chemical composition: $Ca_3Cr_2[SiO_4]_3$ calcium chromium silicate

Refractive index: 1.87
Dispersion: (0.014-0.021)
Absorption: Cannot be evaluated

Named after a Russian statesman. Rarely occurs in gemstone quality. Deposits found in Finland, India, Canada, Poland, Russia (the Urals), and California. Can be confused with demantoid (page 122) and emerald (page 106).

1 Hessonite crystal in host rock
2 Hessonite, 2 cabochons
3 Hessonite, 3 faceted stones
4 Grossularite, rhombic dodecahedron
5 Grossularite, green and copper-brown
6 Leuco garnet, marquise, 1.97ct
7 Demantoid, 3 rough and 3 cut stones
8 Uvarovite crystals, partly rolled
9 Melanite, 2 crystals
10 Topazolite, rough and faceted

1
2
3
4
5
6
7
8
9
10

Zircon

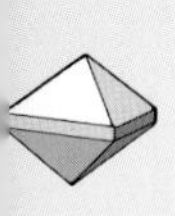

Color: Colorless, yellow, brown, orange, red, violet, blue, green
Color of streak: White
Mohs' hardness: 6½–7½
Density: 3.93–4.73
Cleavage: Indistinct
Fracture: Conchoidal, very brittle
Crystal system: (Tetragonal) short, stocky four-sided prisms with pyramidal ends
Chemical composition: $Zr[SiO_4]$ zirconium silicate
Transparency: Transparent to translucent
Refractive index: 1.810–2.024
Double refraction: +0.002 to +0.059
Low-z: None
Dispersion: 0.039 (0.022), Low z.: None
Pleochroism: Yellow z. very weak: honey-yellow, brown-yellow;
Blue z. distinct: blue, yellow-gray to colorless
Absorption: (High z.) 691, 689, 662, 660, 653, 621, 615, 589, 562, 537, 516, 484, 460, 433
Fluorescence: Blue: very weak; light orange
Red and brown: weak; dark yellow

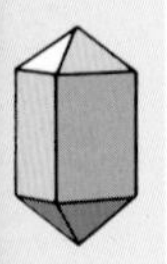

Zircon has been known since antiquity, albeit under various names. Today's name is most likely derived from the Persian language ("golden colored"). Because of its high refractive index and strong dispersion, it has great brilliance and intensive fire. It is brittle and therefore sensitive to knocks and pressure; the edges are easily damaged. The luster is vitreous to a brilliant sheen. A content of radioactive elements (uranium, thorium) causes large variations of physical properties. Zircons with the highest values in optical properties are designated as high zircons, those with the lowest values as low zircons. In between are the medium zircons. The alteration caused by radioactive elements in green (low) zircons is so advanced that these stones can be nearly amorphous, even though their outer appearance seems unchanged. These green, slightly radioactive zircons are rarely found in the gemstone trade, but are highly prized by collectors. Zircons with a cat's-eye effect are also known.

Hyacinth Old term for yellow, yellow-red to red-brown zircon. Also used for hessonite (page 122).

Jargon Old term for straw-yellow to almost colorless zircon.

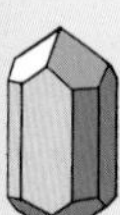

Starlight Trade term for blue zircon variety, created through heating other zircons.

Deposits are mainly alluvial; found in Burma (Myanmar), Cambodia, Sri Lanka, Thailand, as well as Australia, Brazil, Korea, Madagascar, Mozambique, Nigeria, Tanzania, and Vietnam.

In nature the gray-brown and red-brown zircons are the most common. Colorless specimens are rare. In the South Asian countries where found, the brown varieties are heat-treated at temperatures of 1472–1832 degrees F (800–1000 degrees C), producing colorless and blue zircons. These colors do not necessarily remain constant; ultraviolet rays or sunlight can produce changes. Colorless stones are brilliant cut; colored ones are given a brilliant or step (emerald) cut. Synthetic zircons are only of scientific interest.

Possibilities for Confusion With aquamarine (page 110), chrysoberyl (page 114), demantoid (page 122), diamond (page 86), hessonite (page 122), idocrase (page 202), sapphire (page 102), sinhalite (page 202), topaz (page 118), and tourmaline (page 126). Colorless heat-treated zircon is fraudulently offered for diamond as "matura" ("matara") diamond (see page 270).

1 Zircon, rectangular, 9.81ct
2 Zircon, pear and brilliant
3 Zircon, brilliant, 14.35ct
4 Zircon, 2 brilliants
5 Zircon, oval, 5.11ct
6 Zircon, emerald cut, 7.92ct
7 Zircon, emerald cut, 4.02ct
8 Zircon, 4 brilliants
9 Zircon, 3 faceted stones
10 Zircon, rough stones

The photos are 20 percent larger than the originals.

1
2
3
4
5
6
7
8
9
10

Tourmaline Group

Color: Colorless, pink, red, yellow, brown, green, blue, violet, black, multicolored
Color of streak: White
Mohs' hardness: 7–7½
Density: 2.82–3.32
Cleavage: Indistinct
Fracture' Uneven, small conchoidal, brittle
Crystal system: (Trigonal), long crystals with triangular cross section and rounded sides, definite striaton parallel to main axis
Transparency: Transparent to opaque
Refractive index: 1.614–1.666
Double refraction: –0.014 to –0.032
Disperson: 0.017 (0.009–0.011)
Pleochroism: Red t.: definite: dark red, light red;
Brown t.: definite: dark brown, light brown;
Green t.: strong: dark green, yellow-green;
Blue t.: strong: dark blue, light blue
Absorption spectrum: Often extremely faint
Fluorescence: Weak or none

The tourmaline group refers to a number of related species and varieties.

Even though tourmaline has been known since antiquity in the Mediterranean region, the Dutch imported it only in 1703 from Sri Lanka to Western and Central Europe. They gave the new gems a Sinhalese name, *Turamali*, which is thought to mean "stone with mixed colors."

According to color, the following varieties are recognized in the trade:

Achroite (Greek—without color) colorless or almost colorless, quite rare.

Dravite [page 129, nos. 1, 7, 8] yellow-brown to dark brown, sometimes used for stones not of the dravite species (see below).

Indicolite (indigolite) [page 129, nos. 3, 5, 11, 15] named (after color) blue in all shades.

Rubellite [page 129, nos. 2, 4] (Latin—reddish) pink to red, sometimes with a violet tint; ruby color is the most valuable.

Schorl [4, 5] black, very common; used for mourning jewelry. Name derived from an old mining term, sometimes applied to tourmaline that is not actually schorl (see below).

Siberite (after finds in Urals) lilac to violet blue.

Verdelite [page 129, nos. 6, 13] ("green stone") green in all shades.

Recently, instead of variety names, more and more frequently color names are simply added to the word tourmaline, e.g., yellow tourmaline, pink tourmaline.

Mineralogy distinguishes tourmalines according to their chemical composition. The individual members include the following:

Buergerite (after U.S. scholar) $NaFe_3^{3+}Al_6[F|O_3|(BO_3)_3|Si_6O_{18}]$ = iron tourmaline

Dravite (after a deposit near the river Drave, Carinthia/Austria) $NaMg_3Al_6[(OH)_4|(BO_3)_3|Si_6O_{18}]$ = magnesium tourmaline

Elbaite (after the island of Elba/Italy) $Na(Li,Al)_3Al_6[(OH)_4|(BO_3)_3|Si_6O_{18}]$ = lithium tourmaline. Paraiba Tourmaline see pg. 280.

Liddicoatite (after U.S. gemologist) $Ca(Li,Al)_3Al_6[(OH,F)_4|(BO_3)_3|Si_6O_{18}]$ = calcium tourmaline

Schörl (after an old mining expression for "false ore") $NaFe_3^{2+}(Al,Fe^{3+})_6[(OH_4|BO_3)_3|Si_6O_{18}]$ = iron tourmaline

Tsilaisite (after a local name in Madagascar) $NaMn_3^{2+}Al_6[(OH)_4|(BO_3)|Si_6O_{18}]$ = manganese tourmaline

Uvite (after a province in Sri Lanka) $Ca(Mg,Fe^{2+})_3(Al_5Mg)[(OH,F)_4|(BO_3)_3|Si_6O_{18}]$ = magnesium tourmaline

1 Tourmaline, 8 polished cross sections
2 Rubellite cat's-eye, 1.87ct
3 Tourmaline, crystals stem-like on quartz
4 Schorl, crystals, opaque
5 Schorl in quartz, partly polished
6 Tourmaline "watermelon"
7 Verdelite, 2 crystals
8 Multicolored tourmaline, crystal

1
2
3
8
4
6
7
5

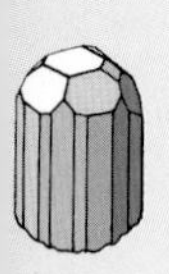

Unicolored tourmaline crystals are quite rare. Most show various tones in the same crystal or even different colors [page 127, nos. 6, 8]. There are colorless tourmalines with black crystal ends, green ones with red crystal ends, and some with different-colored layers. There are stones whose core is red, the inner layer white, and the outer layer green—called "watermelon" [page 127, no. 6].

Tourmaline cat's-eyes exist in various colors, but only in the green and pink varieties [page 127, no. 2] is the chatoyancy usually strong, caused by thin tube-like inclusions. Some tourmalines show a slight change of color in artificial light. They have a vitreous sheen on crystal surfaces, a greasy sheen on fractured surfaces.

By heating and subsequent cooling, as well as by applying pressure, i.e., by rubbing, a tourmaline crystal will become electrically charged. It will then attract dust particles as well as small pieces of paper (pyro- and piezo-electricity). The Dutch, who first imported tourmaline into Europe, knew of this effect. They used a heated stone to pull ash out of their meerschaum pipes and thus called this strange stone *aschentrekker* (ash puller). For a long time this was the popular name for a tourmaline. Due to this pyroelectrical effect, tourmaline has to be cleaned more often than other gemstones.

Deposits are found in pegmatites and alluvial deposits. The most important tourmaline supplier is Brazil (Minas Gerais, Paraiba). Other deposits are in Afghanistan, Australia, Burma (Myanmar), India, Madagascar, Malawi, Mozambique, Namibia, Nepal, Nigeria, Pakistan, Rwanda, Russia, Zambia, Zimbabwe, Sri Lanka, Tanzania, the United States (California, Maine), and Zaire. In Europe, there are tourmaline deposits on Elba (Italy) and in Switzerland (Tessin).

The most desired colors are intense pink and green. It is used in different cuts. Because of the strong pleochroism, dark stones must be cut so that the table lies parallel to the main axis. In the case of pale stones, the table should be perpendicular to the long axis in order to obtain the deeper color.

By heating to 842–1202 degrees F (450–650 degrees C), color changes can be produced in some tourmalines. Some green tourmaline becomes emerald green; others are lightened. The color of tourmalines which have been changed with gamma-irradiation may fade. Synthetic tourmalines are used only for research purposes. The stones, offered as synthetic tourmaline, are really tourmaline-colored synthetic spinels.

Possibilities for Confusion With many gemstones, due to the large variety of colors, especially amethyst (page 134), andalusite (page 194), chrysoberyl (page 114), citrine (page 136), demantoid (page 122), emerald (page 106), hiddenite (page 130), idocrase (page 202), kunzite (page 130), morganite (page 112), peridot (page 174), prasiolite (page 136), ruby (page 98), topaz (page 118), zircon (page 124), and glass imitations.

1 Dravite, 2 faceted stones
2 Rubellite, oval, 1.73ct
3 Indicolite, emerald cut, 6.98ct
4 Rubellite, 2 faceted stones, together 4.55ct
5 Indicolite, emerald cut and antique cut
6 Verdelite, oval, 19.88ct
7 Dravite, cabochon, 19.97ct
8 Dravite, 3 faceted stones
9 Rubellite, oval, 6.16ct
10 Indicolite, crystal
11 Indicolite, faceted, 2.97ct
12 Indicolite crystal
13 Verdelite, 2 cabochons, 9.77ct
14 Tourmaline, 2 faceted yellow-green stones
15 Indicolite, 3 cabochons
16 Tourmaline, multicolored, 24ct
17 Rubellite, 3 cabochons
18 Tourmaline, 3 crystals

The illustrations are 20 percent larger than the originals.

1
2
3
4
5
6
7
8
9
10
11
12
13
14
15
16
17
18

Spodumene Species

The name refers to the mineral spodumene (Greek—ash-colored) because the common non-gem crystals are mostly opaque, white to yellowish. For a long time gem varieties have been known as hiddenite and kunzite; since the 1970s some isolated transparent colorless varieties have been found. Most recently light yellow and green varieties have also been known. Rarely displays the cat's-eye effect.

Hiddenite [1–3, 8]

Spodumene Species

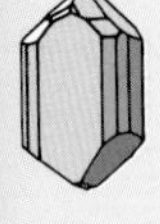

Color: Yellow-green, green-yellow, emerald-green
Color of streak: White
Mohs' hardness: 61/2[fraction]–7
Density: 3.15–3.21
Cleavage: Perfect
Fracture: Uneven, brittle
Crystal system: Monoclinic; prismatic, tabular
Chemical composition: $LiAl[Si_2O_6]$ lithium aluminum silicate

Transparency: Transparent
Refractive index: 1.660–1.681
Double refraction: +0.014 to +0.016
Dispersion: 0.017 (0.010)
Pleochroism: Definite; blue-green, emerald-green, yellow-green
Absorption spectrum: 690, 686, 669, 646, 620, 437, 433
Fluorescence: Very weak; red-yellow

Named after A. E. Hidden, who discovered this stone in 1879 in North Carolina. The coloring agent is chromium. Colors can gradually fade; has a strong vitreous luster. Usually used with the step cut (emerald). In order to display strong colors (due to pleochroism), the table facet must be perpendicular to the main axis of the stone. Deposits occur in granite pegmatite. Deposits found in Burma (Myanmar), Brazil (Minas Gerais), Madagascar, North Carolina, and California. Can be confused with chrysoberyl (page 114), diopside (page 206), emerald (page 106), peridot (page 174), precious beryl (page 112), and verdelite (page 126).

Kunzite [4–7]

Spodumene Species

Color: Pink-violet, light violet
Color of streak: White
Mohs' hardness: 61/2[fraction]–7
Density: 3.15–3.21
Cleavage: Perfect
Fracture: Uneven, brittle
Crystal system: Monoclinic; prismatic, tabular
Chemical composition: $LiAl[Si_2O_6]$ lithium aluminum silicate

Transparency: Transparent
Refractive index: 1.660–1.681
Double refraction: +0.014 to +0.016
Dispersion: 0.017 (0.010)
Pleochroism: Definite; amethyst color, pale red, colorless
Absorption spectrum: Not diagnostic
Fluorescence: Strong, yellow-red, orange

Named after the U.S. mineralogist G. F. Kunz, who first described this gem in 1902. The coloring agent is manganese. Stones are mostly light colored; colors can fade. Brownish and green-violet types can be improved in color by heating to about 300 degrees F (150 degrees C). There are frequently aligned inclusions such as tubes or fractures. The stone has a vivid vitreous luster. The table facet must be perpendicular to the main axis of the stone. Deposits occur in granite pegmatite. The main producer is Brazil (Minas Gerais). Other deposits are found in Afghanistan, Burma (Myanmar), Madagascar, Pakistan, and the United States. Can be confused with many pink-colored stones, especially amethyst (page 134), morganite (page 112), petalite (page 204), rose quartz (page 138), rubellite (page 126), sapphire (page 102), and topaz (page 118), as well as colored glass.

1 Hiddenite, emerald cut, 22.03ct
2 Hiddenite, pear-shaped, 9.30ct
3 Hiddenite, emerald cut, 19.14ct
4 Kunzite, emerald cut, 16.32ct
5 Kunzite, oval, 3.13ct
6 Kunzite, antique cut, 6.11ct
7 Kunzite, 2 crystals
8 Hiddenite, crystal and broken piece

1
2
3
4
5
6

7

8

Quartz

Quartz (named after a Slavic word for "hard") is the name for a group of minerals of the same chemical composition (SiO_2) and similar physical properties.
Macrocrystalline quartz (crystals recognizable with the naked eye) includes stones gemologists classify as varieties of the quartz species: amethyst, aventurine, rock crystal, blue quartz, citrine, hawk's-eye, prasiolite, quartz cat's-eye, smoky quartz, rose quartz, and tiger's-eye.
Cryptocrystalline quartz (microscopically small crystals) generally known as chalcedony, includes agate, petrified wood, chrysoprase, bloodstone, jasper, carnelian, moss agate, and sard.

Rock Crystal [8–11]

Quartz Species

Color: Colorless
Color of streak: White
Mohs' hardness: 7
Density: 2.65
Cleavage: None
Fracture: Conchoidal, very brittle
Crystal system: (Trigonal), hexagonal prisms
Chemical composition: SiO_2 silicon dioxide
Transparency: Transparent
Refractive index: 1.544–1.553
Double refraction: +0.009
Dispersion: 0.013 (0.008)
Pleochroism: Absent
Absorption spectrum: None
Fluorescence: None

The name crystal comes from the Greek for "ice," as it was believed that rock crystal was eternally frozen. Rock crystals weighing many tons have been found. Cuttable material is rare. Inclusions are of goethite [star quartz, no. 12], gold, pyrite, rutile, and tourmaline [page 127, no. 5]; the luster is vitreous. Important deposits, among others all over the world, are found in Brazil, Madagascar, the United States, and the Alps. They are used for costume jewelry and delicate bowls and to imitate diamonds. Rock crystal alters to a smoky color with gamma irradiation. Can be confused with many colorless gems as well as glass. Synthetic rock crystal is used only for industrial purposes.

Smoky Quartz [1–7]

Quartz Species

Color: Brown to black, smoky gray
Color of streak: White
Mohs' hardness: 7
Density. 2.65
Cleavage: None
Fracture: Conchoidal, very brittle
Crystal system: (Trigonal), hexagonal prisms
Chemical composition: SiO_2 silicon dioxide
Transparency: Transparent
Refractive index: 1.544–1.553
Double refraction: +0.009
Dispersion: 0.013 (0.008)
Pleochroism: Dark: definite; brown, reddish-brown
Absorption spectrum: Cannot be evaluated
Fluorescence: Usually none

Named after its smoky color. Very dark stones are called "morion" and "caingorm" [6,7]. Coloring is caused by (natural and artificial) gamma rays. The name *smoky topaz* is improper and no longer acceptable in the trade. Frequent inclusions are rutile needles [nos. 1, 2]. Deposits in Brazil, Madagascar, Russia, Scotland, Switzerland, and Ukraine. It is used just as rock crystal. Can be confused with andalusite (page 194), axinite (page 198), idocrase (page 202), sanidine (page 220), and tourmaline (page 126).

1 Smoky quartz with rutile inclusions
2 Smoky quartz with rutile, cabochon
3 Smoky quartz, oval, 3.8g
4 Smoky quartz, 2 crystals
5 Smoky quartz, emerald cut, 5.6g
6 Smoky quartz, oval, 6.2g
7 Smoky quartz crystal
8 Rock crystal, 4 stones, faceted and cabochon
9 Rock crystal, crystals and twins
10 Rock crystal, brilliant cut, 5g
11 Rock crystal, baguette, 1.8g
12 Star quartz, 15g

2
1
3
5
4
7
6
8
9
10
9
11
12

Amethyst [4–8]

Quartz Species

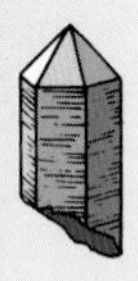

Color: Purple, violet, pale red-violet	Transparency: Transparent
Color of streak: White	Refractive index: 1.544–1.553
Mohs' hardness: 7	Double refraction: +0.009
Density: 2.65	Dispersion: 0.013 (0.008)
Cleavage: None	Pleochroism: Weak: reddish-violet, gray-violet
Fracture: Conchoidal, very brittle	Absorption spectrum: (550–520)
Crystal system: (Trigonal), hexagonal prisms	Fluorescence: Weak; bluish
Chemical composition: SiO_2 silicon dioxide	

Amethyst is the most highly valued stone in the quartz group. The name means "not drunken" (Greek), as amethyst was worn as an amulet against drunkenness. Crystals are always grown onto a base. Prisms are usually not well developed, therefore are often found as crystal points (pointy amethyst) with the deepest color. These parts are broken off at the base for further treatment.

Heat treatment between 878 and 1382 degrees F (470–750 degrees C) produces light yellow, red-brown, green, or colorless varieties. There are some amethysts that lose some color in daylight. The original color can be restored by X-ray radiation. The coloring agent is iron. In artificial light, amethyst does not display as desirable qualities.

Found in geodes in alluvial deposits. The most important deposits are in Brazil ("Palmeira" amethyst of Rio Grande do Sul, "Maraba" amethyst of Para), Madagascar, Zambia, Uruguay, as well as in Burma (Myanmar), India, Canada, Mexico, Namibia, Russia, Sri Lanka, and the United States (Arizona). The best stones are faceted; others are tumbled or worked into ornaments. Formerly amethyst was a favorite gemstone of high officials of the Christian church. Can be confused with precious beryl (page 112), fluorite (page 214), kunzite (page 130), spinel (page 116), topaz (page 118), tourmaline (page 126), and tinted glass. Synthetic amethyst is abundant on the gemstone market.

Ametrine (also called trystine) Color-zoned quartz variety, which consists half of amethyst and other half of citrine. Deposits are found in Brazil (Rio Grande do Sul) and Bolivia.

Amethyst Quartz [1–3]

Quartz Species

Color: Violet with whitish stripes	Transparency: Translucent, opaque
Color of streak: White	Refractive index: 1.54–1.55
Mohs' hardness: 7	Double refraction: +0.009
Density: 2.65	Dispersion: 0.013 (0.008)
Cleavage: None	Pleochroism: Absent
Fracture: Conchoidal, brittle	Absorption spectrum: Cannot be evaluated
Crystal system: (Trigonal) compact	Fluorescence: None
Chemical composition: SiO_2 silicon dioxide	

Amethyst quartz is the rougher, more compact formation of amethyst, layered and striped with milky quartz. Occurs together with amethyst. It is used for beads, baroque stones, cabochons, and ornamental objects. Can be confused with striped fluorite (page 214).

1 Amethyst quartz, rough
2 Amethyst quartz, 7 cabochons
3 Amethyst quartz, polished slice
4 Amethyst, marquise, 3.94ct
5 Amethyst, 4 faceted stones
6 Amethyst, double-ended crystal
7 Amethyst, brilliant cut, 4.16ct
8 Amethyst, geode on agateuf Achat

1
2
3
4
5
6
7
8

Citrine [1–6]

Quartz Species

Color: Light yellow to dark yellow, gold-brown
Color of streak: White
Mohs' hardness: 7
Density: 2.65
Cleavage: None
Fracture: Conchoidal, very brittle
Crystal system: Hexagonal (trigonal); hexagonal prisms with pyramids
Chemical composition: SiO_2 silicon dioxide

Transparency: Transparent
Refractive index: 1.544–1.553
Double refraction: +0.009
Dispersion: 0.013 (0.008)
Pleochroism: Natural: weak; yellow-light yellow
Heat-treated: none
Absorption spectrum: Not diagnostic
Fluorescence: None

The name is derived from its lemon yellow color. The coloring agent is iron. Natural citrines are rare. Most commercial citrines are heat-treated amethysts (page 134) or smoky quartzes (page 132). Brazilian amethyst turns light yellow at 878 degrees F (470 degrees C) and dark yellow to red-brown at 1022–1040 degrees F (550–560 degrees C). Some smoky quartzes turn into citrine color already at about 390 degrees F (200 degrees C).

Almost all heat-treated citrines have a reddish tint. The natural citrines are mostly pale yellow. Names for citrine such as Bahia, Madeira, and Rio Grande topaz are improper and no longer accepted in the trade as they are deceptive. On the other hand, when one, for example, speaks of Madeira color and/or Madeira citrine, this is a correct usage; the expert properly connects a certain color with the locality name.

Deposits of natural-colored citrines are found in Brazil, Madagascar, and the United States, as well as in Argentina, Burma (Myanmar), Namibia, Russia, Scotland, and Spain. Well-colored citrines are used as ring stones and pendants; less attractive stones are made into necklaces or ornaments.

Can be confused with many yellow gemstones, especially apatite (page 210), golden beryl (page 112), orthoclase (page 180), topaz (page 118), and tourmaline (page 126), as well as tinted glass.

Prasiolite [7, 8]

Quartz Species

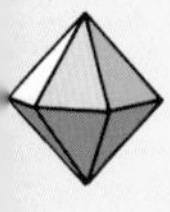

Color: Leek-green
Color of streak: White
Mohs' hardness: 7
Density: 2.65
Cleavage: None
Fracture; Conchoidal, very brittle
Crystal system: Hexagonal (trigonal); hexagonal prisms
Chemical composition: SiO_2 silicon dioxide

Transparency: Transparent
Refractive index: 1.544–1.553
Double refraction: +0.009
Dispersion: 0.013 (0.008)
Pleochroism: Very weak; light green, pale green
Absorption spectrum: Not diagnostic
Fluorescence: None

Prasiolite (Greek—leek-green stone) is not found in nature. It is produced by heating violet amethyst or yellowish quartz from the deposit Montezuma in Minas Gerais, Brazil, to a temperature of about 930 degrees F (500 degrees C). Other deposits of heatable amethyst have recently been reported in Arizona. In sunlight, the color commonly fades.

Can be confused with precious beryl (page 112), peridot (page 174), tourmaline (page 126), and other green gemstones. See also page 278.

1 Citrine, heat-treated, rough
2 Citrine, heat-treated, faceted
3 Citrine, heat-treated, rectangle
4 Citrine, natural, rough
5 Citrine, natural, oval
6 Citrine, 2 emerald cuts
7 Prasiolite, rough
8 Prasiolite, 2 faceted stones

1
2
3
4
5
6
7
8

Rose Quartz [3-7]

Quartz Species

Color: Strong pink, pale pink	Chemical composition: SiO_2 silicon dioxide
Color of streak: White	Transparency: Semitransparent, translucent
Mohs' hardness: 7	Refractive index: 1.544–1.553
Density: 2.65	Double refraction: +0.009
Cleavage: None	Dispersion: None
Fracture: Conchoidal, very brittle	Pleochroism: Absent
Crystal system: (Trigonal) prisms, mostly compact	Absorption: Cannot be evaluated
	Fluorescence: Weak; dark violet

Rose quartz (named after its pink color) is often crackled, usually a little turbid. Coloring agent is titanium. Color can fade. Traces of included rutile needles cause six-rayed stars when cut en cabochon [no. 4]. Deposits are found in Brazil and Madagascar, as well as India, Mozambique, Namibia, Sri Lanka, and the United States. Worked into cabochons, bead necklaces, and ornamental pieces; only the larger clear stones can be faceted [no. 5]. Can be confused with kunzite (page 130), morganite (page 112), and topaz (page 118).

Aventurine [1, 2, 8] Also called Aventurine Quartz

Quartz Species

Color: Green, red-brown, gold-brown, aventurescent	Chemical composition: SiO_2 silicon dioxide
Color of streak: White	Transparency: Translucent, opaque
Mohs' hardness: 7	Refractive index: 1.544–1.553
Density: 2.64–2.69	Double refraction: +0.009
Cleavage: None	Dispersion: None
Fracture: Conchoidal, brittle	Pleochroism: Absent
Crystal system: (Trigonal) massive	Absorption spectrum: Green a.: 682, 649
	Fluorescence: Green a.: Reddish

A type of glass discovered by chance (Italian—*a ventura*) around 1700 gave the same name to the similar-looking stone. Mostly dark green with metallic glittery appearance caused by included fuchsite (green mica) or red- to gold-brown caused by hematite leaves. Deposits are found in Brazil, India, Austria (Steiermark), Russia (Urals, Siberia), and Tanzania. Used for ornamental objects and cabochons. Can be confused with iridescent analcite (page 228), aventurine feldspar (page 182), emerald (page 106), jade (page 170), as well as aventurine glass.

Prase [9]

Quartz Species

Prase is a leek-green (Greek—*prason*) quartz aggregate, usually classified as a chalcedony, whose color is caused by chlorite inclusions. Deposits occur in Saxony (Germany), Finland, Austria (Salzburg), and Scotland. Can be confused with amazonite (page 180) and jade (page 170).

Blue Quartz [10]

Quartz Species

This is a turbid-blue quartz aggregate (quartzite). The inclusions of crocidolite fibers cause the color. Deposits are found in Brazil, Austria (Salzburg), Scandinavia, South Africa, and Virginia. Used for ornaments. See also page 278. Can be confused with dumortierite quartz (page 198) and lapis lazuli (page 188).

1 Aventurine, 5 cabochons
2 Aventurine, rough, partly polished
3 Rose quartz, rough
4 Star rose quartz, 20.23ct
5 Rose quartz, octagon, 8.16ct
6 Rose quartz, 6 cabochons
7 Rose quartz, baroque neclace
8 Aventurine, rough, partly polished
9 Prase, 2 cabochons
10 Blue quartz, rough, partly polished

1
2
3
4
5
6
7
8
9
10

Quartz Cat's-Eye [1, 2]

Quartz Species

Color: White, gray, green, yellow, brown
Color of streak: White
Mohs' hardness: 7
Density: 2.58–2.64
Cleavage: None
Fracture: Irregular
Crystal system: (Trigonal) usually massive
Chemical composition: SiO_2 silicon dioxide

Transparency: Semitransparent translucent
Refractive index: 1.534–1.540
Double refraction: None
Dispersion: None
Pleochroism: Absent
Absorption: Not diagnostic
Fluorescence: None

Quartz cat's-eye is quartz in which numerous fiber-like inclusions of rutile create chatoyancy. Sensitive to some acids. Deposits found in Sri Lanka, as well as Brazil and India. Cut en cabochon, shows chatoyancy like a cat's-eye (hence the name).

Can be confused with chrysoberyl cat's-eye (page 114). Decolorized hawk's- and tiger's-eye are sometimes substituted for quartz cat's-eye. Syntheses are known. The name "cat's-eye" (without the word "quartz") is taken to mean only chrysoberyl cat's-eye (page 114).

Hawk's-Eye [3, 4]

Quartz Species

Finely fibrous, opaque aggregate formed when quartz replaces the mineral crocidolite (type of asbestos), blue-gray to blue-green. Iridescence of planes; fractures have silky luster. It is sensitive to some acids. Found associated with tiger's-eye (see below). Used for ornamental objects and costume jewelry. Cabochons show chatoyancy (small ray of light on surface) which is reminiscent of the eye of a bird of prey. Even flat pieces show a similar effect.

Tiger's-Eye [5, 6]

Quartz Species

Color: Gold-yellow, gold-brown
Color of streak: Yellow-brown
Mohs' hardness: 61/2—7
Density: 2.58–2.64
Cleavage: None
Fracture: Fibrous
Crystal system: (Trigonal) fibrous aggregate
Chemical composition: SiO_2 silicon dioxide

Transparency: Opaque
Refractive index: 1.534–1.540
Double refraction: None
Dispersion: None
Pleochroism: Absent
Absorption: Not diagnostic
Fluorescence: None

Formed from hawk's-eye (see above) where the iron from the decomposed crocidolite has oxidized to a brown color, keeping the fibrous structure. The luster is silky on fractures. Typically displays chatoyant stripes, because structural fibers are crooked or bent. It is sensitive to some acids. Found together with hawk's-eye in slabs of a few inches' thickness. Most important deposits are found in South Africa (where the export of raw material is forbidden), also in Australia, Burma (Myanmar), India, Namibia, and the United States (California). Used for necklaces, costume jewelry, and objets d'art. When cut en cabochon, the surface shows chatoyancy reminiscent of the eyes of a cat. Reddish tiger's-eye (a.k.a. Ox's eye) is artificially dyed. (For tiger's-iron matrix, see page 218.)

1 Quartz cat's-eye
2 Quartz, cat's-eye, 3.96ct
3 Hawk's-eye, rough
4 Hawk's-eye, 2 cabochons
5 Tiger's-eye, partly polished
6 Tiger's-eye, 7 cabochons

1
2
3
4
5
6

Chalcedony

Chalcedony is used by gemologists as a species name for all cryptocrystalline quartzes (compare page 132, e.g., agate, petrified wood, chrysoprase, bloodstone, jasper, carnelian, moss agate, onyx, and sard) as well as specifically only the bluish-white-gray variety, the actual chalcedony (i.e., chalcedony in the narrow sense). Some scientists ascribe only the fibrous varieties to the chalcedonies; the grainy jasper (page 162) then forms its own group.

Whereas the crystal quartzes (rock crystal, amethyst, etc.) have a vitreous luster, chalcedonies are, in their natural state, waxy or dull. Syntheses which can be used economically are not known.

The following data refers to actual chalcedony (i.e., in the narrow sense).

Color: Bluish, white, gray
Color of streak: White
Mohs' hardness: 6½–7
Density: 2.58–2.64
Cleavage: None
Fracture: Uneven, shell-like
Crystal system: (Trigonal) fibrous aggregates
Chemical composition: SiO_2 silicon dioxide

Transparency: Dull, translucent
Refractive index: 1.530–1.540
Double refraction: up to 0.004
Dispersion: None
Pleochroism: Absent
Absorption spectrum: Dyed blue: 690-660, 627
Fluorescence: Blue-white

Chalcedony [nos. 4–6], probably named after an ancient town at the Bosporus, consists of microscopic fibers, which are parallel to each other. Chalcedony shows macroscopically radiating, stalactitic, grape-like or kidney shapes [no. 4]. Always porous; can therefore be dyed (compare to page 152). Natural chalcedony normally has no banding. The trade also offers parallel layered, artificial blue-colored agate as chalcedony [no. 5]. Deposits are found in Brazil, India, Madagascar, Malawi, Namibia, Zimbabwe, Sri Lanka, Uruguay, and California. In ancient times used for cameos; today used in the arts and crafts for rings and necklaces. Can be confused with tanzanite (page 176).

Chrome chalcedony (also called mtorodite or mtorolite) Trade name for a chalcedony from Zimbabwe, which has a natural green color from chromium.

Carnelian [2, 3] — Chalcedony Species

Carnelian is probably named after the color of the kornel cherry because of its color. It is a brownish red to orange, translucent to opaque chalcedony variety. The coloring agent is iron; the color can be enhanced by heating. Deposits are found in Brazil, India, and Uruguay. Most carnelians offered today are agates which are dyed and then heat-treated. When held against the light, the color variety shows stripes, natural carnelian shows a cloudy distribution of color. Used similarly to other chalcedonies. Can be confused with jasper (page 162).

Sard [1] — Chalcedony Species

Red-brown to brown variety of chalcedony (named after town in Asia Minor). No strict separation from carnelian (darker and browner stones are usually called sard), with common deposits, and uses. Many of the "sards" on the market are dyed.

1 Sard, 8 faceted and cabochon stones
2 Carnelian, rough
3 Carnelian, 7 tablet and cabochon stones
4 Chalcedony nodule, partly polished
5 Chalcedony, 3 banded stones
6 Chalcedony, 7 cabochons

1
2
3
4
5
5
6
5

Chrysoprase [1-4]

Color: Green, apple-green
Color of streak: White
Mohs' hardness: 6½–7
Density: 2.58–2.64
Cleavage: None
Fracture: Rough, brittle
Crystal system: (Trigonal) microcrystalline aggregates
Chemical composition: SiO_2 silicon dioxide
Transparency: Translucent, opaque
Refractive index: 1.530–1.540
Double refraction: up to 0.004
Dispersion: None
Pleochroism: Absent
Absorption: Natural ch.: 444; Dyed with nickel: 632, 444
Fluorescence: None

Chrysoprase is considered the most valuable stone in the chalcedony group. The name (Greek—gold-leek) seems misapplied today. The microscopic fine quartz fibers have a radial structure. The coloring agent is nickel. Large broken pieces are often full of fissures with irregular colors. Color can fade in sunlight and when heated (be careful when soldering). Colors may recover under moist storage.

Occurs as nodules or fillings of clefts in serpentine rocks and in weathered materials of nickel ore deposits. Long ago, the deposit of Frankenstein (Zabkowice) in Upper Silesia, Poland, was the most important mine, but it has been worked out since the 14th century. Today's deposits include Australia (New South Wales), Brazil, India, Kazakhstan, Madagascar, Russia (the Urals), Zimbabwe, South Africa, Tanzania, and California.

Used as cabochons, for necklaces, and for ornamental objects. In earlier centuries, it was used as a luxurious decorative stone for interior decoration, such as in the Wenceslaus Chapel in Prague and in Sanssouci Castle in Potsdam (near Berlin).

Possibilities for Confusion With chrome chalcedony (page 142), jade (page 170), prase opal (page 168), prehnite (page 204), smithsonite (page 214), variscite (page 212), and artificially colored green chalcedony.

Chrysoprase Matrix [3, 4] Chrysoprase with brown or white matrix rock. Used for ornamental objects and in jewelry when cut en cabochon.

Bloodstone [5, 6]

Chalcedony Species

Bloodstone is an opaque, dark-green chalcedony with red spots. An old name still used in Europe is heliotrope (Greek—sun turner). Particles of chlorite or included hornblende needles cause the green color. Red spots are caused by iron oxide. The colors are not always constant. The most important deposits are in India, also in Australia, Brazil, China, and the United States. Used often as seals for men's rings and for other ornamental objects.

In the trade, the term *blood jasper* is sometimes used. Bloodstone, however, is not a jasper at all (page 162), even though a radial structure with spherical aggregates can simulate a grainy appearance. Other data as for chrysoprase.

1 Chrysoprase, 2 pieces, partly polished
2 Chrysoprase, 4 cabochons
3 Chrysoprase, 2 stones with matrix
4 Chrysoprase-matrix, partly polished
5 Heliotrope, rough, partly polished
6 Heliotrope, 7 tablet and cabochon stones

1
2
3
4
5
6

Dendritic Agate [1-4] Also called Mocha Stone Chalcedony Species

Dendritic agate is a colorless or whitish-gray, translucent chalcedony with tree- or fern-like markings, the dendrites (Greek—tree-like). According to some authorities, the term *agate* (compare page 148) is not strictly correct, as there are no bandings, but the term is not generally considered deceptive. The dendrites are iron or manganese inclusions of brown or black color. Despite their looks, they have nothing to do with the organic world, but rather they resemble ice crystals on windows in the winter. They are formed at very fine fracture surfaces through crystallization of weathered solutions of neighboring rock.

They are found with other chalcedonies. Important deposits are in Brazil (Rio Grande do Sul); also in India and the United States. Because the Indian stones formerly came via the Arabian harbor of Mocha, these stones are also called mocha stones.

Scenic Agate [2] A dendritic agate where the included dedrites resemble landscape-like images in brown or reddish color tones.

Mosquito Stone [4] (also called midge stone) A dendritic agate where the dendrites do not hang together but show ball-like growths, reminiscent of swarms of mosquitoes.

Used to make rings, brooches, and pendants. Because the colorful inclusions are at various depths of the rough stone, the cutter must try to bring the image-like markings closer to the surface by taking off the upper layers. Inevitably, a not quite even surface-form may result. Imitations have been attempted with silver nitrate. Sometimes petrified wood (page 164) is also referred to as dendritic agate.

Moss Agate [5, 6] Chalcedony Species

Color: Colorless with green, brown, or red inclusions
Color of streak: White
Mohs' hardness: 6½–7
Density: 2.58–2.64
Cleavage: None
Fracture: Rough
Crystal system: (Trigonal) microcrystalline
Chemical composition: SiO_2 silicon dioxide

Transparency: Translucent
Refractive index: 1.530–1.540
Double refraction: up to 0.004
Dispersion: None
Pleochroism: Absent
Absorption: Cannot be evaluated
Fluorescence: Variable, cannot be used for a diagnosis

Moss agate is a colorless, translucent chalcedony with inclusions of green hornblende (or chlorite) in moss-like patterns (therefore the name). Moss agate colors are brown and red through oxidation of the iron hornblende. The name agate (page 148) is generally accepted even though the stone is not banded.

It occurs as filler in fissures or as pebbles. India supplies the best quality specimens. Other deposits are in China, Russia (the Urals), and Colorado.

It is used in thin slabs, so that the moss-like image can be seen effectively, especially as plates, cabochons for rings, brooches, and pendants, as well as for ornamental objects.

Possibilities for Confusion With other natural gemstones hardly exist due to their characteristic looks. But there are good imitations by doublets.

1 Dendritic agate, fern-like
2 Scenic agate
3 Dendritic agate, 2 pieces, radial inclusions
4 Mosquito agate
5 Moss agate, 10 cabochons
6 Moss agate, 2 pieces, partly polished

1
2
3
4
5
6
6

Agate

Color: All colors, banded
Color of streak: White
Mohs' hardness: 6½–7
Density: 2.60–2.64
Cleavage: None
Fracture: Uneven
Crystal system: (Trigonal) microcrystalline aggregates
Chemical composition: SiO_2 silicon dioxide
Transparency: Translucent, opaque
Refractive index: 1.530–1.540
Double refraction: up to 0.004
Dispersion: None
Pleochroism: Absent
Absorption spectrum: Dyed green: 700, (665), (634)
Fluorescence: Varies with bands: partly strong; yellow, blue-white

Agate is a banded, concentric shell-like chalcedony, sometimes containing opal substance (see page 166). The fine quartz fibers are oriented vertically to the surface of the individual band layers. The bands can be multicolored or of the same color. The agates of the exhausted German mines had soft to strong colors (especially pink, red, or brownish) and were separated by bright gray bands. The South American agates are mostly dull gray and without special markings; only through dyeing (page 152) do they receive their lively colors. Transparency of agates varies from nearly transparent to opaque. In thin slabs, even the opaque agates are mostly translucent. The name agate is supposedly derived from the river Achates (now called the Drillo) in Sicily.

Origin Agates are found as ball- or almond-shaped nodules with sizes ranging from a fraction of an inch to a circumference of several yards; more rarely they are found as fillings of crevices in volcanic rocks (such as rhyolite and dacite). The bands are thought to be formed by rhythmic crystallization, but scientific opinions vary as to how. It was thought that the agate bands crystallize gradually in hollows formed by gas bubbles from a siliceous solution. Recently, the theory that their formation is simultaneous with that of the matrix rock has won support. According to this idea, the liquid drops of the silicic acid cool with the cooling rock and produce a layered crystallization from the outside. A new theory postulates that rather than the liquids penetrating the agate walls, colloid solutions, i.e., substances with very fine sizes of grains, flow into the agate hollows. The various agate bands vary in thickness, but normally their thicknesses remain constant throughout the nodule. Where the inner cavity of the nodule is not filled with an agate mass, well-developed crystals may have formed: rock crystal (page 132), amethyst (page 134), and smoky quartz (page 132); sometimes accompanied by anhydritspat (page 206), ankerit (page 82), baryite (page 222), calcite (page 224), Goethit, hematite (page 178), siderite (page 222), and zeolite. A nodule with crystals in the central cavity if called a druse [no. 5]; if the inside is completely filled, one speaks of a geode [2].

For further information, see the following:

Agate varieties, page 150
Agate coloring, page 152
Agate industry, page 154
History of agate polishing, page 156
Usage of agate, page 156
Layer stones, page 158

1 Agate nodule in cross section, banded agate, ⅓ natural size, Uruguay
2 Agate nodule in cross section, orbicular agate with concentric eye (eye agate), ⅔ natural size, India
3 Agate nodule in cross section, orbicular, ⅔ natural size, India
4 Agate nodule in cross section, orbicular agate with eccentric eye (eye agate), ⅔ natural size, India
5 Agate geode, ⅔ natural size, Brazil

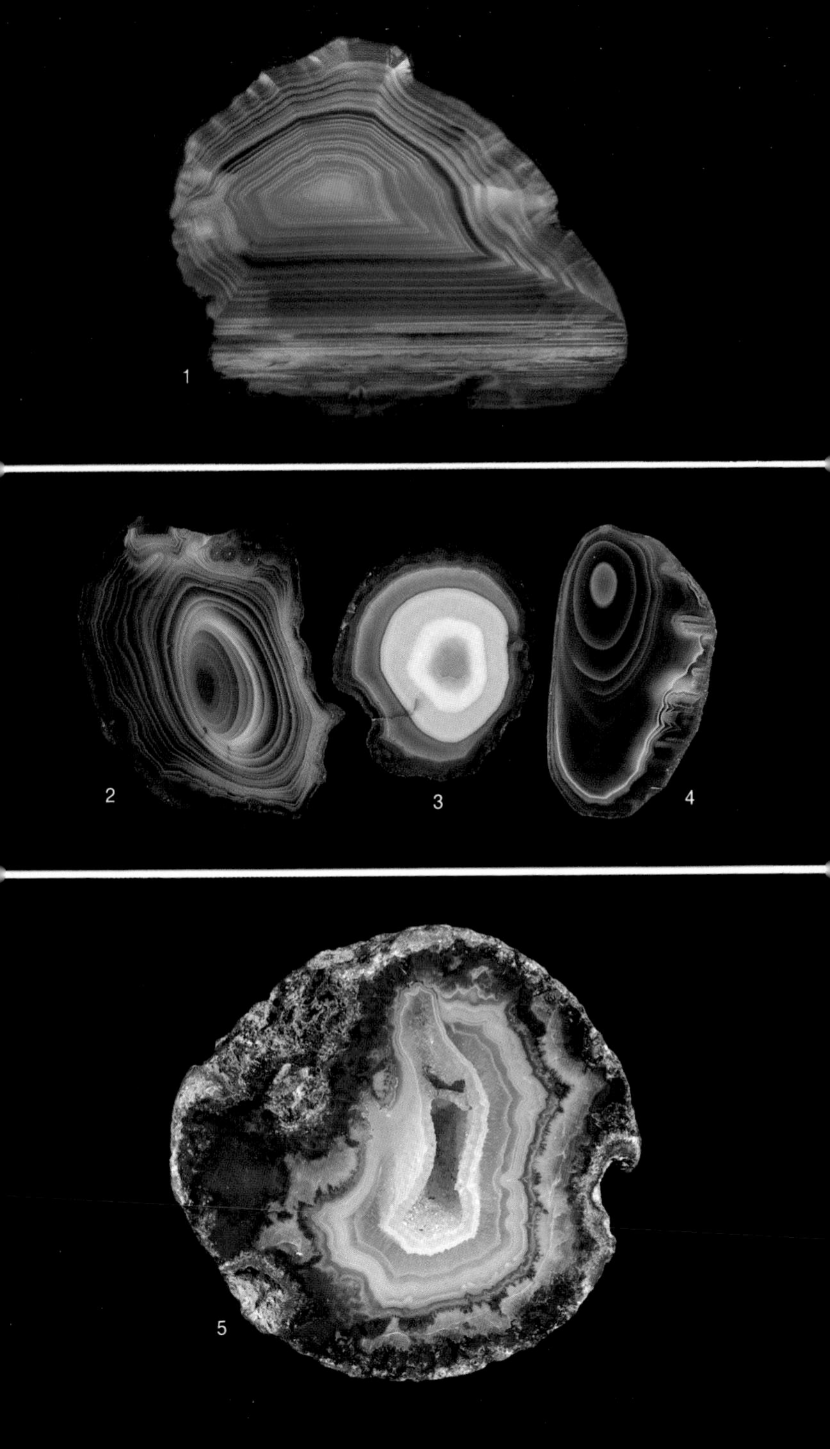
1
2
3
4
5

Agate Varieties According to sample, design, or structure of the agate layer, there are numerous trade names.
Eye agate [page 149, nos. 2, 4] Ring-shaped design with point in the center, similar to an eye. A type of orbicular agate.
Layer agate [page 149, no. 1] Layers/bands of about the same size parallel to the outer wall of the agate nodule.
Dendritic agate [page 147, no. 1-4] Colorless or whitish, translucent chalcedony with dendrites.
Enhydritic agate (also called enhydro or water stone) Agate nodule or mono-colored chalcedony nodule, partly filled with water which can be seen through the walls. After the agate is taken from surrounding rock, the water often dries out.
Fortification agate [page 151, no. 2] Agate bands with jutting-out corners like the bastions of old fortresses.
Fire agate Opaque, limonite-bearing layered chalcedony with iridescence which is created through diffraction of the light by the layered structure.
Orbicular agate [page 149, no 3] Circles of the agate layers, arranged concentrically or excentrically around a centerpoint.
Moss agate [page 147, nos. 5, 6] Translucent chalcedony with moss-like inclusions of hornblende or chlorite.
Scenic agate [page 147, no. 2] Shows landscape-like images through dendrites.
Pseudo-agate (also called polyhedric quartz) [page 153, no. 2] Interior similar to agate with layering and druse opening, although outside not nodule-like, but geometric shape. Formed probably as wedge-filling of crystals, which later were loosened. Occurrences in loose rocks in Brazil. Individual pieces can measure up to 28 in (75 cm). Many fanciful names.
Tubular agate [page 151, no. 3] Agate with numerous tubes (old feeding canals). The mouth-opening of the canals is usually bordered concentrically.
Thunder egg or Sandstone [page 151, no. 1] Layered agate nodule with strongly furrowed outer surface. Deposits in Oregon (United States), Australia, and Mexico.
Brecciated agate [page 153, no. 3] Agate, broken, but cemented together by quartz.

Deposits The most important agate deposits at the beginning of the 19th century in the neighborhood of Idar-Oberstein/Rhineland-Palatinate, Germany. These have now been worked out. Nodules were found as large as a human head with beautiful colors of gray, pink, red, yellow, brown, and pale blue. These could not be dyed.

The most important deposits today are in the south of Brazil (Rio Grande do Sul) and in the north of Uruguay. The deposits are layered in weathered materials and river sediments, and are derived from melaphyric rocks. The color is generally gray; the striations are hardly recognizable. They can be given an attractive appearance by dyeing (see below). Other deposits are in Australia (Queensland), China, India, the Caucasus, Madagascar, Mexico, Mongolia, Namibia, and the United States (Wyoming and Montana).

---

1 Cross section of agate nodule, sard stone with concentric, shell-like outer layers and parallel layers on the inside; ⅓ natural size. Found in Brazil
2 Cross section of agate nodule, fortification agate with agate layers having corners that jut out like bastions; ⅓ natural size. Found in Brazil
3 Agate nodule partially cut, tubular agate with numerous canals, former feeding tubes; ½ natural size. Found in Idar-Oberstein/Rhineland-Palatinate, Germany

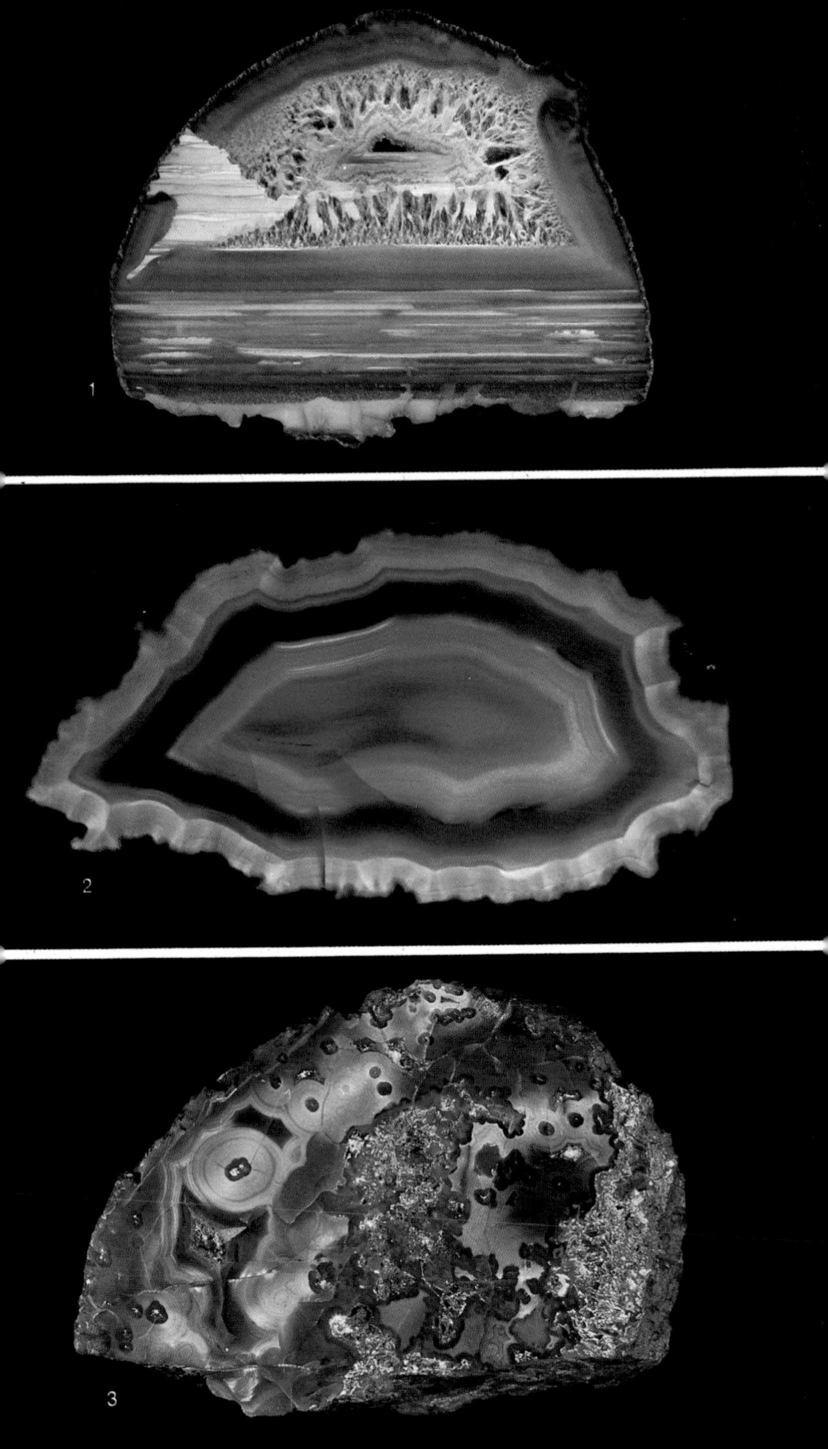
1
2
3

Agate Coloring The South American deposits, the most important suppliers of agate overall, produce agates which normally appear opaque gray and without markings. Only when dyed do they obtain their coloring and lively patterns.

The art of dyeing was known to the Romans. In Idar-Oberstein/Rhineland-Palatinate, it has been practiced since the 1820s and brought to a perfection not achieved anywhere else. This is the reason why the town has developed into the most important center for the cutting of agate and other stones.

The absorption of the dye varies with the porosity and water content of the individual layer of the agate. White layers, consisting of dense quartz aggregates, absorb little or no color. Layers which are easily dyed are called soft, the others hard.

Details of the process are a commercial secret. Generally, inorganic pigments are used, as the organic ones tend to fade in light and are less intense. Dyed agates are not recognizable with the naked eye, unless a certain color, such as deep blue, does not exist in natural agate. Although agate dyeing is considered routine in the trade, US FTC guidelines and many industry organizations say it should be disclosed (see page 33). Since 2007, all banded and single-colored agates, no matter whether they are permanently or non-permanently artificially colored, have to be marked as "treated" or "dyed" for the buyer.

Before they are dyed, the agates are cleaned with a warm acid or a caustic solution, cut into their final form, and sometimes even polished.

Coloring red [1b] Pigment is iron oxide. The agate is laid in a solution of iron nitrate, then strongly heated. By altering techniques, various reds can be obtained. Natural yellow layers turn red by heating alone.
Coloring yellow [1b] Pigment is iron chloride. The agate is first saturated with hydrochloric acid, then slightly warmed, producing a lemon-yellow color.
Coloring brown [1b] By treatment with sugar solution and heating, browns are produced. By use of cobalt nitrate, supposedly the same results can be achieved.
Coloring black [1c] Pigment is carbon. Use of concentrated honey or sugar solution, following treatment with heated sulfuric acid, produces a dark black color in agate. By certain variations, browns can be produced. Cobalt nitrate is said to have the same effect.
Coloring green [1d] Pigment is iron. Saturation of the agate with chromium salt solution and subsequent heat treatment produces green. The use of nickel nitrate solution and heat is said to produce the same result.
Coloring blue [1e] Pigment is iron. Agates are first placed in a solution of yellow potassium ferro-cyanide and subsequently boiled in hydrous iron sulfate. This produces "Berliner" blue.

---

1 Agate plate, a—natural, b–e—colored; ⅓ natural size, Brazil
2 Pseudo-agate, ½ natural size, Brazil (compare with page 150)
3 Brecciated agate, partly polished; ½ natural size, U.S. (compare with page 150)

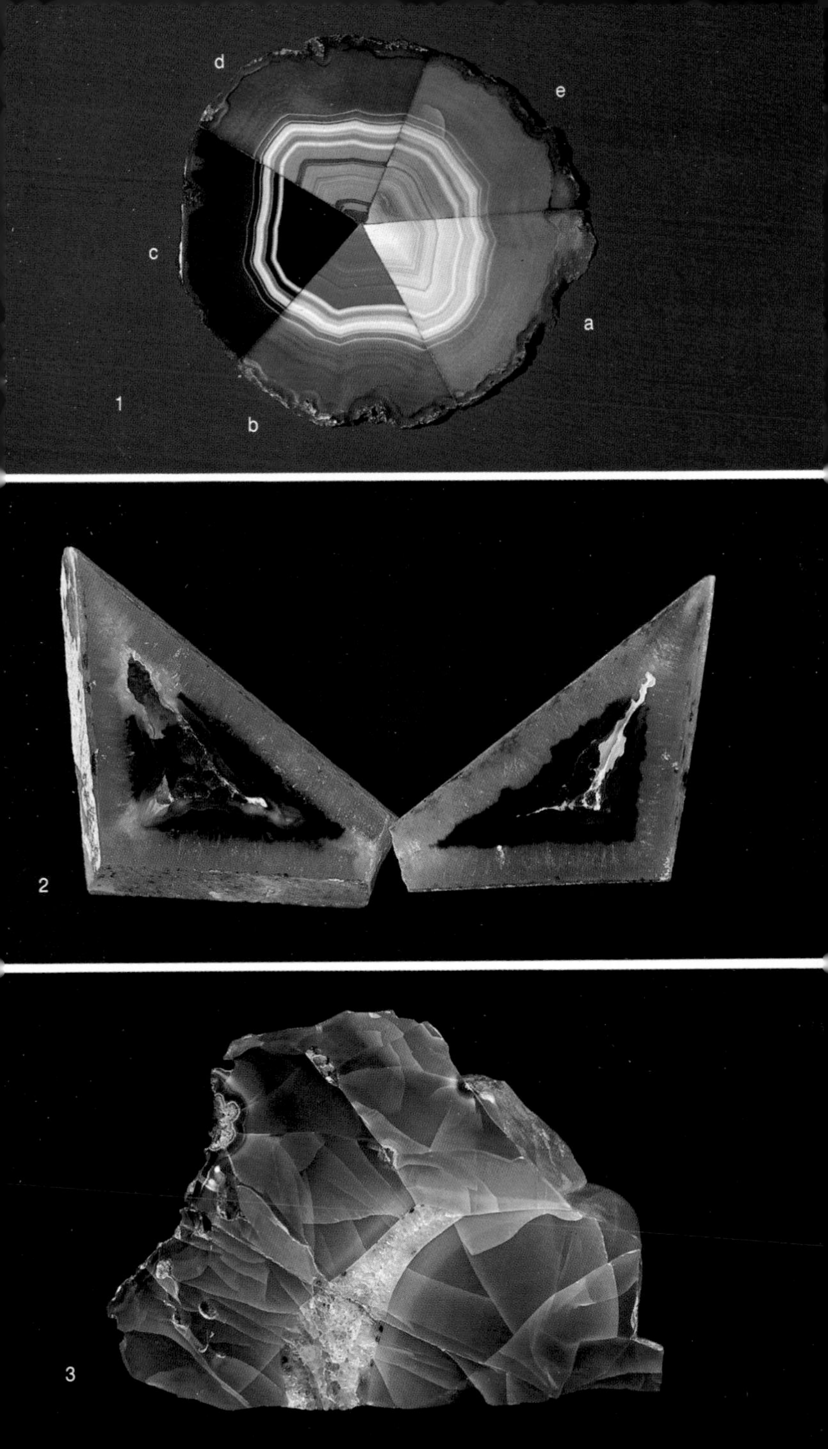
d
e
c
a
1
b
2
3

Old agate mill with waterwheel, Idar-Oberstein/Rhineland-Palatinate, Germany.

Development of the Agate Industry in Idar-Oberstein Agate has a special position among gemstones; there is a unique industry centered around it in Idar-Oberstein in Rhineland-Palatinate, Germany. The bases for this development were favorable natural occurrences, for instance, some finds of agate and jasper, good local sandstone for the production of cutting and polishing wheels, and water power to work the wheels.

Gemstones have been worked in and near Idar-Oberstein since the first part of the 16th century. Agate polishing at the Idar-brook was first mentioned in documents in 1548, but it is known that, one hundred years before that, agate, jasper, and quartz were locally mined, but possibly worked somewhere else.

Toward the end of the 17th century, there were about 15 workshops; and around 1800, 30 workshops were cutting agate and using the river for energy. Toward the beginning of the 19th century, local agate deposits were beginning to be worked out, and many experts left the area. However, new life was brought to the industry when, by chance, some emigrants who were wandering around as musicians discovered large agate deposits in Brazil. By 1834 the first supply of Brazilian agate had reached Idar-Oberstein, and by 1867 there were 153 polishing shops.

With the emergence of steam power and especially since the advent of electric energy, the industry has been decentralized. Today there are numerous workshops in and around the district.

Now that more and more countries around the world have established their own gemstone industries, Idar-Oberstein has found its place in specializing in processing high-quality agates almost exclusively.

1 Bowl of Brazilian agate, diameter 4.2 in/11.2 cm, height 2.6 in/6.5 cm
2 Bowl of Brazilian agate, diameter 51.4 in/14 cm, height 1½ in/4 cm

1
2

Agate polishers at work toward the end of the 19th century; Idar-Oberstein/Rhineland-Palatinate, Germany.

The History of Agate Polishing The oldest way to polish agate was to rub it against a horizontal sandstone. Polishing against a vertically rotating sandstone probably became common during the 14th century. A waterwheel outside the house was turned by a river or dammed pond to drive the axle inside the house. Several sandstone grinding wheels were mounted vertically onto this axle. These wheels were about 56 in (150 cm) high and 15–19 in (40–50 cm) wide. The polishers lay on their stomachs on special chairs and pushed the agate hard against the rotating wheels, which were sprinkled with water constantly. Because the sandstone wheel had an obtuse angle in the middle of its working surface, two workers could share one wheel.

With the increasing use of steam and, later, electricity and also the new method of polishing using carborundum, a sitting posture was adopted. It is more comfortable and does not require as much effort as formerly (see modern polishing on page 70).

Uses At least 3,000 years ago, agate was used as a gemstone by the Egyptians. Today it is used for objets d'art, decorative purposes, beads, rings, brooches, pendants, and as layer stones for cameos (page 158). It is also used by industry because of its toughness and resistance to chemicals.

1 Agate, decorative egg
2 Agate, pendant
3 Agate, ring
4 Agate, mortar
5 Agate, handles for knives and forks
6 Agate, handles for a manicure set
7 Agate, handle for a letter opener
8 Agate, brooch
9 Agate, seal
10 Agate, pill box
11 Agate, dentist's instrument
12 Agate, letter opener

1
4
3
2
5
6
7
8
9
10
11
12

Layer Stones

Layer stones are multilayered materials used in the art of gem carving and engraving, also called glyptography. Usually this material is cut from agates or other chalcedonies with even parallel layers, a lighter layer above a darker one. Brazil supplies the best raw material, usually two-layered, but sometimes three-layered ones are seen. Some masterpieces are cut out of five-layered material. Engravings in multilayered and curved agates are rare.

Some explanations concerning layer stones and engraving work follow.

Onyx Layer stone with the combination of a black base and a white upper layer, also called true onyx or Arabic onyx [page 161, no. 4]. In cabochons and beads, black and white layers may alternate. On the other hand, onyx is also a name sometimes used for unicolored chalcedony (e.g., black onyx). This must not be confused with onyx marble, which in short is also called onyx (compare with page 244). The name onyx has its origin in the Greek language and means "fingernail," probably because of its weak transparency.
Sard onyx Layer stone with brown base, upper layer white [page 159, no. 1].
Cornelian onyx Layer stone with red base, upper layer white [page 161, no. 5].
Niccolo Layer stone whose upper layer is very thin, which results in translucent blue-gray color tones due to diffusion of the light and showing through of the black base. Popular as seal rings for engraving coats-of-arms and monograms.
Intaglio (Italian) Name for engraving with negative image, as used for seals.
Cameo (Italian) Name for a relief which is cut so that it is raised.

The layers in agate or chalcedony required for this work are not often found in nature in the colors of onyx, carnelian, or sard. Therefore, such stones are usually dyed, as described in more detail on page 152. The natural and the dyed stones have the same names.

Recently onyx layer stones have been produced from unlayered, unicolored gray chalcedony, which can be dyed, as follows: A square block is saturated with a solution of cobalt chlorate and chlorammonium, and is thus dyed black. With the help of hydrochloric acid, this color is then removed up to a depth of 0.04 in (1 mm). When the block is sawn in half, the sawn surfaces are black and the reverse sides are white; however, it is said that the dark color tends to fade.

Creation of a Cameo

1 A piece of layer stone with different colors in even parallel layers is cut out or broken out of a block of agate, as seen in the background.
2 Several two-layered stones can be cut out of the first piece.
3 The lower layer is dyed black or red-brown. The upper layer remains white because it does not absorb the pigment.
4 The main features of the cameo are indicated. In mass production, a template is used.
5–8 These illustrate the art of the engraver; his experience, knowledge of the stone, and his technical perfection lend the personal touch.
9 The final result is an example of the masterly precision of the engraver's art.

1
2
3
4
5
6
7
8
9

A gem engraver at work

The Technique of Stone Engraving The main tool of the engraver is a small lathe with a horizontal spindle, to which, according to need, various instruments can be attached. These can be wheels, spheres, cones, or needles, which are kept handy on a rack nearby. The spindle is driven by an electric motor at 3000–5000 r.p.m. The spindle is rigid and the engraver guides the stone by hand. This requires great precision and knowledge of the particular stone. A working period lasts no more than two to three hours.

The rotating tips are prepared with diamond polishing powder and oil; by this means, they are cooled and given an abrasive surface because the tiny diamond particles are pressed into the softer iron of the instruments during the engraving process.

The polishing is performed with wood, leather, or another softer material using water and special polishing pastes. This process also removes any marks made by a metal pencil during preliminary sketching.

Recently, flexible spindles are also available. However, they are usually used only for larger sculptures, where the stone is too heavy to be guided by hand.

Even though nowadays all gemstones are used for engravings, agate, especially layer stones, is worked most of all.

1 Abstract modern engraving, onyx
2 Abstract modern engraving, onyx
3 Shadow cameo, onyx, Brazil
4 Woman's head, Parisian style, onyx
5 Girl with blossom, carnelian onyx
6 Coat-of-arms, onyx, Brazil
7 Shaded engraving, onyx, Brazil
8 Initials, onyx, Brazil

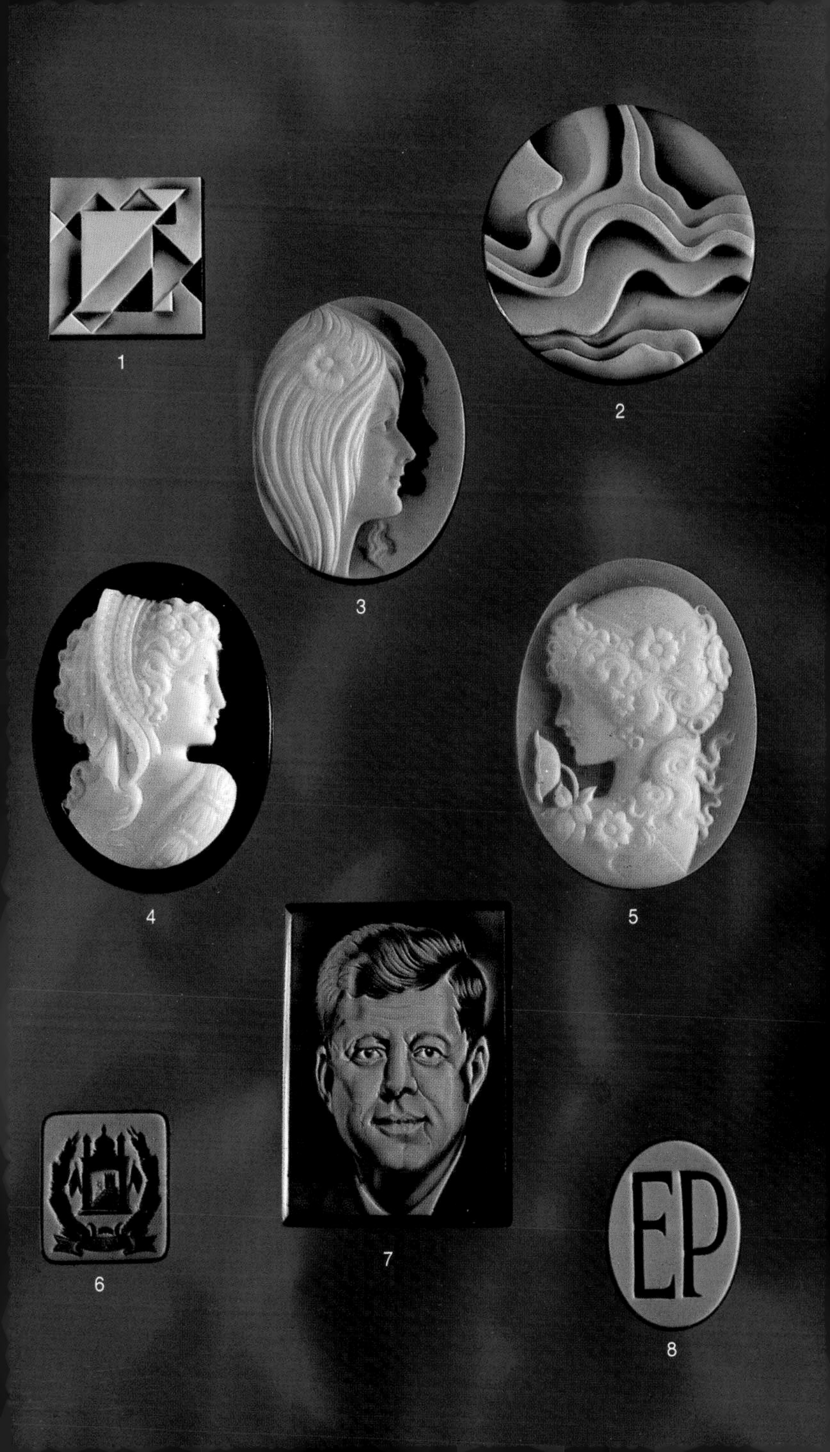
1
2
3
4
5
6
7
EP
8

Jasper

Color: All colors, mostly striped or spotted
Color of streak: White, yellow, brown, red
Mohs' hardness: 61/2[fraction]–7
Density: 2.58–2.91
Cleavage: None
Fracture. Splintery, conchoidal
Crystal system: (Trigonal) microcrystalline aggregate
Chemical composition: SiO_2 silicon dioxide
Transparency: Opaque, even in thin slabs opaque
Refractive index: About 1.54
Double refraction: None
Dispersion: None
Pleochroism: Absent
Absorption: Cannot be evaluated
Fluorescence: None

Jasper is usually considered as chalcedony (page 142); sometimes, however, scientists put it in a group by itself within the quartz group because of its grainy structure.

The name jasper is derived from the Greek and means "spotted stone." In antiquity, however, jasper referred to completely different stones than today, i.e., to green, transparent varieties.

The finely grained, dense jasper contains up to 20 percent foreign materials, which determine its color, streak, and appearance. Uniformly colored jasper is rare; usually it is multicolored, striped, or flamed. Sometimes jasper can be grown together with agate or opal. There is also fossilized material (page 164).

Occurs as fillings of crevices or fissures or in nodules. Deposits are found in Egypt, Australia, Brazil, India, Canada, Kazakhstan, Madagascar, Russia, Uruguay, and the United States. Used for ornamental objects, cabochons, and for stone mosaics. Care must be taken during cutting and polishing; banded jasper tends to separate along the layers.

Varieties According to color, appearance, occurrence, or composition, there are many names used in the trade.

Agate jasper (jaspagate) Yellow, brown, or green blended, grown together with agate.

Egyptian jasper (Nile pebble) Strongly yellow and red.

Banded jasper [13] Layered structure with more or less wide bands.

Basanite (touchstone) Fine-grained, black. Is used by jewelers and goldsmiths for streak-tests of precious metals.

Blood jasper Name sometimes used for bloodstone (page 144).

Hornstone (chert) Very fine grained, gray, brown-red, more rarely green or black. Sometimes hornstone is understood as a general synonym for jasper.

Scenic jasper Brown marking, caused by iron oxide, resembling a landscape.

Moukaite [4] Pink to light red, cloudy. Found in Australia.

Nunkirchner jasper Whitish gray, rarely yellow or brownish-red (named after deposit in Rhineland-Palatinate). Dyed with Berliner blue, misleadingly called "German lapis" or "Swiss lapis" in imitation of lapis lazuli (page 188).

Plasma Dark green, sometimes with white or yellow spots.

Silex [14] Yellow and brown-red spotted or striped.

1 Jasper breccia, Australia
2 Pop jasper, 2 stones, South Africa
3 Pop jasper, 2 stones, Australia
4 Moukaite, Australia
5 Multicolored jasper, 2 stones, India
6 Multicolored jasper, cabochon, Australia
7 Zebra jasper, South Africa
8 Yellow jasper, cabochon, Australia
9 Yellow jasper, Australia
10 Multicolored jasper, India
11 Striped jasper, South Africa
12 Multicolored jasper, India
13 Banded jasper, Australia
14 Silex, Egypt
15 Multicolored jasper, India
16 Multicolored jasper, rough, India

1
2
3
4
5
6
7
8
9
10
11
12
13
14
15
16

Petrified Wood Also called Fossilized Wood

Chalcedony Species

Color: Brown, gray, red
Color of streak: White, partly colored
Mohs' hardness: 61/2[fraction]–7
Density: 2.58–2.91
Cleavage: None
Fracture: Uneven, splintery
Crystal system: (Trigonal) microcrystalline aggregate

Chemical composition: SiO_2 silicon dioxide
Transparency: Opaque, even in thin slabs
Refractive index: About 1.54
Double refraction: Weak or none
Dispersion: None
Pleochroism: Absent
Absorption spectrum: Cannot be evaluated
Fluorescence: None

Petrified wood is fossilized wood with the mineral composition of jasper, chalcedony, and, less frequently, opal; it consists of silicon dioxide only. The wood has not actually become stone as is usually understood by the layman. The organic wood is not changed into stone, but only the shape and structural elements of the wood are preserved. The expert speaks of a pseudomorphosis of chalcedony (or jasper or opal) after wood.

Well-preserved petrification occurs only where trees after their death are quickly covered with fine-grained sedimentary rock. Thus the outer structure of the wood is preserved in a negative form within the enclosing rock. Circulating waters loosen and decompose the organic substances and replace them with mineral substances. Therefore, it is not a *change* that takes place, but rather an *exchange*. Sometimes this process is successive (from removal and adding of substances) so that the inner structural elements of the wood, the annuals rings [no. 5], the structure of the cells, even wormholes, are preserved. It can also happen that the appearance is totally changed by the crystallization process. Either way, the process ensures the creation of petrified wood.

The colors are mostly dull gray or brown, sometimes also red, pink, light brown, yellow, and even blue to violet. The colors become stronger with cutting and polishing.

The most important occurrence is the "petrified wood" near Holbrook in Arizona (United States). There are fossilized tree trunks of up to 213 ft (65 m) long and 10 ft (3 m) thick belonging to the araucaria variety of plants. The tree trunks were deposited there from various parts by water about 200 million years ago, and then covered by several hundred yards of sediment. In the course of time, part of the fossilized wood was exposed by weathering from the enclosing sandstone. Nowhere is the fossilized wood as splendidly colored as in Arizona. In order to preserve this unique natural beauty spot, the "Petrified Forest" was declared a national park in 1962. That also means that no visitor is allowed to take a piece of these petrified materials as souvenirs.

There are smaller deposits of petrified wood on all the continents. Egypt supplies good quality (Dschel Moka Ham near Cairo), as does Argentina (Patagonia), Canada (Alberta), and Wyoming. In Nevada (Virgin Valley), the fossilized wood shows the beautiful color play of opal. Other deposits can be found in Australia (Queensland), Greece (Lesbos), India, Madagascar, Mongolia, and Namibia. See also petrified wood page 282.

It is mostly used for ornamental objects and decorative pieces (tabletops, ashtrays, bookends, paperweights), less frequently for jewelry purposes.

1 Petrified wood, ashtray
2 Petrified wood, bottle cork holder
3 Petrified wood, five fern-tree pieces
4 Petrified wood, partly polished
5 Petrified wood, with year rings
6 Petrified wood, two sections of tree trunk

The illustrations are 50 percent smaller than the originals.

1
2
3
5
6

Opal Species

The name is derived from an Indian (Sanskrit) word for "stone." It is divided into three subgroups: the precious opals, the yellow-red fire opals, and the common opals. Their physical properties vary considerably.

Precious Opal

Color: All colors, partially play-of-color
Color of streak: White
Mohs' hardness: 5½–6½
Density: 1.98–2.50
Cleavage: None
Fracture: Conchoidal, splintery, brittle
Crystal system: Amorphous; kidney- or grape-shaped aggregates
Chemical composition: $SiO_2 \bullet nH_2O$ hydrous silicon dioxide

Transparency: Transparent, opaque
Refractive index: 1.37–1.52
Double refraction: None
Dispersion: None
Pleochroism Absent
Absorption spectrum: Fire opal: 700–640, 590–400
Fluorescence: White o.: white, bluish, brownish, greenish
Fire o.: greenish to brown

The special characteristic of these gems is their play-of-color, a display of rainbow-like hues which (especially in rounded cut forms) changes with the angle of observation. The electron-microscope, using a magnification of 20,000, reveals the cause: tiny spheres (as small as 0.001 millimeter in diameter) of the mineral cristobalite layered in siliceous jelly cause the diffraction and interference patterns.

Opal always contains water (3 to 30 percent). It can happen that in the course of time, the stone loses water, cracks, and the play-of-color diminishes. This can, at least temporarily, be restored by saturation with oil, epoxy resin, or water. The aging process is avoided when stored in moist absorbent cotton wool. Care must be taken during setting. A little heat can evaporate the water. Opal is also sensitive to pressure and knocks as well as being affected by acids and alkalies. Opal is often impregnated with plastic to improve appearance.

White opal [11–16] A precious opal of white or otherwise light basic color with color play.

Black opal [4, 5 and 7–10] Precious opal with dark gray, dark blue, dark green, and gray-black basic color and play-of-color. Deep black is an exception. Black opals are rarer than white opals.

Opal matrix [1, 2, 6] Banded growth or leafed inclusion of precious opal with and/or in the matrix rock.

Boulder opal Precious opal with dark base surface, color play, and high density. Occurs as pebble rock, where opal fills hollows (see picture on page 33).

Harlequin opal Transparent to translucent precious opal with effective mosaic-like color patterns. Counted among the most desirable opals.

Jelly opal Bluish-gray precious opal with little play-of-color.

Crystal opal Transparent with strong color play on colorless, vitreous surface.

1 White opal in matrix
2 Black opal in matrix
3 Opalized snail
4 Black opal, diverse shapes
5 Black opal, 86ct
6 Opal matrix, pendant
7 Black opal, 2 triplets
8 Black opal, 4 cabochons
9 Black opal doublet, 16.90ct
10 Black opal, 2 doublets
11 White opal, 4 cabochons
12 White opal, rough, partly polished
13 White opal, cabochon, 10.39ct
14 White opal, cabochon, 33.75ct
15 White opal, 2 cabochons, 7.78ct
16 White opal, 4 cabochons, 14.21ct

About 20 percent smaller than original. Stones not numbered belong to no. 4.

1
2
3
4
5
6
7
8
9
10
8
11
12
13
14
15
16

Deposits Up to the end of the 19th century, the andesite lavas in the east of Slovakia supplied the best qualities. Then the Australian deposits were discovered. Famous deposits in New South Wales are at Lightning Ridge and White Cliffs; in South Australia at Coober Pedy and Andamooka. Numerous deposits are also found in Queensland. Most of the 0.04–0.08 in (1–2 mm)-thin opal layers are bedded in sandstone. Since 2008 there has been almost no production in Australia due to movement of miners to copper mines. Almost at the same time rich resources were discovered in Ethiopia. Further deposits are found in Brazil, Guatemala, Honduras, Indonesia, Japan, Mexico, Peru, Russia, Nevada, and Idaho.
Possibilities for Confusion With ammonite (page 266), labradorite (page 182), mother-of-pearl (page 265), and moonstone (page 180). Very thin pieces of opal are sometimes mounted on a piece of common opal or black chalcedony to create an opal doublet or layer opal. Triplets are also made with a protective top layer of rock crystal. Several good imitations made from glass or plastic are also known. In 1970, the U.S. succeeded in creating a synthesis of white and black opal, and in 1995, Russia did the same.

Fakes are prepared by coloring black or matrix opal in order to liven up the play of color.

Fire Opal [1-7]

Fire opal (named after its often fire-red color) often shows no play-of-color. It is usually milky and turbid. The best qualities are clear and transparent [nos. 3, 4, 6], which makes them suitable for being faceted. They are very sensitive to every stress. Deposits are found in Mexico, as well as Brazil, Guatemala, the United States, and Western Australia.
Possibilities for Confusion With garnet (page 120), rhodochrosite (page 184).

Common Opal [8-14] (Also called Potch)

Common opal is usually opaque, rarely translucent, and shows no play-of-color. A wide variety of trade names are used. New varieties are found on page 280. Many trade names:
Agate opal (opal agate) Agate with light and dark opal layers.
Angel skin opal Misleading name for palygorskite, an opaque, whitish- to pink-colored silicate mineral.
Wood opal (zeasite) Yellowish or brownish opal as petrified wood (page 164)
Honey opal [8] Honey-yellow, translucent opal.
Hyalite (glass opal, waterstone) Colorless, water-clear opal with strong sheen.
Hydrophane A milk opal, which has turned turbid due to the loss of water. Through absorption of water, it can become translucent again and have color play.
Porcelain opal White, opaque milk opal.
Moss opal [11] Milk opal with dendrites.
Girasol Almost colorless, transparent opal with a bluish opalescence (see page 54).
Prase opal (chrysopal) [10] Apple-green opal. Substitute for chrysoprase (page 144).
Wax opal [12] Yellow-brown opal with wax-like luster.
Andes opal Opals from Peru that come in many colors (see page 280).

---

1 Fire opal, rough, Mexico
2 Fire opal, 5 cabochons, 11.80ct
3 Fire opal, 9 faceted stones, 11.95ct
4 Fire opal, 4 faceted stones, 13.61ct
5 Fire opal, rough, Mexico
6 Fire opal, 3 faceted stones, 5.89ct
7 Fire opal, cabochon and oval, 24.53ct
8 Honey opal, Western Australia
9 Common opal, 3 stones, Mexico
10 Prase opal, Nevada
11 Moss opal, India
12 Wax opal, rough, Hungary
13 Dendrite opal, rough
14 Liver opal or menilite, rough, Hungary
The illustrations are 20 percent smaller than the originals.

1
2
3
4
5
6
7
8
9
10
11
12
13
14

Jade (Jadeite and Nephrite)

The name jade goes back to the time of the Spanish conquest of Central and South America and derives from *piedra de ijada*, i.e., hip stone, as it was seen as a protection against and cure for kidney diseases. The corresponding Chinese word *yu* has not been generally accepted.

In 1863 in France, the gemstone, which had been known for 7,000 years, was proved to consist of two separate, distinct minerals, namely jadeite and nephrite. Differentiation between jadeite and nephrite is based on properties, but the term jade is used as a description of both.

In prehistoric times, jade was used in many parts of the world for arms and tools because of its exceptional toughness. Therefore nephrite is sometimes called "axe stone." For over 2,000 years, jade was part of the religious cult in China and mystic figures and other symbols were carved from it. In pre-Columbian Central America, jade was more highly valued than gold. With the Spanish conquest, the high art of jade carving in America came to a sudden end. In China, however, this art was never interrupted. In former times, only nephrite was worked in China, but since about 1750 jadeite imported from Myanmar has also been used.

Jadeite Species [9–13] — Jade

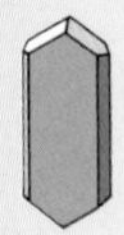

Color: Green, also all other colors
Color of streak: White
Mohs' hardness: 6½–7
Density: 3.30–3.38
Fracture: Splintery, brittle
Crystal system: Monoclinic; intergrown, grainy aggregate
Chemical composition: $Na(Al,Fe^{3+})[Si_2O_6]$ sodium aluminum silicate

Transparency: Opaque, translucent
Refractive index: 1.652–1.688
Double refraction: 0.020
Dispersion: None
Pleochroism: Absent
Absorption spectrum: Green j.: 691, 655, 630, (495), 450, 437, 433
Fluorescence: Greenish j.: very weak: whitish glimmer

Jadeite (name derived from jade) is very tough and resistant because of its tight growth of tiny interlocking grains. It occurs in all colors. Fractures are dull and when polished greasy.

Imperial jade A jadeite from Myanmar, colored emerald-green with chromium translucent to almost transparent. Most desired jade variety of all.

Chloromelanite (jade albite, maw-sit-sit) [nos. 15, 16] A rock composed of kosmochlor (a mineral related to jadeite) combined with varying amounts of jadeite, albite feldspar, and other minerals. Its color is deep green with dark green-black spots or veins caused by chlorite. The source is upper Myanmar.

1 Nephrite, rough, partly polished
2 Jadeite, 2 stones, table cut
3 Jadeite, 6 cabochons
4 Nephrite, 2 navettes, together 7.68ct
5 Nephrite, 3 cabochons
6 Jadeite, cabochon
7 Nephrite, cabochon, Wyoming (U.S.)
8 Nephrite, octagon, cabochon
9 Jadeite, rough
10 Jadeite, flat table
11 Jadeite, 4 qualities
12 Jadeite, 2 drops
13 Jadeite, 3 different cuts
14 Nephrite, cat's eye
15 Chloromelanite, antique cut, 14.32ct
16 Chloromelanite, 4 different cuts

All illustrations are 20 percent smaller than the original.

1
2
3
4
5
6
7
8
10
11
12
13
14
15
16

Yünan-Jade Chinese term for jadeite, named for the province through which it was imported from Burma (Myanmar).

Historically important jadeite deposits are in upper Burma (Myanmar), near Tawmaw, interlayered in serpentine, or in secondary deposits in conglomerates or in river gravel. Other deposits are in China, Japan, Canada, Guatemala, Kazakhstan, Russia (Siberia), and California. (For more about usage, possibilities for confusion, and imitations, see below.)

Much of the jadeite sold today is treated by dyeing or impregnation with wax or plastic-type resins to improve color and appearance.

Nephrite [page 171, nos. 1–8, 14] Jade

Color: Green, also other colors
Color of streak: White
Mohs' hardness: 6–6½
Density: 2.90–3.03
Cleavage: None
Fracture: Splintery, sharp edged, brittle
Crystal system: Monoclinic; intergrown fine fibrous aggregate

Chemical composition: $Ca_2(Mg,Fe^{2+})_5[OH|Si_4O_{11}]_2$ basic calcium mag. iron silicate
Transparency: Opaque
Refractive index: 1.600–1.627
Double refraction: –0.027, often none
Dispersion: None
Pleochroism: Absent
Absorption spectrum: (689), 589, 490, 460
Fluorescence: None

Nephrite (Greek—kidney) is a dense, felt-like fibrous aggregate variety of the actinolite-tremolite mineral series that is even tougher than jadeite. Occurs in all colors, often with a yellow tint. The most valuable is green. The luster is vitreous.

Nephrite is more common than jadeite. No deposits are in New Zealand (greenstone), occuring in serpentine rocks, and as river or beach pebbles. Other deposits are found in Australia, Brazil, China (Sinkiang), Canada, Zimbabwe, Russia, Taiwan, Alaska, and formerly also in Poland.

Uses Jadeite and nephrite are used as cabochons, in other jewelry, and for vases, both decorative and religious. The main cutting centers are in China (particularly in Hong Kong) and in Taiwan.

Possibilities for Confusion (For jadeite and nephrite) with agalmatolite (page 248), amazonite (page 180), aventurine (page 138), californite (page 202), chrysoprase (page 144), hydrogrossular garnet (page 122), pectolite (page 228), plasma (page 162), prase (page 138), prehnite (page 204), serpentine (page 218), emerald (page 106), smaragdite (page 220), smithsonite (page 214), and verdite (page 276).

There are many imitations made from glass and plastic. Glued triplets and colorations in order to improve the color value are known. A large number of greenish stones are falsely offered as jade in the trade.

Indian Jade Misleading trade name for aventurine (page 138) and for aventurine glass.

Russian Jade Trade name for spinach-green nephrite from the region of Lake Baikal, Russia.

Wyoming Jade Trade name for nephrite from Wyoming, also for a green-colored growth of tremoite with albite.

1 Nephrite, elephant, China
2 Jade, necklace, China
3 Jade, cigarette holder
4 Nephrite, necklace, Myanmar
5 Jade, Buddha, China
6 Jade, horse, China
7 Jade 3 symbolic figures
8 Jade, necklace, multicolored
9 Nephrite, pendant
10 Chloromelanite, pendant

The illustrations are 20 percent smaller than the originals.

1
2
3
5
6
7
9
8
10

Peridot Also called Chrysolite, Olivine

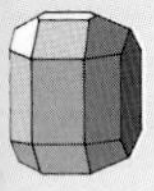

Color: Yellow-green, olive-green, brownish
Color of streak: White
Mohs' hardness: 6½–7
Density: 3.28–3.48
Cleavage: Indistinct
Fracture: Brittle, small conchoidal
Crystal system: Orthorhombic; short, compact prisms, vertically striated
Chemical composition: $(Mg,Fe^{2+})_2[SiO_4]$ magnesium iron silicate
Transparency: Transparent
Refractive index: 1.650–1.703
Double refraction: +0.036 to +0.038
Dispersion: 0.020 (0.012–0.013)
Pleochroism: Very weak; colorless to pale green, lively green, olive-green
Absorption: 497, 425, 493, 473, 453.
Fluorescence: None

The name probably derives from the Arabic word *faridat* (gem). The name *chrysolite* (Greek—gold stone) was formerly applied not only to peridot but also to many similarly colored stones. The name commonly used in mineralogy is olivine (because of its olive-green color).

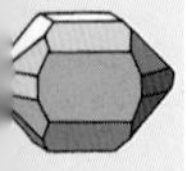

Olivine is a mineral that occurs in the series with the end members forsterite and fayalite. It has a vitreous and greasy luster, and is not resistant to acids. It tends to burst under great stress; therefore, it is sometimes metal-foiled. Rarities are peridot cat's eye and star peridot.

Historically important deposit was on the Red Sea volcanic island Zabargad (St. John), 188 miles (300 km) east of Aswan, Egypt; it was mined for over 3500 years but forgotten for many centuries, and rediscovered only around 1900. Beautiful material is also found in the serpentine quarries in upper Burma (Myanmar). Other deposits have been found in Australia (Queensland), Brazil (Minas Gerais), China, Kenya, Mexico, Pakistan, Sri Lanka, South Africa, Tanzania, and Arizona. In Europe, deposits are found in Norway, north of Bergen.

Peridot was brought to Central Europe by the crusaders in the Middle Ages, and was often used for ecclesiastical purposes. It was the most popular stone during the Baroque period. Popular are table and step cuts; sometimes also brilliant cut, especially when set in gold.

The largest cut peridot weighs 319ct and was found on the island Zabargad; it is in the Smithsonian Institution in Washington, D.C. In Russia, there are some cut peridots which came out of a meteorite that fell in 1749 in eastern Siberia.

Possibilities for Confusion With chrysoberyl (page 114), demantoid (page 122), diopside (page 206), emerald (page 106), idocrase (page 202), moldavite (page 246), prasiolite (page 136), precious beryl (page 112), prehnite (page 204), sinhalite (page 202), and tourmaline (page 126). A green foil sometimes enhances pale stones. Imitations are often constructed of corundum and spinel syntheses. Evergreen bottle glass can be mistaken for peridot. The strong double refraction of peridot is an important distinguishing mark. In thick stones the doubling of the edges of the lower facets can be clearly seen through the polished table with the naked eye.

1 Peridot, 2 emerald cuts, each 4.65ct
2 Peridot, 2 ovals, 5.67 and 6.38ct
3 Peridot, emerald cuts, 4.14ct
4 Peridot, oval, 12.45ct
5 Peridot, 4 different cuts
6 Peridot, antique cut, 24.02ct
7 Peridot, 5 faceted stones
8 Peridot, 4 faceted stones
9 Peridot, 5 cabochons
10 Peridot, crystals, partly tumbled

The illustrations are 20 percent larger than the originals.

1
2
3
4
5
6
7
8
9
10

Zoisite Species

The mineral zoisite (named after the collector Zois) was first found in the Sau-Alp mountains in Kärnten, Austria, in 1805. It was originally called saualpite, and gemstone quality specimens have only recently been found. The gemstone members of the group are tanzanite and thulite as well as anyolite. For the time being, only individual specimens and/or small amounts of other transparent zoisite varieties of gemstone quality (colorless, green, brown) are known. See page 278 for further details.

Tanzanite [1-11] — Zoisite Species

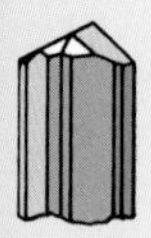

Color: Sapphire blue, amethyst, violet
Color of streak: White
Mohs' hardness: 6½–7
Density: 3.35
Cleavage: Perfect
Fracture: Uneven, brittle
Crystal system: Orthorhombic; multifaced prisms, mostly striated
Chemical composition: $Ca_2Al_3[O|OH|SiO_4|Si_2O_7]$ calcium aluminium silicate
Transparency: Transparent
Refractive index: 1.691–1.700
Double refraction: +0.009
Dispersion: 0.030 (0.011)
Pleochroism: Very strong; purple, blue, brown, or yellow
Absorption: 525, 528, 455
Fluorescence: None

The name tanzanite (after Tanzania) was introduced by the New York jewelers Tiffany & Co. In good quality the color is ultramarine to sapphire blue; in artificial light, it appears more amethyst violet. When heated to 752–932 degrees F (400–500 degrees C), the interfering yellowish and brown tints vanish, and the blue deepens. Green tanzanite (page 278) is rare. Tanzanite cat's-eyes are also rarely found. It is used in facet cut. The only deposit is in Tanzania near Arusha; it occurs in veins or filling of fissures of gneisses.

Possibilities for Confusion With amethyst (page 134), iolite (page 196), lazulite (page 208), sapphire (page 102), spinel (page 116), and synthetic corundum. There are glass imitations and doublets made from glass with tanzanite crown or made from two colorless synthetic spinels glued with tanzanite-colored glue.

Thulite [15-16] — Zoisite Species

Dense, opaque pink zoisite variety. Named after the legendary island Thule. Deposits found in Western Australia, Namibia, Norway, and North Carolina. Used as cabochon or as an ornamental stone.

Possibilities for Confusion With carnelian (page 142), eudialyte (page 228), rhodonite (page 184), and ruby (page 98).

Anyolite [12-14] — Zoisite Species

Green zoisite rock with black hornblende inclusions and large, but mostly opaque, rubies. Named according to the native language of the Massai ("green"). It was first discovered in 1954 in Tanzania. Due to its color contrasts, it is an effective gem and ornamental stone.

Possibilities for Confusion With chloromelanite (page 170) and tourmaline (page 126).

1 Tanzanite, crystal
2 Tanzanite in host rock
3 Tanzanite, 3 broken crystals
4 Tanzanite, pear-shaped, 5.2ct
5 Tanzanite, antique cut, 24.4ct
6 Tanzanite, oval, 3.5ct
7 Tanzanite, brilliant cut, 6.8ct
8 Tanzanite, oval, 3.lct
9 Tanzanite, cabochon, 8.5ct
10 Tanzanite, 5 faceted stones
11 Tanzanite, 5 cabochons
12 Anyolite with ruby
13 Anyolite, 2 cabochons
14 Anyolite with ruby
15 Thulite, 2 pieces, rough
16 Thulite, cabochon

2
1
6
4
3
3
5
7
8
9
10
3
11

12
14
13
16
15

Hematite [1–4]

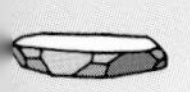

Color: Black, black-gray, brown-red
Color of streak: Brown-red
Mohs' hardness: 5½-6½
Density: 5.12–5.28
Cleavage: None
Fracture: Conchoidal, uneven, fibrous
Crystal system: (Trigonal) mostly platy
Chemical composition: Fe_2O_3, iron oxide
Transparency: Opaque
Refractive index: 2.940–3.220
Double refraction: –0.287
Dispersion: None
Pleochroism: Absent
Absorption spectrum: Not diagnostic
Fluorescence: None

The name hematite (Greek—blood) (in some countries, also called bloodstone) is derived from the fact that, when cut, the saw coolant becomes colored red. The English name bloodstone, however, is applied to a chalcedony variety (page 144). In mineralogy, well-crystallized hematite varieties are called iron luster, finely crystallized ones red iron ore or red ironstone, and radial aggregates are called red glass head. When cut into very thin plates, hematite is red and transparent; when polished, it it metallic and shiny. Deposits with cuttable material are found in Cumberland, England, as well as in Bangladesh, Brazil, China, New Zealand, Czech Republic (Erzgebirge), Minnesota, and occasionally Elba/Italy. Formerly used as mourning jewelry, but today mainly for rings, bead necklaces, and intaglio (deepened engravings).
Possibilities for Confusion With cassiterite (page 200), davidite (page 232), magnetite (page 230), neptunite (page 228), pyrolusite (page 232), and wolframite (page 230).

Hematine Trade name for an imitation of hematite with pressed and sintered iron oxide, from the United States. In contrast to hematite, it is slightly magnetic.

Pyrite [5-10] Also called Fool's Gold

Color: Brass-yellow, gray-yellow
Color of streak: Green-black
Mohs' hardness: 6–6½
Density: 5.00–5.20
Cleavage: Indistinct
Fracture: Conchoidal, uneven, brittle
Crystal system: (Cubic) cubes, pentagonal dodecahedra, octahedra
Chemical composition: FeS_2 iron sulphide
Transparency: Opaque
Refractive index: Cannot be determined
Double refraction: None
Dispersion: None
Pleochroism: Absent
Absorption spectrum: Not diagnostic
Fluorescence: None

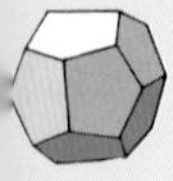

Pyrite (Greek—fire, as it produces sparks when knocked) is wrongly called marcasite in the trade. True marcasite is a mineral, in many ways similar to pyrite, but unsuitable for jewelry, as it sometimes powders in air. Because of its similarity to gold, pyrite is often called fool's gold. Disk-like, radial-fibrous aggregates, so-called pyrite-suns (page 292), are known. Metallic shiny. Deposits in Peru, also in Bolivia, Mexico, Romania, Sweden, and Colorado.
Possibilities for Confusion With chalcopyrite (page 222) and gold (page 224).

---

1 Hematite, radial red glass head
2 Hematite, 2 broken crystals
3 Hematite, 5 cut stones
4 Hematite, tablet cut, truncated corners
5 Pyrite aggregate covered with crystals
6 Pyrite, cabachon cut
7 Pyrite, faceted
8 Pyrite crystals in matrix rock
9 Pyrite, 4 different crystals
10 Pyrite aggregate as brooch
11 Pyrite octahedron crystal

1
2
3
4
5
6
7
8
9
10
11

Feldspar Group

For the feldspars (from German *Feld,* field, and *spalten,* to split), there are tw main subgroups that produce gems: the potassium feldspars and the plagioclases (series from calcium to sodium feldspars). There are numerous gemstone varieties.

Amazonite [1–3] Also called Amazon Stone — Feldspar Grou

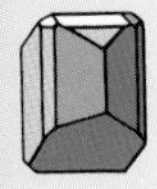

Color: Green, blue-green
Color of streak: White
Mohs' hardness: 6–6½
Density: 2.56–2.58
Cleavage: Perfect
Fracture: Uneven, splintery, brittle
Crystal system: Triclinic; prismatic
Chemical composition: $K[AlSi_3O_8]$ potassium aluminum silicate
Transparency: Translucent, opaque
Refractive index: 1.522–1.530
Double refraction: –0.008
Dispersion: None
Pleochroism: Absent
Absorption spectrum: Not diagnostic
Fluorescence: Weak; olive-green

Amazonite (derived from the Amazon) is a mostly opaque, green sodium feldspa Color distribution is irregular; luster is vitreous; it is sensitive to pressure. Deposit are found in Colorado, Brazil, India, Kenya, Madagascar, Namibia, and Russia.
Possibilities for Confusion With chrysoprase (page 144), jade (page 170 serpentine (page 218), and turquoise (page 186).

Moonstone [7–10] — Feldspar Grou

Color: Colorless, yellow, pale sheen
Color of streak: White
Mohs' hardness: 6–6½
Density: 2.56–2.59
Cleavage: Perfect
Fracture: Uneven, conchoidal
Crystal system: Monoclinic; prismatic
Chemical composition: $K[AlSi_3O_8]$ potassium aluminum silicate
Refractive index: 1.518–1.526
Double refraction: –0.008
Dispersion: None
Pleochroism: Absent
Absorption spectrum: Not diagnostic
Fluorescence: Weak; bluish, orange

A potassium feldspar of the orthoclase (adularia) species with white shimmer, similar to moonshine (therefore the name) the so-called adularescence (see page 52 Moonstone cat's-eye is also known. Vitrous luster; sensitive to pressure. Deposit found in Sri Lanka, Burma (Myanmar), Brazil, India, Madagascar, and the Unite States. Used as cabochons.
Possibilities for Confusion With chalcedony (page 142), synthetic spinel, and glas imitations. Other moonstones from the feldspar group are also known (page 278)

Orthoclase [4–6] — Feldspar Group

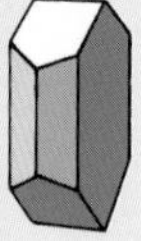

Orthoclase (Greek—to break straight) is transparent to opaque, often colorless o champagne-colored potassium feldspar. Vitreous luster. Deposits are found i Madagascar, Burma (Myanmar), and Kenya.
Possibilities for Confusion With apatite (page 210), chrysoberyl (page 114), citrine (page 136), precious beryl (page 112), prehnite (page 204), topaz (page 118 and zircon (page 124).

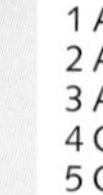

1 Amazonite, broken crystal
2 Amazonite, rough
3 Amazonite, 6 different cuts
4 Orthoclase, broken crystal
5 Orthoclase, 3 faceted stones
6 Orthoclase, rough, Kenya
7 Moonstone, rough, 2 pieces
8 Moonstone, 7 cabochons, India
9 Moonstone, 2 cabochons, Sri Lanka
10 Moonstone, 3 cabochons, 13.23ct
The illustrations are 10 percent smaller than the originals.

1
2
3
5
6
8
9
10
7

Labradorite [1-4]

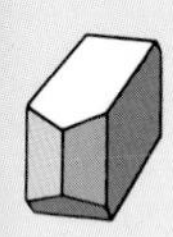

Color: Dark gray to gray-black, with colorful iridescence; also colorless, brownish
Color of streak: White
Mohs' hardness: 6–6½
Density: 2.65–2.75
Cleavage: Perfect
Fracture: Uneven, splintery, brittle
Crystal system: (Triclinic), platy, prismatic
Chemical composition: $NaAlSi_3O_8$ to $CaAl_2Si_2O_8$ sodium calcium aluminum silicate
Transparency: Transparent to opaque
Refractive index: 1.559–1.570
Double refraction: +0.008 to + 0.010
Dispersion: 0.019 (0.010)
Absorption spectrum: Not diagnostic
Fluorescence: Yellow striations

Labradorite (named after peninsula of Labrador in Canada, where it was found) i a plagioclase feldspar. It shows a schiller (labradorescence) in lustrous metalli tints, often blue and green, although specimens with the complete spectrum ar most appreciated. It is mainly caused by interference of light from lattice distor tions resulting from alternating microscopic exsolution lamellae of high- and low calcium plagioclase phases. Vitreous luster; sensitive to pressure. Deposits ar found in Canada (Labrador, Newfoundland), also in Australia (New South Wales) Madagascar, Mexico, Russia, and the United States.

Used for bead necklaces, brooches, rings, and ornamental objects. Colorless an yellowish-brown transparent labradorites [no. 3] are cut with facets.

Spectrolite [2] Trade name for a labradorite from Finland that shows the spectra colors especially effectively.

Madagascar Moonstone Trade name for an almost transparent oligoclase moon stone (Plagioclase) from Madagascar with a strong blue schiller.

Aventurine Feldspar [5-8] Also called Sunstone Feldspar Group

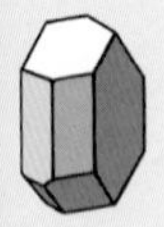

Color: Orange, red-brown, sparkling
Color of streak: White
Mohs' hardness: 6–6½
Density: 2.62–2.65
Cleavage: Perfect
Fracture: Grainy, splintery, brittle
Crystal system: (Triclinic), rare, solid aggregates
Chemical composition: $(Ca,Na)[(Al,Si)_2Si_2O_8]$ sodium calcium aluminum silicate
Transparency: Translucent, opaque
Refractive index: 1.525–1.548
Double refraction: +0.010
Dispersion: None
Pleochroism: Weak or absent
Absorption: Not diagnostic
Fluorescence: Dark brown-red

A metallic shiny plagioclase feldspar; named after a type of glass that wa discovered by chance (Italian—*a ventura)* around 1700. Typically, it has a red more rarely a green or blue, glitter which is caused by light reflections from tiny hematite or goethite platelets. Deposits found in India, Canada Madagascar, Norway, Russia (Siberia), and the United States. Used with flat sur faces or en cabochon.

Possibilities for Confusion With aventurine quartz (page 138) and th artificial glass "goldstone." Other feldspar gemstones: albite (page 240), andesin (page 232), anorthite (page 242), oligoclase (page 226), peristerite (page 220), an sanidine (page 220).

1 Labradorite, rough, Canada
2 Spectrolite, rough, Finland
3 Labradorite, faceted, 4.08ct, U.S.
4 Labradorite, 13 cabochons
5 Aventurine feldspar, 2 rough pieces
6 Aventurine feldspar, 4 cabochons
7 Aventurine feldspar, faceted
8 Aventurine feldspar, cabochon
The illustrations are 10 percent smaller than the originals.

1
2
3
4
5
6
7
8

Rhodochrosite [1–5] Also called Manganesespar, Raspberryspa

Color: Rose-red to yellowish, striped
Color of streak: White
Mohs' hardness: 4
Density: 3.45–3.70
Cleavage: Perfect
Fracture: Uneven, conchoidal
Crystal system: (Trigonal) rhombohedra, usually compact aggregate
Chemical composition: $Mn[CO_3]$ manganese carbonate
Transparency: Transparent to opaque
Refractive index: 1.600–1.820
Double refraction: –0.208 to –0.220
Dispersion: 0.015 (0.010–0.020)
Pleochroism: Absent in aggregate
Absorption spectrum: 551, 449, 415
Fluorescence: Weak; red

Rhodochrosite (Greek—rose colored) has been on the market only since about 1940. Transparent crystals are very rare. The aggregates are light-dark stripes with zigzag bands. Raspberry red and pink are the most common colors. It has a vitreous luster; on cleavage faces, there is a pearly luster. The most important deposits are in Argentina, among others those near San Luis, 144 miles (230 km) east of Mendoza. The rhodochrosite has formed as stalagmites in the silver mines of the Incas since they were abandoned in the 13th century. Other deposits are in Chile, Mexico, Peru, South Africa, and the United States.

Usually used in larger pieces, as then the marking is most distinct, for ornamental objects as well as for cabochons and necklaces. The transparent stones are usually faceted and in demand by collectors. The largest faceted rodochrosite with 59.65ct is from South Africa (private collection).

Possibilities for Confusion With fire opal (page 168), rhodonite (see below), tugtupite (page 220), tourmaline (page 126), and bustamite (page 238).

Rhodonite [6-12] Also called Manganese Gravel

Color: Dark red, flesh red, black dendritic inclusions
Color of streak: White
Mohs' hardness: 5½–6½
Density: 3.40–3.74
Cleavage: Perfect
Fracture: Uneven, conchoidal, tough
Crystal system: (Triclinic), platy, columnar, usually compact aggregate
Chemical composition: $(CaMg)(Mn^{2+}, Fe^{2+})_4[Si_5O_{15}]$ manganese silicate
Transparency: Transparent to opaque
Refractive index: 1.716–1.752
Double refraction: +0.010 to +0.014
Dispersion: None
Pleochroism: Definite; yellow-red, rose-red, red-yellow in transparent
Absorption spectrum: 548, 503, 455, (412), (408)
Fluorescence: None

Rhodonite (Greek–rose), in addition to its red color, usually has black dendritic inclusions of manganese oxide. Transparent varieties are very rare. It has a vitreous luster; on cleavage faces, there is a pearly luster. Deposits are found in Australia (New South Wales), Finland, Japan, Canada, Madagascar, Mexico, Russia (the Urals), Sweden, South Africa, Tanzania, and New Jersey. It is cut with a table or en cabochon for necklaces, ornamental objects, and even wall tiles (e.g., subway in Moscow). Transparent stones are faceted with step or brilliant cut.

Possibilities for Confusion With rhodochrosite (see above), thulite (page 176), transparent varieties with hessonite (page 122), pyroxmangite (deep red manganese silicate with slight brown tint, page 242), spessartite (page 120), spinel (page 116), tourmaline (page 126), and bustamite (page 238).

Fowlerite Rhodonite variety with brownish or yellowish tint.

1 Rhodochrosite, baroque necklace
2 Rhodochrosite, 3 partly polished pieces
3 Rhodochrosite, crystals
4 Rhodochrosite, 4 cabochons
5 Rhodochrosite, 3 cabochons
6 Rhodonite, high cabochon
7 Rhodonite, bead necklace
8 Rhodonite, flat cut
9 Rhodonite, unicolored cabochon
10 Rhodonite, 5 transparent stones
11 Rhodonite, rough, partly polished
12 Rhodonite, 5 cabochons

The illustrations are 40 percent smaller than the originals

1
2
3
4
5
5
5
6
7
8
9
10
11
12

Turquoise

Color: Sky-blue, blue-green, apple-green
Color of streak: White
Mohs' hardness: 5–6
Density: 2.31–2.84
Cleavage: None
Fracture: Conchoidal, uneven
Crystal system: (Triclinic) seldom; grape-shaped aggregates

Chemical composition: $Cu(Al,Fe^{3+})_6[(OH)_4|(PO_4)_2]_2 \cdot 4H_2O$; a copper containing basic aluminum phosphate
Transparency: Translucent, opaque
Refractive index: 1.610–1.650
Double refraction: +0.040
Dispersion: None
Pleochroism: Absent
Absorption spectrum: (460), 432, (422)
Fluorescence: Weak; green-yellow, light blue

The name turquoise means "Turkish stone" because the trade route that brought it to Europe used to come via Turkey. Pure blue color is rare; mostly turquoise is interspersed with brown, dark gray, or black veins of other minerals or the host rock. Such stones are called turquoise matrix. It can also be intergrown with malachite (page 192) and chrysocolla (page 216). It has a waxy luster or mat. Most of the so-called turquoise found in the United States contains Fe (substituting for Al) and is thus really a mixture with chalcosiderite. Iron imparts a greenish color.

The popular sky-blue color changes at 482 degrees F (250 degrees C) into a dull green (be careful when soldering). A negative change in color can also be brought about by the influence of light, perspiration, oils, cosmetics, and household detergents, as well as loss of natural water content. Turquoise rings should be removed before hands are washed.

Occurs in dense form, filling in fissures, as grape-like masses or nodules. Thickness of veins up to 0.8 in (20 mm). The best qualities are found in northeast Iran near Nishapur. Additional deposits are found in Afghanistan, Argentina, Australia, Brazil, China, Israel, Mexico, Tanzania, and the United States (New Mexico, Arizona).

The deposits in Sinai, Egypt, were already worked out by 2000 B.C. In the early Victorian period, sky-blue turquoise was most popular. Today, it is used en cabochon, for brooches, necklaces, and bracelets, as well as ornamental objects.

Possibilities for Confusion With amazonite (page 180), chrysocolla (page 216), hemimorphite (page 214), lazulite (page 208), odontolite (page 252), serpentine (page 218), smithsonite (page 214), and variscite (page 212).

Because the stone is so porous, turquoise is often soaked with artificial resin, which improves color and at the same time hardens the surface. The color can also be improved with oil or paraffin, Berliner blue, aniline colors, or copper salt. It is imitated by dyed chalcedony (page 142), dyed howlite (page 224), powdered turquoise pieces that are baked with a glue mixture, as well as by glass, porcelain, and plastics. Synthetic turquoise, with or without matrix, has been on the market with good qualities since about 1970.

Neolite (also called Reese turquoise) Well-done turquoise imitations with dark matrix.
Neo Turquoise Good turquoise imitations with dark matrix.
Viennese Turquoise Turquoise imitations with good color.

1 Turquoise with matrix, 2 cabochons
2 Turquoise, Chinese carving
3 Turquoise, 3 cabochons, 25.89ct
4 Turquoise, rough
5 Turquoise, 9 cabochons, 26.10ct
6 Turquoise, 3 cabochons, octagon
7 Turquoise, bead necklace
8 Turquoise, 2 matrix cabochons
9 Turquoise, 7 cabochons, 14.30ct
10 Turquoise, baroque necklace
11 Turquoise, rough, partly polished
12 Turquoise, 2 cabochons, 38.53ct
13 Turquoise, 4 cabochons, 42.48ct
14 Turquoise, 3 cabochons

The illustrations are 40 percent smaller than the originals.

1
2
4
3
5
8
6
7
11
13
12
9
14
14
14
10

Lapis Lazuli

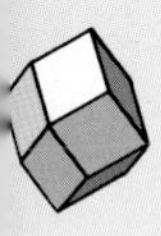

Color: Lazur blue, violet, greenish-blue
Color of streak: Light blue
Mohs' hardness: 5–6
Density: 2.50–3.00
Cleavage: Indistinct
Fracture: Conchoidal, grainy
Crystal system: (Cubic) rare; dense aggregates
Chemical composition:
$Na_6Ca_2[S,SO_4,Cl_2)_2|Al_6Si_6O_{24}]$
sodium calcium aluminum silicate

Transparency: Opaque
Refractive index: About 1.50
Double refraction: None
Dispersion: None
Pleochroism: Absent
Absorption spectrum: Not diagnostic
Fluorescence: Strong: white, also orange, copper-colored

As lapis lazuli (Arabic and Latin—blue stone) is composed of several minerals–in addition to lazurite (25–40 percent), including augite, calcite, diopside, enstatite, mica, haüynite, hornblende, nosean, sodalite, and/or pyrite—it is considered to be not a mineral but a rock. The variation in composition also causes a wide range in the above data.

The coloring agent is sulfur. In the best-quality specimens, the color is evenly distributed, but in general it is spotty or striated. In the lapis lazuli from Chile (Chilean lapis) and from Russia, a strongly protruding whitish or gray calcite diminishes the value. Well-distributed fine pyrite is advantageous and is taken to show genuineness. Too much pyrite, on the other hand, causes a dull, greenish tint. Color can be improved through slight heating and dyeing. It has a vitreous and greasy luster.

Lapis lazuli is sensitive to strong pressure and high temperatures, hot baths, acids and alkalies. Rings should be taken off during household work!

For 6,000 years, the deposit producing the finest material has been in the West Hindu Kush mountains in Afghanistan. Lapis lazuli is present there as an irregular occurrence in limestone in difficult terrain. Russian deposits are at the southwest end of Lake Baikal. The matrix rock is white dolmitic marble. Chile's deposits are north of Santiago. Other deposits are found in Angola, Burma (Myanmar), Canada, Pakistan, California, and Colorado.

Lapis lazuli was already used in prehistoric times for jewelry. During the Middle Ages, it was also used as a pigment to produce ultramarine. Some palaces and churches have wall panels and columns inlaid with lapis lazuli. Today it is used for ring stones and necklaces, as well as sculptures, vases, and other ornamental objects.

Possibilities for Confusion With azurite (page 190), dumortierite (page 198), dyed howlite (page 224), lazulite (page 208), sodalite (page 190), and glass imitations.

In 1954, a synthetic grainy spinel, colored with cobalt oxide, with a good lapis color, made an appearance on the market. Inclusions of thin gold pieces simulated the pyrite. Imitations with lapis lazuli pieces and powder, pressed or bound with artificial resin, are also on the market.

German Lapis (also called Swiss lapis) Misleading trade name for a dyed jasper (see page 162). It has no pyrite inclusions.

1 Lapis lazuli, bowl, Chile
2 Lapis lazuli, rough, Afghanistan
3 Lapis lazuli, broken crystal
4 Lapis lazuli, Afghanistan, partly polished
5 Lapis lazuli, Buddha, Afghanistan
6 Lapis lazuli, bead necklace, Afghanistan
7 Lapis lazuli, cabochon, Afghanistan
8 Lapis lazuli, 3 cabochons, Afghanistan
9 Lapis lazuli ring stone, Siberia, Russia
10 Lapis lazuli, 7 different cuts
11 Lapis lazuli, tablet cut, Chile
12 Lapis lazuli, rough, Siberia, Russia
13 Lapis lazuli, rough, Afghanistan
14 Lapis lazuli, rough, Chile

The illustrations are 40 percent smaller than the originals.

2
7
4
5
6
9
8
10
11
14
12
13

Sodalite [1-4]

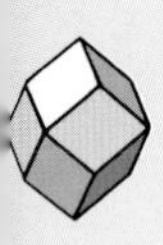

Color: White, blue, gray
Color of streak: White
Mohs' hardness: 5½–6
Density: 2.14–2.40
Cleavage: Indistinct
Fracture: Uneven, conchoidal
Crystal system: (Cubic) rhombic dodecahedra
Chemical composition: $Na_8[Cl,OH)_2|Al_6Si_6O_{24}]$ chloric sodium aluminum silicate
Transparency: Transparent to opaque
Refractive index: 1.48
Double refraction: None
Dispersion: 0.018 (0.009)
Pleochroism: Absent
Absorption spectrum: Not diagnostic
Fluorescence: Strong; orange

The name sodalite refers to its sodium content. For jewelry, only blue tones are used; sometimes they have a violet tint; frequently they are dispersed with white veins from white calcite. It has a vitreous luster; on fractures, there is a greasy luster. The low density is quite noticeable. Occurrence is usually in syenite and trachyte rocks, as well as pegmatite. Deposits are found in Brazil (Bahia), Greenland, India, Canada (Ontario), Namibia (transparent crystals), Russia (Urals), and Montana. Used as cabochon and for necklaces, especially for arts and crafts objects, also as ornamental stone.
Possibilities for Confusion With azurite (see below), dumortierite (page 198), haüynite (page 240), lapis lazuli (page 188), and lazulite (page 208). Synthetic sodalite has been known since 1975.

Hackmanite Pink-colored sodalite variety. It was discovered for the first time in cuttable qualities in 1991, in Quebec, Canada. The color fades in light.

Azurite [5-8] Also called Chessylite

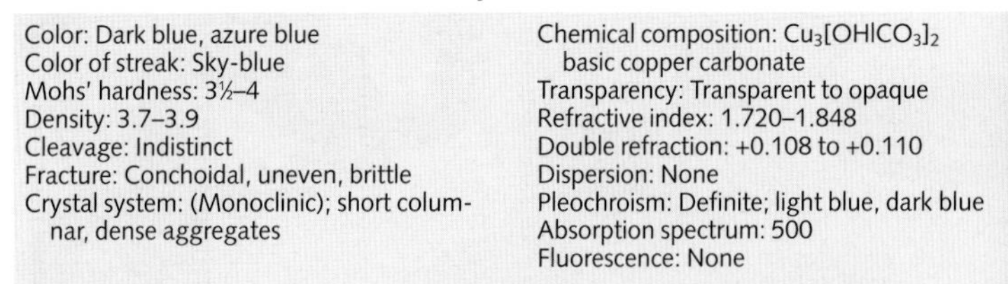

Color: Dark blue, azure blue
Color of streak: Sky-blue
Mohs' hardness: 3½–4
Density: 3.7–3.9
Cleavage: Indistinct
Fracture: Conchoidal, uneven, brittle
Crystal system: (Monoclinic); short columnar, dense aggregates
Chemical composition: $Cu_3[OH|CO_3]_2$ basic copper carbonate
Transparency: Transparent to opaque
Refractive index: 1.720–1.848
Double refraction: +0.108 to +0.110
Dispersion: None
Pleochroism: Definite; light blue, dark blue
Absorption spectrum: 500
Fluorescence: None

Named after its azure-blue color. It has a vitreous luster. Occurs with malachite in or near copper deposits. Exceptional dark blue—almost black—crystals have come from Tsumbeb, Namibia, with similar material now coming from Morocco. Deposits are found in Australia (Queensland), Chile, Mexico, Russia (Urals), and the United States (Arizona, New Mexico). The famous deposit in Chessy near Lyon, France, seems to be worked out.

It was formerly used for azure pigment. Because of its low hardness, it is mainly used for ornamental objects. It is also cut by collectors en cabochon and even faceted.
Possibilities for Confusion With dumortierite (page 198), haüynite (page 240), lapis lazuli (page 188), lazulite (page 208), and sodalite (see above).

Azure-Malachite [8] Ball- or kidney-shaped, striped intergrowth of azurite and malachite (page 192). Frequently cut en cabochon.

Burnite Intergrowth of azurite and cuprite (page 222).

1 Sodalite, rough, partly polished
2 Sodalite, baroque necklace
3 Sodalite, 4 flat cut pieces
4 Sodalite, 2 cabochons
5 Azurite, crystals
6 Azurite, 5 differently cut stones
7 Azurite, part of crystal
8 Azure-malachite, rough
The illustrations are 20 percent smaller than the originals.

1
2
3
4
5
6
7
8

Malachite

Color: Light- to black-green, banded	Chemical composition: $Cu_2[(OH)_2|CO_3]$ basic copper carbonate
Color of streak: Light green	Transparency: Translucent, opaque
Mohs' hardness: 3½–4	Refractive index: 1.655–1.909
Density: 3.25–4.10	Double refraction: –0.254
Cleavage: Perfect	Dispersion: None
Fracture: Splintery, scaly	Pleochroism: Absent
Crystal system: (Monoclinic) small, long prismatic; usually aggregate	Absorption spectrum: Not diagnostic
	Fluorescence: None

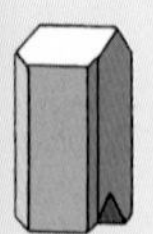

The name is probably derived from the green color (Greek—*malache* = mallow), perhaps from its low hardness (Greek—*malakos* = soft). In fracture or when cut, aggregates show a banding of light and dark layers with concentric rings, straight stripes, or other figurative shapes caused by its shell-like formation. Large monocolored pieces are rare. In thin plates, it is translucent, otherwise opaque. The coloring agent is copper. Crystals are rare, mostly dense, fibrous fine-crystalline aggregates. As rough stone, it has a weak vitreous or mat luster; on fresh fractures and when polished, it has a silky luster. Malachite is sensitive to heat, acids, ammonia, and hot water.

Occurs in rounded nodules, grape shapes, cone shapes, or stalactitic and, rarely, encrusted slabs. Formed from copper-containing solutions in or near copper ore deposits. The most important deposits used to be in the Urals near Yekaterinburg (Sverdlovsk). The quarries delivered blocks of over 20 tons in weight. From there, the Russian tsars obtained the malachite for decorating their palaces, paneling the walls, and for beautiful inlaid works.

Today Shaba (Katanga) in Zaire near the Democratic Republic of the Congo is the most important malachite producer. Other deposits are in Australia (Queensland, New South Wales), Chile, Namibia, Zimbabwe, and Arizona.

Malachite was popular with the ancient Egyptians, Greeks, and Romans for jewelry, amulets, and as a powder for eye shadow. It is used as pigment for mountain green.

Although it is not very hard and not very resistant, malachite is nowadays popular for jewelry and ornaments. Used en cabochon and as slightly rounded table stones for necklaces and especially for objets d'art, such as plates, boxes, ashtrays, and sculptures. The cutter must work the malachite so as to show the decorative marking to its best advantage. Concentric eye-like rings are most popular (called malachite peacock's eye). Because of its low hardness, malachite is easily scratched and sometimes becomes dull. The surface can be hardened with artificial resin.

Possibilities for Confusion Unlikely for larger pieces because of its striped formation; small, unbanded stones, on the other hand, can be confused with opaque green gemstones.

Azure-Malachite Intergrowth of azurite and malachite (see page 190).

Eilat Stone Intergrowth of malachite with turquoise and chrysocolla (see page 216).

1 Malachite, rough, partly polished
2 Malachite, bead necklace
3 Malachite, 2 cabochons
4 Malachite, cabochon, Zimbabwe
5 Malachite, 7 various samples
6 Malachite, rough

The illustrations are 40 percent smaller than the originals.

1
2
3
4
5
6

Lesser-Known Gemstones

Those gemstones that are not commonly known but which are becoming increas-ingly more popular belong to this group. Compare also with page 84.

Andalusite [2-3]

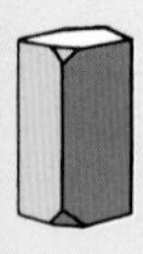

Color: Yellow-green, green, brownish-red
Color of streak: White
Mohs' hardness: 7½
Density: 3.05–3.20
Cleavage: Good
Fracture: Uneven, brittle
Crystal system: (Orthorhombic), thick-columnar
Chemical composition: $Al_2[O|SiO_4]$ aluminum silicate

Transparency: Transparent to opaque
Refractive index: 1.627–1.649
Double refraction: –0.007 to –0.013
Dispersion: 0.016 (0.009)
Pleochroism: Strong; yellow, olive, red-brown to dark red
Absorption spectrum: 553, 550, 547, (525), (518), (495), 455, 447, 436
Fluorescence: Weak; green, yellow-green

Transparent andalusite (named after Andalusia in Spain) of gemstone quality is rare. The luster is vitreous or mat. When cutting, the strong pleochroism has to be taken into consideration. Occurs in schists, gneisses, and in placers. Deposits are found in Australia, Brazil, Canada, Russia, Spain (Andalusia), Sri Lanka, and the United States. Used in table and brilliant cut.
Possibilities for Confusion With chrysoberyl (page 114), smoky quartz (page 132), sin-halite (page 202), sphene (page 210), tourmaline (page 126), and idocrase (page 202).

Chiastolite [1] (Also called Cross Stone) Opaque variety of andalusite; white, gray, yellowish; Mohs' hardness is 5–5½. Occurs in long prisms that show a dark cross in cross section, when viewed perpendicular to the prism axes; caused by carbona-ceous inclusions. Deposits are found in Algeria, South Australia, Bolivia, Chile, France (Brittany), Russia (Kola, Siberia), Spain (Galicia), and California. Worn as an amulet; now of value only to the collector. Cut en cabochon, flat.
Possibilities for Confusion Do not exist due to its distinct marking.

Viridine Dark-green andalusite variety, containing iron and manganese.

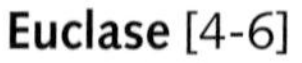

Euclase [4-6]

Color: Colorless, sea-green, light blue, dark blue
Color of streak:
White Mohs' hardness: 7½
Density: 3.10
Cleavage: Perfect
Fracture: Conchoidal, brittle
Crystal system: Monoclinic; prisms
Chemical composition: $BeAl[OH|SiO_4]$ basic beryllium aluminum silicate

Transparency: Transparent
Refractive index: 1.650–1.677
Double refraction: +0.019 to +0.025
Dispersion: 0.016 (0.009)
Pleochroism: Very weak; white-green, yel-low-green, blue-green
Absorption spectrum: 706, 704, 650, 639, 468, 455
Fluorescence: Weak or none

Euclase (Greek—breaks well) is difficult to cut and polish because of its perfect cleavage. It has a bright, vitreous luster. Occurs in pegmatites, in placers, and in druses. Deposits are found in Brazil (Minas Gerais), the Democratic Republic of the Congo, Russia (Urals), Zimbabwe, Tanzania, and Zaire. Used with a step cut.
Possibilities for Confusion With aquamarine (page 110), precious beryl (page 112), hiddenite (page 130), and sapphire (page 102). With radiation, colorless euclase can be changed to blue.

1 Chiastolite, 4 pieces, partly polished
2 Andalusite, 2 broken crystals
3 Andalusite, 4 faceted stones
4 Euclase, 2 colorless faceted stones
5 Euclase, light blue, faceted
6 Euclase, crystal in host rock

1
2
2
3
4
5
6

Hambergite [1, 2]

Color: Colorless, gray-white, yellow-white
Color of streak: White
Mohs' hardness: 7½
Density: 2.35
Cleavage: Perfect
Fracture: Conchoidal, brittle
Crystal system: Orthorhombic; prisms
Chemical composition: $Be_2[(OH,F)|BO_3]$

Transparency: Transparent, translucent
Refractive index: 1.553–1.628
Double refraction: +0.072
Dispersion: (0.015 (0.009–0.010)
Pleochroism: Absent
Absorption spectrum: Cannot be evaluated
Fluorescence: Usually none, sometimes orange

Hambergite (after a Swedish mineralogist) has the same vitreous luster as glass when cut. The strong double refraction can be recognized through the table along the lower facet edges.

Possibilities for Confusion With rock crystal (page 132), danburite (page 198), euclase (page 194), leuco garnet (page 122), and zircon (page 124).

Iolite [3-6] Also called Cordierite and Dichroite

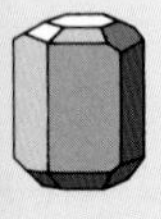

Color: Blue, violet, brownish
Color of streak: White
Mohs' hardness: 7–71/2
Density: 2.58–2.66
Cleavage: Good
Fracture: Conchoidal, uneven, brittle
Crystal system: Orthorhombic; short prisms
Chemical composition: $Mg_2Al_3[AlSi_5O_{18}]$ magnesium aluminum silicate

Transparency: Transparent, translucent
Refractive index: 1.542–1.578
Double refraction: –0.008 to –0.012
Dispersion: 0.017 (0.009)
Pleochroism: Very strong; yellow, dark blue-violet, pale blue
Absorption spectrum: 645, 593, 585, 535, 492, 456, 436, 426
Fluorescence: None

The color of iolite (Greek—violet) is usually blue. Inclusions of hematite and goethite sometimes cause reddish sheen or aventurescence. It has a greasy vitreous luster. Deposits found in Myanmar, Brazil, India, Madagascar, Sri Lanka, and the United States.

Possibilities for Confusion With benitoite (page 200), kyanite (page 212), sapphire (page 102), and tanzanite (page 176). Glass imitations are known.

Water Sapphire Misleading trade name for blue iolite.

Phenakite [7, 8]

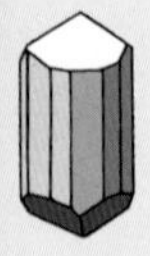

Color: Colorless, wine-yellow, pink
Color of streak: White
Mohs' hardness: 7½–8
Density: 2.95–2.97
Cleavage: Good
Fracture: Conchoidal
Crystal system: (Trigonal) short columnar
Chemical composition: $Be_2[SiO_4]$ beryllium silicate

Transparency: Transparent
Refractive index: 1.650–1.670
Double refraction: +0.016
Dispersion: 0.015 (0.009)
Pleochroism: Definite; colorless, orange-yellow
Absorption spectrum: Not diagnostic
Fluorescence: Pale greenish, blue

Phenakite (Greek—deceiver) is mostly water-clear. It has a strong vitreous luster; when polished, a greasy luster. Colored stones can fade. Deposits are found in Brazil, Mexico, Namibia, Norway, Zimbabwe, Sri Lanka, Tanzania, and the United States.

Possibilities for Confusion With rock crystal (page 132), beryllonite (page 206), cerussite (page 216), danburite (page 198), precious beryl (page 112), and topaz (page 118).

1 Hambergite, 3 faceted stones
2 Hambergite, 2 broken crystals
3 Iolite, 6 faceted stones
4 Iolite, 2 rough pieces
5 Iolite, 2 cut cubes
6 Iolite, 3 ovals
7 Phenakite, 3 rough pieces
8 Phenakite, 2 faceted stones

1
2
3
4
5
6
7
8
7

Dumortierite [1, 2]

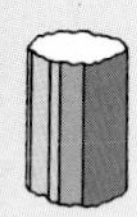

Color: Dark blue, violet-blue, red-brown, colorless
Color of streak: White
Mohs' hardness: 7–8½
Density: 3.26–3.41
Cleavage: Good
Fracture: Conchoidal
Crystal system: (Orthorhombic) very rare; fibrous or radial aggregates
Chemical composition: $Al_7[(O,OH_3)|BO_3|(SiO_4)_3]$ aluminum borate silicate
Transparency: Transparent to opaque
Refractive index: 1.678–1.689
Double refraction: –0.015 to –0.037
Dispersion: None
Pleochroism: Strong; black, red-brown, brown
Absorption: Not diagnostic
Fluorescence: Weak; blue, blue-white, violet

Aggregates of dumortierite, mainly cut, have a Mohs' hardness of 7; crystals have 8½. Named after a French palaeontologist. Deposits are found in Brazil, France, India, Canada, Madagascar, Mozambique, Namibia, Sri Lanka, and the United States.
Possibilities for Confusion With azurite (page 190), blue quartz (page 138), lapis lazuli (page 188), and sodalite (page 190).

Dumortierite Quartz Quartz which is intergrown with dumortierite [1].

Danburite [3, 4]

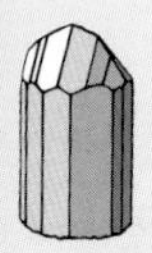

Color: Colorless, wine-yellow, brown, pink
Color of streak: White
Mohs' hardness: 7–7½
Density: 2.97–3.03
Cleavage: Imperfect
Fracture: Uneven, conchoidal
Crystal system: Orthorhombic; prismatic
Chemical composition: $Ca[B_2Si_2O_8]$ calcium boric silicate
Transparency: Transparent
Refractive index: 1.630–1.636
Double refraction: –0.006 to –0.008
Dispersion: 0.017 (0.009)
Pleochroism: Weak; light yellow
Absorption spectrum: 590, 586, 585, 584, 583, 582, 580, 578, 573, 571, 658, 566, 564
Fluorescence: Sky-blue

Named after the first place of discovery in Danbury, Connecticut. It has a greasy, vitreous luster. It can be well faceted because of its hardness and little cleavage. Deposits are found in Burma (Myanmar), Japan, Madagascar, Mexico, Russia, and Connecticut (page 280).
Possibilities for Confusion With citrine (page 136), hambergite (page 196), phenakite (page 196), and topaz (page 118).

Axinite (group) [5, 6]

Color: Brown, violet, blue
Color of streak: White
Mohs' hardness: 6½–7
Density: 3.26–3.36
Cleavage: Good
Fracture: Conchoidal, brittle
Crystal system: Triclinic; platy crystals
Chemical composition: $Ca_2Fe^{2+}Al_2B[O|OH|(Si_2O_7)_2]$ calcium aluminum borate silicate
Transparency: Translucent, transparent
Refractive index: 1.656–1.704
Double refraction: –0.010 to –0.012
Dispersion: 0.018–0.020 (0.011)
Pleochroism: Strong; olive-green, red-brown, violet-blue
Absorption spectrum: 532, 512, 492, 466, 444, 415
Fluorescence: Red, orange

The axinite group includes ferroaxinite, manganaxinite, magnesioaxenite, and tinzenite. Named (Greek—axe) because of its sharp-edged crystals. It has a strong, vitreous luster. As it is pyro- and piezo-electric, axinite attracts dust and must be cleaned frequently. Deposits are found in Brazil, England (Cornwall), France (Pyrenees, Dep. Isère), Mexico (Baja California), Russia (Urals), Sri Lanka, Tanzania, and California.
Possibilities for Confusion With andalusite (page 194), baryite (page 222), smoky quartz (page 132), and sphene (page 210).

1 Dumortierite, quartz, California
2 Dumortierite, 2 cabochons
3 Danburite, 9 different cuts
4 Danburite, 3 broken crystals
5 Axinite, 5 different cuts
6 Axinite, rough

1
2
4
3
4
6
5

Benitoite [1, 2]

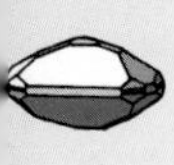

Color: Blue, purple, pink, colorless
Color of streak: White
Mohs' hardness: 6–61/2
Density: 3.64–3.68
Cleavage: Indistinct
Fracture: Conchoidal, brittle
Crystal system: (Hexagonal) bipyramidal
Chemical composition: $BaTi[Si_3O_9]$ barium titanium silicate
Transparency: Transparent
Refractive index: 1.757–1.804
Double refraction: +0.047
Dispersion: 0.046 (0.026)
Pleochroism: Very strong; colorless, blue
Absorption spectrum: Not diagnostic
Fluorescence: Strong; blue

Named after discovery in San Benito County, California. Only small crystals are of gemstone quality. A vitreous to adamantine luster. Deposits only in California.
Possibilities for Confusion With iolite (page 196), kyanite (page 212), sapphire (page 102), spinel (page 116), tanzanite (page 176), tourmaline (page 126), zircon (page 124).

Cassiterite [3–5] Also called Tin Stone

Color: Various browns, colorless
Color of streak: White to light yellow
Mohs' hardness: 6–7
Density: 6.7–7.1
Cleavage: Indistinct
Fracture: Conchoidal, brittle
Crystal system: Tetragonal; short columnar
Chemical composition: SnO_2 tin oxide
Transparency: Transparent to opaque
Refractive index: 1.997–2.098
Double refraction: +0.096 to +0.098
Dispersion: 0.071 (0.035)
Pleochroism: Weak to strong; green-yellow, brown, red-brown
Absorption spectrum: Not diagnostic
Fluorescence: None

Cassiterite (Greek—tin) has anadamantine luster. Deposits in Australia, Bolivia, England, Malaysia, Mexico, Namibia, Spain, and California.
Possibilities for Confusion With diamond (page 86), smoky quartz (page 132), scheelite (page 212), sinhalite (page 202), sphene (page 210), idocrase (page 202), and zircon (page 124).

Epidote [6–8] Also called Pistacite

Color: Pistachio green
Color of streak: Gray
Mohs' hardness: 6–7
Density: 3.3–3.5
Cleavage: Perfect
Fracture: Conchoidal, splintery
Crystal system: Monoclinic; prisms
Chemical composition: $Ca_2(Fe^{3+},Al)Al_2[O|OH|SiO_4|Si_2O_7]$ calcium aluminum iron silicate
Transparency: Transparent to opaque
Refractive index: 1.729–1.768
Double refraction: +0.015 to –0.049
Dispersion: 0.030 (0.012–0.027)
Pleochroism: Strong; green, brown, yellow
Absorption spectrum: 475, <u>455</u>, 435
Fluorescence: None

Named (Greek—addition) after the numerous crystal faces. It has a bright, vitreous luster. Deposits are found in Brazil, Kenya, Mexico, Mozambique, Norway, Austria (Untersulzbach Valley/Salzburg), Sri Lanka, and California.
Possibilities for Confusion With diopside (page 206), dravite (page 126), and idocrase (page 202).

Clinozoisite Epidote mineral, low on iron. (Compare with page 228, picture on page 229, no. 17.)

Piemontite (piedmontite) Manganian-epidote, cherry red, opaque, Mohs' hardness 6, density 3.45–3.52. Locality: Italy (Piedmont), France (Bretagne), Japan.

Tawmawite Dark green, chrome-containing epidote variety, transparent to opaque. Locality: Finland (Outokumpu), Myanmar.

1 Benitoite, 2 crystal formations
2 Benitoite, 8 faceted stones
3 Cassiterite, crystals, Cornwall, England
4 Cassiterite, 3 light brown stones, Malaysia
5 Cassiterite, crystal, Cornwall
6 Epidote, 3 faceted stones
7 Epidote, 2 broken crystals
8 Epidote, twinned aggregate
The illustrations are 30 percent larger than the originals.

1
2
3
4
5
6
7
7
8

Idocrase [1–4] Also called Vesuvianite

Color: Olive-green, yellow-brown, pale blue
Color of streak: White
Mohs' hardness: 6½
Density: 3.32–3.47
Cleavage: Indistinct
Fracture: Uneven, splintery
Crystal system: Tetragonal; thick columnar crystals
Chemical composition: $Ca_{19}(Mg,Fe^{2+},Ti)_4Al_9[(OH,F)_{10}|(SiO_4)_{10}|(Si_2O_7)_4]$ complicated calcium aluminum silicate
Transparency: Transparent, translucent
Refractive index: 1.700–1.723
Double refraction: +0.002 to –0.012
Dispersion: 0.019–0.025 (0.014)
Pleochroism: Weak; present color lighter and darker
Absorption spectrum: Green: (528), 461; Brown: 591, 588, 584, 582, 577, 574
Fluorescence: None

Due to wide variations in composition, there is a range of physical properties displayed. The luster is greasy. Deposits are found in Brazil, Mexico, Kenya, Russia, Switzerland, Sri Lanka, and the United States.

Possibilities for Confusion With demantoid (page 122), diopside (page 206), epidote (page 200), zircon (page 124), peridot (page 174), and sinhalite (see below).

Californite Green variety; falsely called California jade (page 279).

Cyprine Sky-blue variety from Norway.

Sinhalite [5, 6]

Color: Yellow-brown, green-brown
Color of streak: White
Mohs' hardness: 6½–7
Density: 3.46–3.50
Cleavage: None
Fracture: Conchoidal
Crystal system: (Orthorhombic) very rare; grains
Chemical composition: $MgAl[BO_4]$ magnesium aluminum iron borate
Transparency: Transparent, translucent
Refractive index: 1.665–1.712
Double refraction: –0.036 to –0.042
Dispersion: 0.018 (0.010)
Pleochroism: Definite; green, light brown, dark brown
Absorption spectrum: 526, 492, 475, 463, 452
Fluorescence: None

First recognized as an individual mineral in 1952 in Sri Lanka, then Ceylon (Sanskrit—*Sinbala*). It has a vitreous luster. Deposits found in Burma (Myanmar), Russia, Sri Lanka, and Tanzania.

Possibilities for Confusion With chrysoberyl (page 114), peridot (page 174), tourmaline (page 126), idocrase (see above), and zircon (page 124).

Kornerupine [7, 8] Also called Prismatine

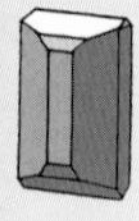

Color: Green, green-brown
Color of streak: White
Mohs' hardness: 6½–7
Density: 3.27–3.45
Cleavage: Good
Fracture: Conchoidal
Crystal system: Orthorhombic; long prisms
Chemical composition: $MgAl_6[(O,OH)_2|BO_4|(SiO_4)_4]$ magnesium aluminum borate silicate
Transparency: Transparent, translucent
Refractive index: 1.660–1.699
Double refraction: –0.012 to –0.017
Dispersion: 0.018 (0.010)
Pleochroism: Strong; green, yellow, reddish-brown
Absorption spectrum: 540, 503, 463, 446, 430
Fluorescence: Usually none; green k. from

Named after a Danish geologist and explorer of Greenland. It has a vitreous luster. Deposits are found in Burma (Myanmar), Canada (Quebec), Kenya, Madagascar, Sri Lanka, Tanzania, and South Africa.

Possibilities for Confusion With enstatite (page 208), epidote (page 200), and tourmaline (page 126).

1 Idocrase, crystal
2 Idocrase, 3 faceted stones, 6.25ct
3 Idocrase, cabochon, 4.19ct
4 Idocrase, 4 faceted stones
5 Sinhalite, rough
6 Sinhalite, 2 faceted stones
7 Kornerupine, 3 faceted stones
8 Kornerupine, aggregate, Sri Lanka

The illustrations are 20 percent larger than the originals.

1
2
3
4
5
6
7
8

Prehnite [1–3]

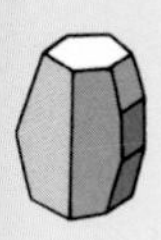

Color: Yellow-green, brown-yellow
Color of streak: White
Mohs' hardness: 6–6½
Density: 2.82–2.94
Cleavage: Good
Fracture: Uneven
Crystal system: Orthorhombic; columnar, tabular crystals and aggregates
Chemical composition: $Ca_2Al[(OH)_2|AlSi_3O_{10}]$ basic calcium aluminum silicate
Transparency: Transparent, translucent
Refractive index: 1.611–1.669
Double refraction: +0.021 to +0.039
Dispersion: None
Pleochroism: Absent
Absorption spectrum: 438
Fluorescence: None

Named after Dutch colonel. It has a vitreous to mother-of-pearl luster. Prehnite cat's-eye occurs. Deposits are found in Australia, China, Scotland, South Africa, and the United States.

Possibilities for Confusion With apatite (page 210), brazilianite (page 206), chrysoprase (page 144), jade (page 170), peridot (page 174), periclase (page 222), and serpentine (page 218).

Petalite [4, 5]

Color: Colorless, pink, yellowish
Color of streak: White
Mohs' hardness: 6–6½
Density: 2.40
Cleavage: Perfect
Fracture: Conchoidal, brittle
Crystal system: (Monoclinic) thick tabular, columnar and aggregates
Chemical composition: $Li[AlSi_4O_{10}]$ lithium aluminum silicate
Refractive index: 1.502–1.519
Double refraction: +0.012 to +0.017
Dispersion: Unknown
Pleochroism: Absent
Absorption spectrum: (454)
Fluorescence: Weak; orange

Named (Greek—leaf) because of its perfect cleavage. It has a vitreous luster; pearly luster on cleavage planes. Crystals are rare; mostly coarse aggregates occur; and petalite cat's-eye is known. Deposits are found in Western Australia, Brazil (Minas Gerais), Italy (Elba), Namibia, Sweden, Zimbabwe, and the United States.

Possibilities for Confusion With other colorless gemstones and glass.

Scapolite [6–8] Also called Wernerite

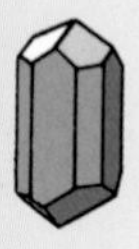

Color: Yellow, brownish pink, violet, colorless
Color of streak: White
Mohs' hardness: 5½–6
Density: 2.57–2.74
Cleavage: Good
Fracture: Conchoidal, brittle
Crystal system: Tetragonal; columnar
Chemical composition: $(Ca,Na)_4[(CO_3Cl)|(SiAl)_6Si_6O_{24}]$ sodium calcium aluminum silicate
Transparency: Transparent, translucent
Refractive index: 1.540–1.579
Double refraction: –0.006 to –0.037
Dispersion: 0.017
Pleochroism: Definite; yellow s.: colorless, yellow
Absorption spectrum: 663, 652
Fluorescence: Pink: orange, pink; Yellow: violet, blue-red

Scapolite is named (Greek—stick) after its crystal habit. It has a vitreous luster. Pink and violet scapolite cat's-eyes are known. Deposits are found in Burma (Myanmar), Brazil, India, Canada, Madagascar, and Tanzania.

Possibilities for Confusion With amblygonite (page 208), chrysoberyl (page 114), golden beryl (page 112), rose quartz (page 138), sphene (page 210), and tourmaline (page 126).

Petschite A violet scapolite variety from Tanzania, discovered in 1975.

---

1 Prehnite, 2 cabochons, 31.91ct
2 Prehnite, 2 faceted stones, Australia
3 Prehnite, with apophyllite crystals
4 Petalite, rough
5 Petalite, 3 faceted stones
6 Scapolite, 5 faceted stones
7 Scapolite cat's-eye, 5 stones
8 Scapolite, 4 crystal pieces

1
2
3
4
5
6
7
8

Diopside [1–3]

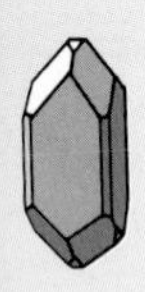

Color: Green, yellow, colorless, brown, black
Color of streak: White
Mohs' hardness: 5–6
Density: 3.22–3.38
Cleavage: Good
Fracture: Uneven, rough
Crystal system: Monoclinic; columnar crystals
Chemical composition: $CaMg[Si_2O_6]$ calcium magnesium silicate
Transparency: Transparent, translucent
Refractive index: 1.664–1.730
Double refraction: +0.024 to +0.031
Dispersion: 0.017–0.020 (0.012)
Pleochroism: Weak; yellow-green, dark green
Absorption spectrum: (505), (493), (446); Chrome diopside: (690), (670), (655), (635), 508, 505, 490
Fluorescence: Violet, orange, yellow, green

Named (Greek—double appearance) because of its crystal shape. Deposits are found in Burma (Myanmar), Finland, India, Madagascar, Austria, Sri Lanka, South Africa, and the United States. Star diopside [3] and diopside cat's-eye are known.
Possibilities for Confusion With hiddenite (page 130), moldavite (page 246), peridot (page 174), emerald (page 106), and idocrase (page 202).

Chrome Diopside Diopside variety with strong emerald green color.

Violane Coarse, violet-blue diopside variety, translucent to opaque, from Piedmont, Italy. Used for ornamental objects.

Beryllonite [4]

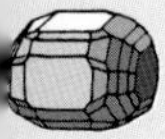

Color: Colorless, white, weak yellow
Color of streak: White
Mohs' hardness: 5½–6
Density: 2.80–2.87
Cleavage: Perfect
Fracture: Conchoidal, brittle
Crystal system: Monoclinic; short prisms
Chemical composition: $NaBe[PO_4]$ sodium beryllium phosphate
Transparency: Transparent
Refractive index: 1.552–1.561
Double refraction: –0.009
Dispersion: 0.010 (0.007)
Pleochroism: Absent
Absorption spectrum: Cannot be evaluated
Fluorescence: None

So named because of its beryllium content. It has a vitreous luster; on fracture planes, pearly luster. Deposits are found in Brazil, Finland, Zimbabwe, and Maine.
Possibilities for Confusion With other colorless gemstones and glass.

Brazilianite [5, 6]

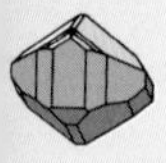

Color: Yellow, green-yellow
Color of streak: White
Mohs' hardness: 5½
Density: 2.98–2.99
Cleavage: Good
Fracture: Small conchoidal, brittle
Crystals: Monoclinic; short prisms
Chemical composition: $NaAl_3[(OH)_2|PO_4]_2$ sodium aluminum phosphate
Transparency: Transparent, translucent
Refractive index: 1.602–1.623
Double refraction: +0.019 to +0.021
Dispersion: 0.014 (0.008)
Pleochroism: Very weak
Absorption spectrum: Not diagnostic
Fluorescence: None

Named after Brazil, the first country of discovery (1944). Deposits are found in Brazil (Minas Gerais) and the United States (New Hampshire).
Possibilities for Confusion With amblygonite (page 208), apatite (page 210), chrysoberyl (page 114), precious beryl (page 112), and topaz (page 118).

1 Diopside, 10 different cuts
2 Diopside, 2 broken crystals
3 Diopside, 4-rayed asterism
4 Beryllonite, 3 faceted stones
5 Brazilianite, rough
6 Brazilianite, 5 faceted stones
The illustrations are 15 percent larger than the originals.

1
2
2
3
4
5
6

Amblygonite [1, 2]

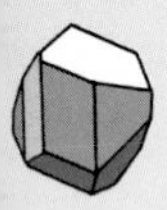

Color: Golden-yellow to colorless, purple
Color of streak: White
Mohs' hardness: 6
Density: 3.01–3.11
Cleavage: Perfect
Fracture: Uneven, brittle
Crystal system: (Triclinic) prismatic
Chemical composition: $(Li,Na)Al[(F,OH)|PO_4]$ basic lithium aluminum phosphorate
Transparency: Transparent, translucent
Refractive index: 1.578–1.646
Double refraction: +0.024 to +0.030
Dispersion: 0.014–0.015 (0.008)
Pleochroism: Absent
Absorption spectrum: Not diagnostic
Fluorescence: Very weak; green

So named (Greek—crooked, angled) because of its crystal shape. Vitreous luster, pearly luster on cleavage planes. Deposits: Burma, Brazil (Minas Gerais, Sao Paulo), Sweden, and California. A purple variety from Namibia is known.
Possibilities for Confusion With apatite (page 210), brazilianite (page 206), citrine (page 136), golden beryl (page 112), hiddenite (page 130), and scapolite (page 204).

Enstatite [3, 4]

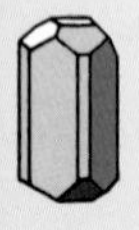

Color: Brown-green, green, colorless, yellowish
Color of streak: White
Mohs' hardness: 5½
Density: 3.20–3.30
Cleavage: Good
Fracture: Scaly
Crystal system: Orthorhombic; prismatic
Chemical composition: $Mg_2[Si_2O_6]$ magnesium silicate
Transparency: Transparent to opaque
Refractive index: 1.650–1.680
Double refraction: +0.009 to + 0.012
Dispersion: Little to none (0.010)
Pleochroism: Definite; green, yellow-green
Absorption spectrum: 547, 509, 505, 502, 483, 459, 449; Chrome e.: 688, 669, 506
Fluorescence: None

So named (Greek—resistor) because it does not melt easily. It has a vitreous luster. Deposits are found in Burma (Myanmar), Brazil, India, Kenya, Mexico, Sri Lanka, South Africa, Tanzania, and the United States. Greenish-gray enstatite cat's-eye and star enstatite are known (page 278).
Possibilities for Confusion With andalusite (page 194), kornerupine (page 202), sphalerite (page 216), idocrase (page 202), and zircon (page 124).

Bronzite Green-brown enstatite with metallic luster; rich in iron (page 234).

Lazulite [5, 6] Also called Bluespar

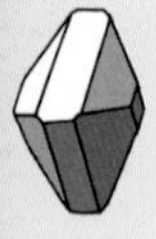

Color: Dark blue to blue-white, green-blue
Color of streak: White
Mohs' hardness: 5–6
Density: 3.04–3 14
Cleavage: Indistinct
Fracture: Uneven, splintery, brittle
Crystal system: Monoclinic; pointed pyramids
Chemical composition: $(Mg,Fe^{24})Al_2[OH|PO_4]_2$ basic magnesium aluminum phosphate
Transparency: Transparent to opaque
Refractive index: 1.612–1.646
Double refraction: –0.031 to –0.036
Dispersion: Unknown
Pleochroism: Strong; colorless, dark blue
Absorption spectrum: Not diagnostic
Fluorescence: None

Named (Persian and Greek—blue stone) because of its color. It has a vitreous luster (page 280). Deposits are found in Afghanistan, Angola, Bolivia, Brazil (Minas Gerais), India, Madagascar, Austria, Sweden, and North Carolina.
Possibilities for Confusion With azurite (page 190), iolite (page 196), indicolite (page 126), lapis lazuli (page 188), sodalite (page 190), topaz (page 118), and turquoise (page 186).

1 Amblygonite, 2 rough pieces
2 Amblygonite, 6 different cuts
3 Enstatite, 2 faceted stones and Star enstatite
4 Enstatite, 3 rough pieces
5 Lazulite, 9 different cuts
6 Lazulite, rough
The illustrations are 20 percent larger than the originals.

1
2
3
4
4
5
6

Dioptase [1, 2]

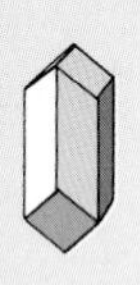

Color: Emerald green, blue-green
Color of streak: Greenish
Mohs' hardness: 5
Density: 3.28–3.35
Cleavage: Perfect
Fracture: Conchoidal, brittle
Crystal system: (Trigonal) short columnar
Chemical composition: $Cu_6[Si_6O_{18}] \cdot 6H_2O$ hydrous copper silicate

Transparency: transparent, translucent
Refractive index: 1.644–1.709
Double refraction: +0.051 to +0.053
Dispersion: 0.036 (0.021)
Pleochroism: Weak; dark emerald green, light emerald green
Absorption spectrum: 550, 465
Fluorescence: None

So named (Greek—view through) because of its crystal structure. Deposits are found in Chile, Kyrgyzstan, the Democratic Republic of the Congo, Namibia, Peru, Russia, Arizona, and Zaire.

Possibilities for Confusion With demantoid (page 122), diopside (page 206), fluorite (page 214), emerald (page 106), uvarovite (page 122), and verdelite (page 126).

Apatite (group) [3–6]

Color: Colorless, pink, yellow, green, blue, violet
Color of streak: White to yellow-gray
Mohs' hardness: 5
Density: 3.16–3.23
Cleavage: Indistinct
Fracture: Conchoidal, brittle
Crystal system: (Hexagonal), columnar, thick tabular
Chemical composition: $Ca_5(F,Cl,OH)[PO_4]_3$ basic fluoro- and chloro-calcium phosphate

Transparency: Transparent
Refractive index: 1.628–1.649
Double refraction: –0.002 to –0.006
Dispersion: 0.013 (0.010)
Pleochroism: Green a.: yellow, green; Blue a.: very strong; blue, yellow
Absorption spectrum: Yellow and green a.: 597, 585, 577, 533, 529, 527, 525, 521, 514, 469; Blue a.: 512, 507, 491, 464
Fluorescence: Yellow a.: purple to pink

Name (Greek—cheat) because it can be easily confused. Vitreous luster; sensitive to acids. Deposits in Myanmar, Brazil, India, Kenya, Madagascar, Mexico, Norway, Sri Lanka, South Africa, and the United States. Apatite cat's-eye is known.

Possibilities for Confusion With amblygonite (page 208), andalusite (page 194), brazilianite (page 206), precious beryl (page 112), sphene (see below), topaz (page 118), and tourmaline (page 126). Synthetic apatite is known.

Asparagus Stone Trade name for a light green apatite variety.

Sphene [7, 8] Also called Titanite

Color: Yellow, brown, green, reddish
Color of streak: White
Mohs' hardness: 5–5½
Density: 3.52–3.54
Cleavage: Good
Fracture: Conchoidal, brittle
Crystal system: (Monoclinic) platy
Chemical composition: $CaTi[O|SiO_4]$ calcium titanium silicate

Transparency: Transparent to opaque
Refractive index: 1.843–2.110
Double refraction: +0.100 to +0.192
Dispersion: 0.051 (0.019–0.038)
Pleochroism: Strong; colorless, greenish-yellow, reddish
Absorption spectrum: 586, 582
Fluorescence: None

The mineralogical name (titanite) derives from the titanium contents. It has an adamantine luster; with brilliant cut, it has an intensive fire. Deposits are found in Myanmar, Brazil, Mexico, Austria, Sri Lanka, and the United States. Titanite is changed to red or orange through heating.

Possibilities for Confusion With chrysoberyl (page 114), dravite (page 126), golden beryl (page 112), scheelite (page 212), topaz (page 118), zircon (page 124), and idocrase (page 202).

1 Dioptase, 2 crystals
2 Dioptase, 12 faceted stones
3 Apatite, crystal
4 Apatite cat's-eye, Brazil
5 Apatite, 8 different cuts
6 Apatite, 3 crystals
7 Sphene, 8 different cuts
8 Sphene, rough

1
1
2
3
6
5
4
7
8

Kyanite [1–3] Also called Disthene

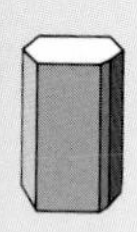

Color: Blue to colorless, blue-green, brown
Color of streak: White
Mohs' hardness: Along axes 4–4½, across 6–7
Density: 3.53–3.70
Cleavage: Perfect
Fracture: Fibrous, brittle
Crystal system: Triclinic; long, flat prisms
Chemical composition: $Al_2O[SiO_4]$ aluminum silicate
Transparency: Transparent, translucent
Refractive index: 1.710–1.734
Double refraction: –0.015
Dispersion: 0.020 (0.011)
Pleochroism: strong; colorless, blue, dark blue
Absorption spectrum: (706), (689), (671), (652), 446, 433
Fluorescence: Weak; red

Name (Greek—blue) because of its color. It has a vitreous luster, often with irregular streaks. Difficult to cut because of variable hardness. Deposits are found in Burma (Myanmar), Brazil, Kenya, Austria, Switzerland, Zimbabwe, and the United States.

Possibilities for Confusion With aquamarine (page 110), benitoite (page 200), iolite (page 196), dumortierite (page 198), sapphire (page 102), and tourmaline (page 126). Green variety found on page 280.

Scheelite [4, 5]

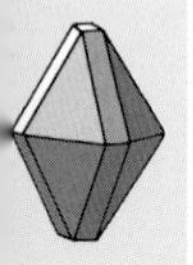

Color: Yellow, brown, orange, colorless
Color of streak: White
Mohs' hardness: 4½–5
Density: 5.9–6.3
Cleavage: Good
Fracture: Conchoidal, splintery, brittle
Crystal system: Tetragonal; dipyramids
Chemical composition: $Ca[WO_4]$ calcium tungstate
Transparency: Transparent, translucent
Refractive index: 1.918–1.937
Double refraction: +0.010 to +0.018
Dispersion: 0.038 (0.026)
Pleochroism: Variable
Absorption spectrum: 584
Fluorescence: Strong; light blue

Named after a Swedish chemist. It has an adamantine luster. Deposits are found in Japan, Korea, Mexico, Sri Lanka, and Arizona.

Possibilities for Confusion With chrysoberyl (page 114), diamond (page 86), golden beryl (page 112), and zircon (page 124). Synthetic scheelite has been known since 1963.

Variscite [6–8] Also called Utahlite

Color: Yellow-green, bluish
Color of streak: White
Mohs' hardness: 4–5
Density: 2.42–2.58
Cleavage: Perfect
Fracture: Conchoidal, brittle
Crystal system: Orthorhombic; short needles
Chemical composition: $Al[PO_4] \cdot 2H_2O$ hydrous aluminum phosphate
Transparency: Translucent to opaque
Refractive index: 1.563–1.594
Double refraction: –0.031
Dispersion: Unknown
Pleochroism: Missing
Absorption spectrum: 688, (650)
Fluorescence: Strong; pale green, green

So named (Latin—*variscia*) after old name for the Vogtland, Germany. Mostly exists as bulbous, coarse aggregates, frequently interspersed with brown matrix; is commonly cut en cabochon.

Possibilities for Confusion With chrysocolla (page 216), chrysoprase (page 144), jade (page 170), smaragdite (page 220), turquoise (page 186), and verdite (page 276).

Amatrix ("American matrix," also spelled amatrice and called variscite quartz) Intergrowth of variscite with quartz or chalcedony, Nevada.

1 Kyanite, 3 broken crystals
2 Kyanite, 5 different cuts
3 Kyanite, crystal
4 Scheelite, broken crystal
5 Scheelite, 3 faceted stones
6 Variscite, 3 cabochons
7 Variscite in host rock, 2 pieces
8 Variscite with host rock, cabochon

3
1
2
1
4
5
6
7
8

Fluorite [1–3] Also called Fluorspar

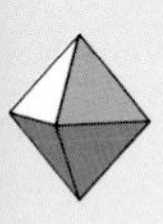

Color: Colorless, all colors
Color of streak: White
Mohs' hardness: 4
Density: 3.00–3.25
Cleavage: Perfect
Fracture: Even to conchoidal, brittle
Crystal system: (Cubic) cubes, octahedra
Chemical composition: CaF_2 calcium fluoride

Transparency: Transparent, translucent
Refractive index: 1.434
Double refraction: None
Dispersion: 0.007 (0.004)
Pleochroism: Absent
Absorption spectrum: Green f.: 634, 610, 582, 445, 427
Fluorescence: Strong; blue-violet

Name (Latin—to flow) follows its being used as flux. It has a vitreous luster; color distribution is often zonal or spotty. Deposits are found in Oberpfak, Bavaria (Germany), Argentina, Burma (Myanmar), England, France, Namibia, Austria, Switzerland, and Illinois.
Possibilities for Confusion With many gemstones, due to its richness of color. Color can be changed with gamma rays. Synthetic fluorite is known in all colors.

Blue John Fluorite variety, banded with colors and white, Derbyshire, England.

Hemimorphite [4–6] Also called Calamine

Color: Blue, green, colorless
Color of streak: White
Mohs' hardness: 5
Density: 3.30–3.50
Cleavage: Perfect
Fracture: Conchoidal, uneven, brittle
Crystal system: Orthorhombic; tabular
Chemical composition: $Zn_4[(OH)_2|Si_2O_7] \cdot H_2O$ hydrous basic zinc silicate

Transparency: Transparent to opaque
Refractive index: 1.614–1.636
Double refraction: +0.022
Dispersion: 0.020 (0.013)
Pleochroism: Absent
Absorption spectrum: Cannot be evaluated
Fluorescence: Weak; not characteristic

Named (Greek—half shape) because of its crystal formation. Aggregates are often blue-white banded, also mixed with dark matrix [no. 5]. Deposits are found in Algeria, Australia, Italy, Mexico, Namibia, Austria, and the United States.
Possibilities for Confusion With chrysocolla (page 216), smithsonite (see below), and turquoise (page 186).

Smithsonite [7, 8] Also called Bonamite

Color: Light green, light blue, pink
Color of streak: White
Mohs' hardness: 5
Density: 4.00–4.65
Cleavage: Perfect
Fracture: Uneven, brittle
Crystal system: (Trigonal) rhombohedral
Chemical composition: $Zn[CO_3]$ zinc carbonate

Transparency: Translucent, opaque
Refractive index: 1.621–1.849
Double refraction: –0.228
Dispersion: 0.014–0.031 (0.008–0.017)
Pleochroism: Absent
Absorption spectrum: Cannot be evaluated
Fluorescence: Blue-white, pink, brown

Named after an American mineralogist. It has a vitreous luster; grape-like clusters have a pearly luster, often slightly banded. Deposits are found in Australia, Greece, Italy (Sardinia), Mexico, Namibia, Spain, and New Mexico.
Possibilities for Confusion With chrysoprase (page 144), hemimorphite (see above), jade (page 170), and turquoise (page 186).

1 Fluorite, 2 cleaved octahedrons
2 Fluorite, 2 rough pieces
3 Fluorite, 9 different cuts
4 Hemimorphite, crystal and 2 stones
5 Hemimorphite, 3 cabochons
6 Hemimorphite, radial aggregate
7 Smithsonite, 2 aggregates
8 Smithsonite, 3 cabochons

1
2
2
3
4
5
6
8
7
7

Sphalerite [1–3] Also called Zinc Blende

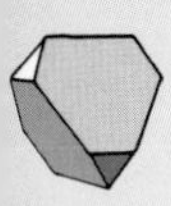

Color: Yellow, reddish, greenish, colorless
Color of streak: Yellowish to light brown
Mohs' hardness: 3½–4
Density: 3.90–4.10
Cleavage: Perfect
Fracture: Uneven, brittle
Crystal system: Cubic; tetrahedral
Chemical composition: ZnS zinc sulphide
Transparency: Transparent, translucent
Refractive index: 2.368–2.371
Double refraction: None
Dispersion: 0.156
Pleochraism: Absent
Absorption spectrum: 690, 667, 651
Fluorescence: Yellow-orange, red

Name (Greek—deceitful) due to the fact that it is used as an ore. It has a greasy and adamantine luster. Dispersion is three times as high as for diamond. Deposits are found in Canada, the Democratic Republic of the Congo, Mexico, Namibia, Spain, Wisconsin, and Zaire (see also page 280).
Possibilities for Confusion With chrysoberyl (page 114), cassiterite (page 200), scheelite (page 212), sinhalite (page 202), topaz (page 118), tourmaline (page 126), idocrase (page 202), zircon (page 124), and colorless diamond (page 86). Synthetics are available in the trade.

Cerussite [4–5] Also called White-Lead Ore

Color: Colorless, gray, brownish
Color of streak: White
Mohs' hardness: 3–3½
Density: 6.46–6.57
Cleavage: Good
Fracture: Conchoidal, uneven, very brittte
Crystal system: Orthorhombic; tabular, columnar
Chemical composition: $Pb[CO_3]$ lead carbonate
Transparency: Transparent, translucent
Refractive index: 1.804–2.079
Double refraction: –0.274
Dispersion: 0.055 (0.033–0.050)
Pleochroism: Absent
Absorption spectrum: Cannot be evaluated
Fluorescence: Yellow, pink, green, bluish

Named (Latin—lead white) after its composition. It has an adamantine luster. It is very difficult to cut because of its high brittleness. Deposits are found in Australia, Italy, Austria, Madagascar, Namibia, Zambia, Scotland, and the United States.
Possibilities for Confusion The same as for sphalerite (see above).

Chrysocolla [6–8]

Color: Green, blue
Color of streak: Green-white
Mohs' hardness: 2–4
Density: 2.00–2.40
Cleavage: None
Fracture: Conchoidal
Crystal system: Monoclinic; compact grape-like aggregates (botryoidal)
Chemical composition: $(Cu,Al)_2H_2[(OH)_4|Si_2O_5] \bullet nH_2O$ hydrous copper silicate
Transparency: Opaque, sometimes just translucent
Refractive index: 1.460–1.570
Double refraction: –0.023 to –0.040
Dispersion: None
Pleochroism: Absent
Absorption spectrum: Cannot be evaluated
Fluorescence: None

It has a greasy vitreous luster. Deposits are found in Chile, Israel, the Democratic Republic of the Congo, Mexico, Peru, Russia, Nevada, and Zaire.
Possibilities for Confusion With azurite (page 190), dyed chalcedony (page 142), malachite (page 192), turquoise (page 186), and variscite (page 212).

Chrysocolla Quartz Intergrowth of chrysocolla with quartz.

Eilat Stone [8] Intergrowth of chrysocolla with turquoise and malachite; found near Eilat, Israel.

1 Sphalerite, 3 rough pieces
2 Sphalerite, faceted, 47.97ct
3 Sphalerite, 3 faceted stones
4 Cerussite, twinned crystal
5 Cerussite, 5 faceted stones
6 Chrysocolla, 4 cabochons
7 Chrysocolla, 2 rough pieces
8 Eilat stone, 2 cabochons

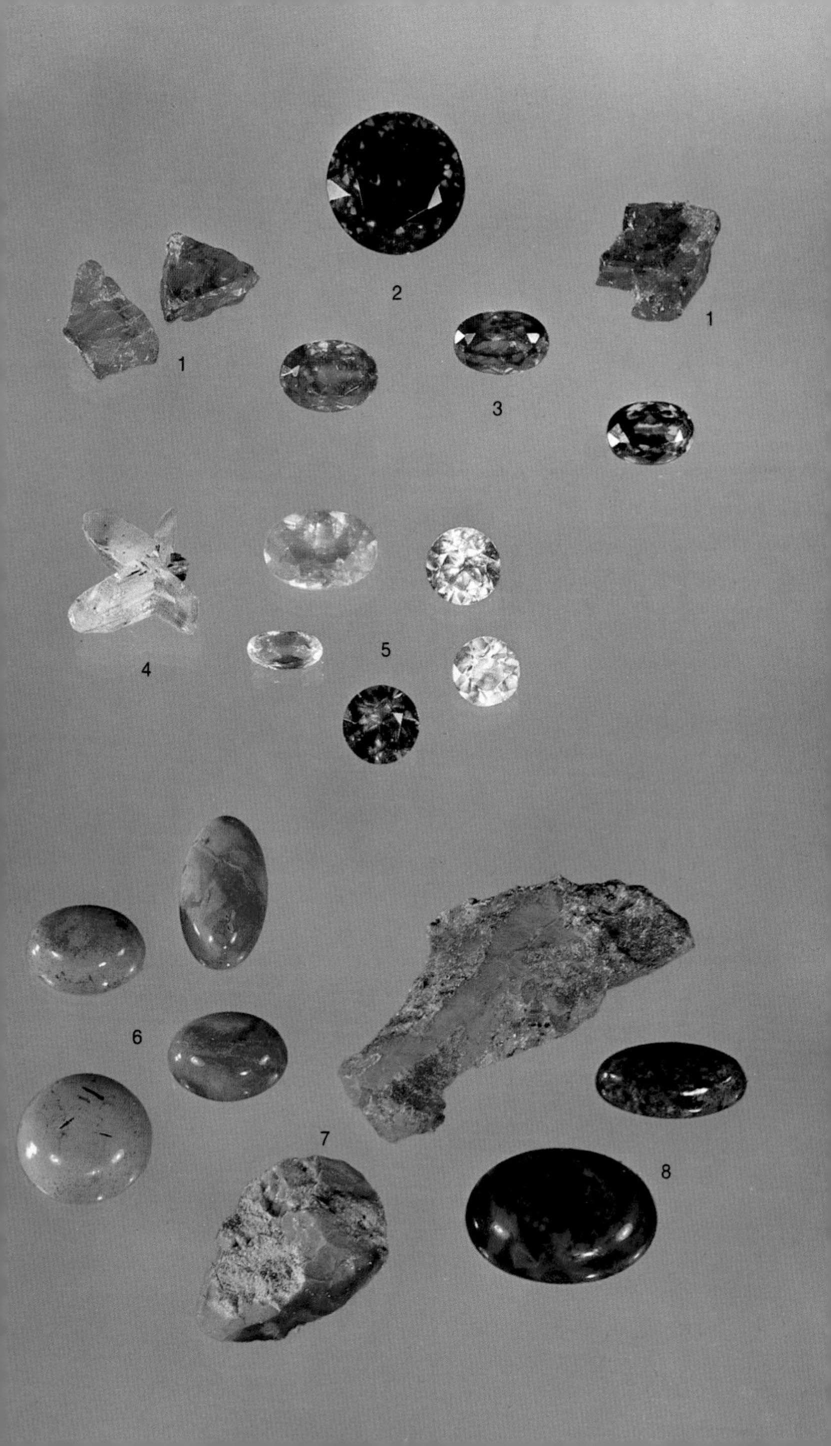
2
1
1
3
4
5
6
7
8

Serpentine [1–3]

Color: Green, yellowish, brown
Color of streak: White
Mohs' hardness: 2½–5½
Density: 2.44–2.62
Cleavage: None
Fracture: Uneven, splintery, tough
Crystal system: (Monoclinic) microcrystalline
Chemical composition: $Mg_6[(OH)_8|Si_4O_{10}]$ basic magnesium silicate
Transparency: semitransparent to opaque
Refractive index: 1.560–1.571
Double refraction: +0.008 to +0.014
Dispersion: None
Pleochroism: Absent
Absorption spectrum: Bowenite: 492, 464
Fluorescence: Williamsite weak; greenish

There are two aggregate structures for serpentine (Latin—snake): leafy antigorite (leafy serpentine) and fibrous chrysotile (fibrous serpentine). Very finely fibrous varieties are called asbestos. It has a greasy to silky luster, and is sensitive to acids. Colors are often spotty. Deposits are found in Afghanistan, China, New Zealand, and the United States. Used mainly as decorative stone or for ornamental objects. There are many trade names for the "precious serpentine," which is sometimes misleadingly sold as jade.

Possibilities for Confusion With jade (page 170), onyx marble (page 244), turquoise (page 186), and verdite (page 276).

Bastite Silky shiny serpentine in the crystal shape of enstatite (page 208).

Bowenite Apple-green serpentine variety; frequently interspersed with light spots.

Connemara Green rock, intergrowth of serpentine with marble.

Verd-Antique Green rock, interspersed with white calcite or dolomite veins; a serpentinite breccia.

Williamsite [3] Oil-green serpentine variety; often with black inclusions.

Stichtite [4]

It is pink-red to purple, formed through the decomposition of chrome-containing serpentinite. It has a vitreous luster. Deposits are found in Australia, Canada, Zimbabwe, and South Africa.

Ulexite [5, 6]

Color: White
Color of streak: White
Mohs' hardness: 2–2½
Density: 1.65–1.95
Cleavage: Perfect
Fracture: Fibrous
Crystal system: (Monoclinic) small; fibrous aggregates
Chemical composition: $NaCa[B_5O_6(OH)_6] \cdot 5H_2O$ hydrous sodium calcium borate
Transparency: Transparent, translucent
Refractive index: 1.491–1.520
Double refraction: +0.029
Dispersion: None
Pleochroism: Absent
Absorption spectrum: Cannot be evaluated
Fluorescence: Green-yellow, blue

Named after a German chemist. It has a silky luster. A piece of writing placed underneath the stone appears on the surface (therefore also called TV stone). Shows a cat's-eye effect [no. 5] when cut en cabochon. Deposits are found in Argentina, Chile, Canada, Kazakhstan, Peru, Russia, and the United States.

Tiger's-Eye Matrix [7, 8]

Trade name for a mineral aggregate, in which tiger's-eye-like structures (page 140) alternate with iron oxide layers.

1 Serpentine, rough
2 Chrysotile, 2 cabochons
3 Williamsite, 2 faceted stones
4 Stichtite, rough and cabochon
5 Ulexite, 3 cabochons
6 Ulexite, 3 rough pieces
7 Tiger's-eye matrix, 2 pieces partly polished
8 Tiger's-eye matrix, rough

1
2
3
4
5
6
7
8

Gemstones for Collectors

Many minerals are not worn as jewelry because they are either too soft, too brittle, endangered, or too rare. They are faceted, though, or offered en cabochon, for collectors and others who are interested in them. Compare also with the description on p. 84.

1 **Gahnite** (also called zinc spinel) Red-violet (0.92 ct), also blue, green, blackish. Color of streak gray-white. Transparent. Mohs' hardness 7½-8. Density 4.00-4.62. Refractive Index 1.791-1.818, double refraction none. Cubic, $ZnAl_2O_4$. Cleavage indistinct.

2 **Binghamite** Gray-brown. Trade name for a quartz (p. 132) with goethite inclusions (p. 238). As cabochon, it shows a fine shimmer.

3 **Willemite** Orange (0.18 ct), green (0.18 ct), light brown (0.20 ct), also white, yellow. Color of streak white. Transparent to opaque. Mohs' hardness 5½. Density 3.89-4.18. Refractive index 1.690-1.723, double refraction 0.028-0.033. Trigonal, $Zn_2[SiO_4]$. Cleavage indistinct.

4 **Sanidine** Light brownish, also colorless, white. Color of streak white. Transparent. Mohs' hardness 6-6½ . Density 2.56-2.58. Refractive index 1.518-1.530, double refraction 0.008. Monoclinic, $(K,Na)\ [AlSi_3O_8]$. Cleavage perfect. Sanidine is a variety of orthoclase (p. 180).

5 **Natrolite** Colorless, white, yellowish, also reddish. Color of streak white. Transparent. Mohs' hardness 5-5½. Density 2.20-2.26. Refractive index 1.480-1.493, double refraction 0.013. Orthorhombic, $Na_2[Al_2Si_3O_{10}] \cdot 2H_2O$.

6 **Tantalite** Red-brown, also black. Color of streak black-brown. Transparent. Mohs' hardness 6-6½. Density 5.18-8.20. Refractive index 2.26-2.43, double refraction 0.160. Orthorhombic, $(Fe^{2+},Mn^{2+})\ (Ta,Nb)_2O_6$.

7 **Smaragdite** Grass to emerald green. Color of streak white. Transparent to translucent. Mohs' hardness 6-6½. Density 3.24-3.50. Refractive index 1.608-1.630, double refraction 0.022. Variety of actinolite (p. 236).

8 **Rutile** Reddish brown, also bloodred, black. Color of streak yellow to light brown. Transparent. Mohs' hardness 6-6½. Density 4.20-4.30. Refractive index 2.616-2.903, double refraction 0.287. Tetragonal, TiO_2. Cleavage good. Synthetic rutile (titania, page 267) is a widespread diamond imitation.

9 **Leucite** Yellowish (0.32 and 0.14 ct), also colorless, white. Color of streak white. Transparent. Mohs' hardness 5½-6. Density 2.45-2.50. Refractive index 1.504-1.509. Double refraction 0.001. Tetragonal, $K[AlSi_2O_6]$.

10 **Peristerite** Brown basic color with bluish shimmer. Color of streak white. Opaque. Mohs' hardness 6-6½. Density 2.59-2.68. Refractive index 1.534-1.543, double refraction 0.009. Triclinic, $(Na,Ca)[(Si,Al)_4O_8]$. Mixed crystal between albite (p. 240) and oligoclase (p. 226)

11 **Hypersthene** Black-green, brown. Color of streak white, gray. Transparent. Mohs' hardness 5-6. Density 3.4-3.5. Refractive index 1.673-1.731, double refraction 0.010-0.016. Orthorhombic, $(Fe,Mg)_2[Si_2O_6]$.

12 **Tugtupite** (also called reindeer stone) Dark red with violet tint. Color of streak white. Opaque. Mohs' hardness 5½-6. Density 2.36-2.57. Refractive index 1.496-1.502, double refraction 0.006. Tetragonal $Na_4[Cl,S]|[BeAlSi_4O_{12}]$. Cleavage good.

13 **Datolite** Colorless (round 8.74 ct.), yellow, greenish. Color of streak white. Transparent. Mohs' hardness 5-5½. Density 2.90-3.00. Refractive index 1.621-1.675, double refraction 0.040-0.050. Monoclinic, $Ca_2B_2[OH|SiO_4]_2$.

1
2
3
4
5
6
7
8
9
10
11
12
13

1 **Synthetic periclase** Yellowish, gray-green, also colorless. Color of streak white. Transparent. Mohs' hardness 5½-6. Density 3.7-3.9. Refractive index 1.74, double refraction none. Cubic, MgO. Cleavage perfect. Since 2001, even natural periclase has been known to have gemstone quality.

2 **Cuprite** (also called red-copper ore) Carmine red (36.03 ct). Color of streak brown-red. Translucent. Mohs' hardness 3½-4. Density 5.85-6.15. Refractive index 2.849, double refraction none. Cubic, Cu_2O.

3 **Baryite** (also called heavy spar) Yellow-brown (octagon 7.62 ct.), yellow (octagon 2.72 ct), colorless (3.38 ct), also red, green, blue. Color of streak white. Transparent. Mohs' hardness 3-3½. Density 4.43-4.46. Refractive index 1.636-1.648, double refraction 0.012. Orthorhombic, $Ba[SO_4]$.

4 **Apophyllite** Colorless (7.47 ct), also reddish, yellowish, greenish or bluish. Color of streak white. Transparent. Mohs' hardness 4½-5. Density 2.30-2.50. Refractive index 1.535-1.537, double refraction 0.002. Tetragonal, $KCa_4[(F,OH) \mid (Si_4O_{10})_2\ 8H_2O$. Apophyllite crystals see p. 205, no. 3.

5 **Zincite** (also called red-zinc ore) Red to orange-red (octagon 6.46 and 0.62 ct, round 0.51 ct). Color of streak orange-yellow. Translucent. Mohs' hardness 4-5. Density 5.66. Refractive index 2.013-2.029, double refraction 0.016. Hexagonal, (Zn,Mn)O. Cleavage perfect.

6 **Dolomite** (also called dolostone) Colorless (octagon 5.91 ct, oval 2.81 and 1.96 ct), also pastel colored. Color of streak white. Transparent. Mohs' hardness 3½-4. Density 2.80-2.95. Refractive index 1.502-1.698, double refraction 0.185. Trigonal, $CaMg[CO_3]_2$. Cleavage perfect.

7 **Kurnakovite** Colorless (round 3.85 ct, octagon 0.81), also pink. Color of streak white. Transparent. Mohs' hardness 3. Density 1.86. Refractive index 1.488-1.525, double refraction 0.036. Triclinic, $Mg[B_3O_3(OH)_5] \cdot 5H_2O$.

8 **Chalcopyrite** (also called copper pyrite) Brass yellow, golden yellow, with a green tint. Color of streak greenish black. Opaque. Mohs' hardness 3½-4. Density 4.10-4.30. Tetragonal, $CuFeS_2$. Cleavage none.

9 **Siderite** (also called chalybite and ironspar) Red-brown and golden brown. Color of streak white. Transparent. Mohs' hardness 3½ -4½. Density 3.83-3.96. Refractive index 1.633-1.875, double refraction 0.242. Trigonal, $Fe[CO_3]$. Cleavage perfect. Radial, globular aggregates (spherosiderites) are cut en cabochon.

10 **Witherite** Yellowish (6.46 ct), golden yellow, also colorless. Color of streak white. Transparent. Mohs' hardness 3-3½. Density 4.27-4.79. Refractive index 1.529-1.677, double refraction 0.148. Orthorhombic, $Ba[CO_3]$. Cleavage good. Witherite dust is toxic; do not inhale.

11 **Colemanite** Colorless, even gray-white. Color of streak white. Transparent. Mohs' hardness 4½. Density 2.40-2.42. Refractive index 1.586-1.615, double refraction 0.028-0.030. Monocline, $Ca[B_3O_4(OH)_3] \cdot H_2O$.

12 **Anhydrite** Colorless, even bluish, reddish. Color of streak white. Translucent. Mohs' hardness 3½. Density 2.90-2.98. Refractive index 1.570-1.614, double refraction 0.044. Orthorhombic, $Ca[SO_4]$. Cleavage perfect.

13 **Magnetite-Jade** Trade name for opaque, black jade with galvanic gilded, originally also black, magnetite inclusions. Mohs' hardness 5½ -7. Density 3.4-4.4. Deposits in California/USA. For jade, see p. 170; for magnetite, see p. 230.

1
2
3
4
5
6
7
8
9
10
11
12
13

1 **Calcite** (also called limespar) Colorless (quadrangle), yellow-gold (octagon 0.49 ct), yellow (antique 20.16 ct), and various other colors. Color of streak white. Transparent to translucent. Mohs' hardness 3. Density 2.69-2.71. Refractive index 1.486-1.658, double refraction 0.172. Trigonal, $Ca[CO_3]$.

2 **Howlite** Milky white, frequently dispersed with black or dark brown veins. Color of streak white. Opaque. Mohs' hardness 3-3½. Density 2.45-2.58. Refractive index 1.586-1.605, double refraction 0.019. Monocline, $Ca_2[SiB_5O_9(OH)_5]$. Cleavage indistinct. Blue color under the turquoise (p. 186).

3 **Crocoite** (also called red-lead ore) Red-orange, also yellow. Color of streak orange-yellow. Transparent to translucent. Mohs' hardness 21/2-3. Density 5.9-6.1. Refractive index 2.29-2.66, double refraction 0.270. Monocline, $Pb[CrO_4]$. Cleavage good.

4 **Gaylussite** White, also colorless, yellow. Color of streak white. Transparent. Mohs' hardness 2½-3. Density 1.99. Refractive index 1.443-1.523, double refraction 0.080. Monocline, $Na_2Ca[CO_3]_2 \cdot 5H_2O$. Cleavage good.

5 **Phosgenite** (also called lead-horn ore) Colorless, white, yellowish, light pink (3.10 ct), even greenish. Color of streak white. Transparent. Mohs' hardness 2-3. Density 6.13. Refractive index 2.114-2.145, double refraction 0.028. Tetragonal, $Pb_2[Cl_2|CO_3]$. Cleavage good.

6 **Sphaerocobaltite** (also called cobaltocalcite) Rose red. Color of streak light red. Translucent. Mohs' hardness 4. Density 4.13. Refractive index 1.601-1.855, double refraction 0.254. Trigonal, $Co[CO_3]$. Cleavage perfect.

7 **Silver** The originally silver-white dendritic inclusions in quartz that have turned deep black. Color of streak white. Opaque. Mohs' hardness 2½-3. Density 9.6-12.0. Refractive index none. Cubic, Ag.

8 **Celestite** Colorless (round 16.77), bluish white (octagon), Pale pink (round), also greenish. Color of streak white. Transparent. Mohs' hardness 3-3½. Density 3.97-4.00. Refractive index 1.619-1.635, double refraction 0.010-0.012. Orthorhombic, $Sr[SO_4]$. Cleavage perfect.

9 **Gold** Pure gold included in quartz. Color gold-yellow. Color of streak yellow. Opaque. Mohs' hardness 2½-3. Density 15.5-19.3. Refractive index none. Cubic, Au. Cleavage none.

10 **Vivianite** Blue-green, also colorless, deep blue. Color of streak colorless or blue. Transparent to translucent. Mohs' hardness 1½-2. Density 2.64-2.70. Refractive index 1.560-1.640, double refraction 0.050-0.075. Monocline, $Fe_3^{2+}[PO_4]_2 \cdot 8H_2O$. Cleavage perfect. The originally white color gradually turns into bluish shades.

11 **Sulfur** Yellow, also brownish. Color of streak white. Translucent. Mohs' hardness 1½-2½ . Density 2.05-2.08. Refractive index 1.958-2.245, double refraction 0.291. Orthorhombic, S. Cleavage indistinct. Extremely sensitive to heat, bursts when warmed in one's hand.

12 **Aragonite** Colorless (0.49 ct), greenish, also other colors. Color of streak white. Transparent. Mohs' hardness 3½-4. Density 2.94. Refractive index 1.530-1.685, double refraction 0.155. Orthorhombic, $Ca[CO_3]$.

13 **Proustite** Cinnabar. Color of streak light red. Translucent. Mohs' hardness 2½. Density 5.51-5.64. Refractive index 2.881-3.084, double refraction 0.203. Trigonal, Ag_3AsS_3. Cleavage good. The color gradually darkens when exposed to light.

3
1
2
4
5
6
7
9
8
10
11
12
2
13

1 **Boleite** Indigo-blue (1.75 ct), also deep blue. Color of streak blue. Transparent to translucent. Mohs' hardness 3-3½. Density 5.05. Refractive index 2.03-2.05, double refraction none. Cubic, $KPb_{26}Ag_9Cu_{24}(OH)_{48}Cl_{62}$.

2 **Oligoclase** Colorless (1.74 ct, 4.14 ct, 2.06 ct). Color of streak white. Transparent. Mohs' hardness 6-6½. Density 2.62-2.67. Refractive index 1.542-1.549, double refraction 0.007. Triclinic, $(Na,Ca)[(Si,Al)_2Si_2O_8]$.

3 **Ludlamite** Light to apple green (1.06 ct), also colorless. Color of streak white. Transparent to translucent. Mohs' hardness 3-4. Density 3.1-3.2. Refractive index 1.650-1.697, double refraction 0.038-0.044. Monoclinic, $(Fe^{2+},Mg,Mn^{2+})_3[PO_4]_2 \cdot 4H_2O$. Cleavage perfect.

4 **Adamite** Brown (0.91 ct), yellow-green (0.45 ct), also colorless, pink, violet. Color of streak white. Transparent to translucent. Mohs' hardness 3½. Density 4.30-4.68. Refractive index 1.708-1.760, double refraction 0.048-0.050. Orthorhombic. $Zn_2[OH|AsO_4]$. Cleavage good.

5 **Augelite** Colorless, also white, pink, yellowish, light blue. Color of streak white. Transparent. Mohs' hardness 4½-5. Density 2.70-2.75. Refractive index 1.570-1.590, double refraction 0.014-0.020. Monoclinic, $Al_2[(OH)_3|PO_4]$.

6 **Friedelite** Red (15.38 ct), also yellow, brown. Color of streak pale red. Translucent to opaque. Mohs' hardness 4-5. Density 3.06-3.19. Refractive index 1.625-1.664, double refraction 0.030. Trigonal, $Mn_8{}^{2+}[(OH,Cl)_{10}[Si_6O_{15}]$. Cleavage perfect.

7 **Talc** Gray-green, also pearl white, yellowish, blue-green. Color of streak white. Translucent to opaque. Mohs' hardness 1. Density 2.55-2.80. Refractive index 1.54-1.59, double refraction 0.050. Monoclinic, $Mg_3[(OH)_2|Si_4O_{10}]$. Dense aggregates are called soapstone or steatite.

8 **Manganotantalite** Scarlet red. Color of streak dark red. Transparent. Mohs' hardness 5½-6½. Density 7.73-7.97. Refractive index 2.19-2.34, double refraction 0.150. Orthorhombic, $(Mn^{2+},Fe^{2+})(Ta,Nb)_2O_6$.

9 **Gadolinite** Black (8.51 ct), also brown, pale green, green-black. Color of streak white, greenish gray. Transparent to opaque. Mohs' hardness 6½-7. Density 4.00-4.65. Refractive index 1.77-1.82, double refraction 0.01-0.04. Monoclinic, $Y_2Fe^{2+}Be_2[O|SiO_4]_2$. Cleavage none.

10 **Anglesite** Colorless, yellow, also white, greenish. Color of streak white. Transparent to translucent. Mohs' hardness 3-3½. Density 6.30-6.39. Refractive index 1.878-1.895, double refraction 0.017. Orthorhombic, $Pb[SO_4]$.

11 **Whewellite** Colorless, yolk color, light yellow, also white. Color of streak white. Transparent. Mohs' hardness 2½-3. Density 2.19-2.25. Refractive index 1.489-1.651, double refraction 0.159-0.163. Monoclinic, $Ca[C_2O_4] \cdot H_2O$.

12 **Ekanite** Dark green (1.34 ct), also yellow, brown. Color of streak white. Transparent to translucent. Mohs' hardness 4½-6½. Density 3.28-3.32. Refractive index 1.572-1.573, double refraction 0.001. Tetragonal, $Ca_2Th[Si_8O_{20}]$. Radioactive!

13 **Phosphophyllite** Blue-green, also colorless. Color of streak white. Transparent. Mohs' hardness 3-3½. Density 3.07-3.13. Refractive index 1.594-1.621, double refraction 0.021-0.033. Monoclinic, $Zn_2(Fe^{2+},Mn^{2+})[PO_4]_2 \cdot 4H_2O$.

14 **Gypsum** (also called selenite) White, also colorless, pink, bluish. Color of streak white. Transparent to opaque. Mohs' hardness 2. Density 2.20-2.40. Refractive index 1.520-1.529, double refraction 0.009. Monoclinic, $Ca[SO_4] \cdot 2H_2O$.

The illustrations are 40 percent larger than the originals.

1
2
3
4
5
6
7
8
9
10
11
12
13
14

1, 2 **Analcite** Colorless, white, pink, cat's-eye, greenish. Color of streak white. Transparent to opaque. Mohs' hardness 5-5½. Density 2.22-2.29. Refractive index 1.479-1.489, double refraction none. Cubic, $Na_2[Al_2Si_4O_{12}] \cdot 2H_2O$.

3 **Triphylite** Brownish, also greenish, bluish. Color of streak white. Transparent to translucent. Mohs' hardness 4-5. Density 3.34-3.58. Refractive index 1.689-1.702, double refraction 0.006-0.008. Orthorhombic, $LiFe^{2+}[PO_4]$.

4 **Staurolite** Brown, yellow. Color of streak white. Transparent to opaque. Mohs' hardness 7-7½. Density 3.65-3.77. Refractive index 1.736-1.762, double refraction 0.010-0.015. Monoclinic, $(Fe^{2+},Mg,Zn)_2Al_9[O_6|(OH,O)_2|(SiO_4)_4]$.

5 **Hornblende** Brown, green, black. Color of streak gray-green, brown. Translucent to opaque. Mohs' hardness 5-6. Density 2.9-3.4. Refractive index 1.675-1.695, double refraction 0.02. Monoclinic.

6 **Pectolite** White, green-blue, also colorless, blue. Color of streak white. Transparent to translucent. Mohs' hardness 4½-5. Density 2.74-2.88. Refractive index 1.595-1.645, double refraction 0.038. Triclinic, $NaCa_2[Si_3O_8(OH)]$.

7 **Zektzerite** Colorless, also light pink. Color of streak white. Transparent to translucent. Mohs' hardness 6. Density 2.79. Refractive index 1.582-1.585, double refraction 0.003. Orthorhombic, $Na_2Li_2Zr_2[Si_{12}O_{30}]$.

8 **Nepheline** Pink, also colorless, white, green. Color of streak white. Mohs' hardness 5-6. Density 2.55-2.65. Refractive index 1.526-1.546, double refraction 0.004. Hexagonal.

9 **Greenockite** Orange, also yellow, brown. Color of streak orange to red. Transparent to translucent. Mohs' hardness 3-3½. Density 4.73-4.79. Refractive index 2.506-2.529, double refraction 0.023. Hexagonal, CdS.

10 **Anatase** Brown, also colorless, reddish, yellow, blue, black. Color of streak white. Transparent to translucent. Mohs' hardness 5½-6. Density 3.82-3.97. Refractive index 2.488-2.564, double refraction 0.046-0.067. Tetragonal, TiO_2.

11 **Milarite** Yellow, colorless, white, green. Color of streak white. Transparent. Mohs' hardness 5½-6. Density 2.46-2.61. Refractive index 1.529-1.551, double refraction 0.003. Hexagonal.

12 **Descloizite** Red-brown, also brown-black. Color of streak light-brown. Transparent to opaque. Mohs' hardness 3-3½. Density 5.5-6.2. Refractive index 2.185-2.350, double refraction 0.165. Orthorhombic, $Pb(Zn,Cu)[OH|VO_4]$.

13 **Lithiophilite** Brown, also yellow, blue. Color of streak gray-white. Transparent to translucent. Mohs' hardness 4-5. Density 3.4-3.6. Refractive index 1.68-1.70, double refraction 0.01. Orthorhombic, $Li(Mn^{2+},Fe^{2+})[PO_4]$.

14 **Jeremejevite** Bluish, also colorless, yellowish. Color of streak white. Transparent. Mohs' hardness 6½-7½. Density 3.28-3.31. Refractive index 1.637-1.653, double refraction 0.007-0.013. Hexagonal, $Al_6[(F,OH)_3|(BO_3)_5]$.

15 **Clinohumite** Golden yellow, also white, brown. Color of streak white. Transparent to opaque. Mohs' hardness 6. Density 3.13-3.75. Refractive index 1.629-1.674, double refraction 0.028-0.041. Monoclinic, $(Mg,Fe^{2+})_9[(F,OH)_2|(SiO_4)_4]$.

16 **Neptunite** Black. Color of streak reddish brown. Translucent to opaque. Mohs' hardness 5-6. Density 3.19-3.23. Refractive index 1.690-1.736, double refraction 0.029-0.045. Monoclinic. $KNa_2Li(Fe^{2+},Mg,Mn^{2+})_2Ti_2{}^{4+}[O|Si_4O_{11}]_2$.

17 **Eudialyte** Brown-red, red-green. Color of streak white. Translucent to opaque. Mohs' hardness 5-5½. Density 2.74-2.98. Refractive index 1.591-1.633, double refraction 0.003-0,010. Trigonal.

1
2
3
4
5
7
8
6
9
10
12
11
13
14
15
16
17

1 **Montebrasite** Colorless, yellowish, light green, light blue, also white. C streak white. Transparent. Mohs' hardness 5½-6. Density 2.98-3.11. Ref index 1.594-1.633, double refraction 0.22. Triclinic, $LiAl[(OH,F)|PO_4]$.

2 **Cinnabar** Red (1.53 and 1.13 ct), also pale gray. Color of streak cinnab Translucent to opaque. Mohs' hardness 2-2½. Density 8.0-8.2. Ref index 2.905-3.256, double refraction 0.351. Trigonal, HgS. Cleavage pe

3 **Boracite** Light green (0.62 ct), also colorless, white, yellow, bluish. C streak white. Transparent to translucent. Mohs' hardness 7-7½. Density 2.96. Refractive index 1.658-1.673, double refraction 0.010-0.011. rhombic, $Mg_3[Cl|BO_3|B_6O_{10}]$. Cleavage none.

4 **Magnesite** Colorless (1.99 ct), also white, yellow, brown. Color of white. Transparent to translucent. Mohs' hardness 3½-4½. Density 2.96 Refractive index 1.509-1.717, double refraction 0.208. Trigonal, Mg[C

5 **Wolframite** Black (6.37ct), also dark-brown. Color of streak brown to Translucent to opaque. Mohs' hardness 5-5½. Density 7.1-7.6. Ref index about 2.21, double refraction 0.14. Monoclinic, $(Fe^{2+}, Mn^{2+})WO_4$ vage perfect.

6 **Herderite** Grayish blue (0.65 and 1.05 ct), also colorless, yellowis green. Color of streak white. Transparent to translucent. Mohs' hardness Density 2.95-3.02. Refractive index 1.587-1.627, double refraction 0.032. Monoclinic, $CaBe[(F,OH)|PO_4]$. Cleavage indistinct.

7 **Leucophanite** Yellow with needle-like aegirine inclusions (0.63 ct), also ish. Color of streak white. Mohs' hardness 4. Density 3.0. Refractive 1.556-1.598, double refraction 0.025. Triclinic, $(Na,Ca)_2Be[Si_2O_6(F,OH$

8 **Pyrargyrite** Dark red (8.13 ct). Color of streak crimson. Translucent. hardness 2½-3. Density 5.85. Refractive index 2.88-3.08, double refr 0.200. Trigonal, Ag_3SbS_3. Cleavage good.

9 **Magnetite** Black (3.10 ct). Color of streak black. Opaque. Mohs' hardn 6½. Density 5.2. Refractive index 2.42, double refraction none. $Fe^{2+}Fe_2^{3+}O_4$. Cleavage indistinct.

10 **Strontianite** Light yellow, also colorless, white, brown, reddish. Color of white. Transparent to translucent. Mohs' hardness 3½. Density 3.63 Refractive index 1.52-1.67, double refraction 0.150. Orthorhombic, Sr[

11 **Parisite** Yellow-brown, also reddish. Color of streak white. Transpar translucent. Mohs' hardness 4½. Density 4.33-4.42. Refractive index 1.772, double refraction 0.081-0.101. Trigonal, $Ca(Ce,La)_2[F_2|(CO_3)_3]$.

12 **Taaffeite** Violet, also colorless, red, green, bluish. Color of streak white. parent. Mohs' hardness 8-8½. Density 3.60-3.62. Refractive index 1.730, double refraction 0.004-0.009. Hexagonal, $Mg_3Al_8BeO_{16}$.

13 **Simpsonite** Orange, also colorless, white, brownish. Color of streak Translucent. Mohs' hardness 7-7½. Density 5.92-6.84. Refractive index 2.040, double refraction 0.058. Hexagonal, $Al_4(Ta,Nb)_3O_{13}(OH)$.

14 **Diaspore** Greenish brown (7.73 ct), also colorless, white, pink, yellow, Color of streak white. Transparent to translucent. Mohs' hardness 6½-7 sity 3.30-3.39. Refractive index 1.702-1.750, double refraction 0.048. rhombic, AlO(OH). Cleavage perfect.

The illustrations are 40 percent larger than the originals.

1
2
3
4
5
6
7
8
9
10
11
12
13
14

1 **Thaumasite** Colorless, also white, light yellow. Color of streak white. Translucent. Mohs' hardness 3½. Density 1.91. Refractive index 1.464-1.507, double refraction 0.036. Hexagonal, $Ca_6Si_2[(OH)_{12}|(CO_3)_2|(SO_4)_2] \cdot 24H_2O$.

2 **Cancrinite** Yellow to orange, also colorless, white, pink, bluish. Color of streak white. Transparent to translucent. Mohs' hardness 5-6. Density 2.42-2.51. Refractive index 1.495-1.528, double refraction 0.024-0.029. Hexagonal, $Na_6Ca_2[(CO_3)_2|Al_6Si_6O_{24}] \cdot 2H_2O$.

3 **Tremolite** (also called grammatite) Gray-brown, green, also colorless, white, pink. Color of streak white. Transparent to opaque. Mohs' hardness 5-6. Density 2.95-3.07. Refractive index 1.560-1.643, double refraction 0.017-0.027. Monoclinic, $Ca_2Mg_5[(OH,F)|Si_4O_{11}]_2$. Cleavage good.

4 **Yugawaralite** Colorless (0.51 ct), dull white (0.64 ct). Color of streak white. Transparent to translucent. Mohs' hardness 4½. Density 2.19-2.23. Refractive index 1.490-1.509, double refraction 0.011-0.014. Monoclinic, $Ca[Al_2Si_6O_{16}] \cdot 4H_2O$. Cleavage indistinct.

5 **Sapphirine** Dark blue, also colorless, pink, greenish, violet. Color of streak white. Transparent. Mohs' hardness 7½. Density 3.40-3.58. Refractive index 1.701-1.734, double refraction 0.004-0.007. Monoclinic, $(Mg,Al)_8[O_2|(Al,Si)_6O_{18}]$.

6 **Aegirine-Augite** Black (2.53 ct), also green-black. Color of streak greenish gray. Opaque. Mohs' hardness 6. Density 3.40-3.55. Refractive index 1.700-1.800, double refraction 0.030-0.050. Monoclinic, $(NaCa)(Mg,Fe^{3+})[Si_2O_6]$. Cleavage good.

7 **Meliphanite** (also called melinophane) Honey yellow (1.35 ct), orange (0.50 ct), also colorless. Color of streak white. Translucent. Mohs' hardness 5-5½. Density 3.00-3.03. Refractive index 1.593-1.612, double refraction 0.019. Tetragonal, $(Ca,Na)_2(Be,Al)[Si_2O_6(OH,F)]$. Cleavage perfect.

8 **Pollucite** (also called pollux) Colorless, slightly brownish (0.99 ct), also white, bluish, violet. Color of streak white. Transparent to translucent. Mohs' hardness 6½-7. Density 2.85-2.94. Refractive index 1.517-1.525, double refraction none. Cubic, $(Cs,Na)_2[Al_2Si_4O_{12}] \cdot ½ - 1H_2O$.

9 **Andesine** Light pink (0.89 ct) also red, white, gray, yellowish, green. Color of streak white. Transparent to opaque. Mohs' hardness 6-6½. Density 2.65-2.69. Refractive index 1.543-1.551, double refraction 0.008. Triclinic, $(Na,Ca)[(Al,Si)_2Si_2O_8]$. Cleavage perfect.

10 **Muscovite** Pink (2.57 ct), also colorless, silver-white, yellowish, greenish. Color of streak white. Transparent to translucent. Mohs' hardness 2-3. Density 2.78-2.88. Refractive index 1.552-1.618, double refraction 0.036-0.043. Monoclinic, $KAl_2[(OH,F)_2 \cdot AlSi_3O_{10}]$.

11 **Davidite** Black (2.04 ct). Color of streak gray-black. Opaque. Mohs' hardness 5-6. Density 4.5. Refractive index cca. 2.3, double refraction none. Trigonal, $(Ce,La,Y,U)_2Fe_2^{3+}(Ti,Fe^{3+})_{18}O_{38}$. Radioactive!

12 **Mesolite** Colorless (0.98 ct), also white. Color of streak white. Transparent to translucent. Mohs' hardness 5-5½. Density 2.26-2.40. Refractive index 1.504-1.508, double refraction 0.001. Monoclinic, $Na_2Ca_2[Al_2Si_3O_{10}]_3 \cdot 8H_2O$.

13 **Pyrolusite** Black (24.46 ct), also dark gray. Color of streak black. Opaque. Mohs' hardness 6-7. Density 4.5-5.0. Refractive index indeterminable since aggregate. Tetragonal, MnO_2. Cleavage perfect.

The illustrations are 50 percent larger than the originals.

1
2
3
4
5
6
7
8
9
10
11
12
13

1 **Cordierite** Pale blue, colorless (together 14.07 ct). Transparent. For rare varieties of the otherwise strong blue-colored cordierite, see p. 196.

2 **Legrandite** Yellow (0.97 ct), also colorless. Color of streak white. Transparent. Mohs' hardness 4½-5. Density 3.98-4.04. Refractive index 1.675-1.740. Double refraction 0.060. Monoclinic, $Zn_2[OH|AsO_4] \cdot H_2O$.

3 **Eosphorite** Orange pink (0.67 ct), also colorless, brown. Color of streak white. Transparent to translucent. Mohs' hardness 5. Density 3.05-3.08. Refractive index 1.638-1.671. Double refraction 0.028-0.035. Orthorhombic, $(Mn^{2+}, Fe^{2+})Al[(OH)_2|PO_4] \cdot H_2O$. Cleavage indistinct.

4 **Cabochon quartz** Colorless with brass to golden colored inclusion of pyrite fibers, cat's-eye effect. Quartz p. 132, pyrite p. 178.

5 **Palygorskite** Gray-yellowish (2.34 ct) also white, pink. Color of streak white. Translucent. Mohs' hardness 1-2. Density 2.21. Refractive index 1.522-1.548. Double refraction none. Monoclinic or orthorhombic, $(Mg,Al)_2[OH|Si_4O_{10}] \cdot 4H_2O$. Due to the opal-like appearance, also falsely called Angel Skin Opal (compare with p. 168).

6 **Ceruleite** Turquoise blue, also sky blue. Color of streak bluish white. Translucent to opaque. Mohs' hardness 5-6. Density 2.70-2.80. Refractive index about 1.60. Triclinic, $Cu_2Al_7[(OH)_{13}|(AsO_4)_4] \cdot 12H_2O$. Tuberous aggregates made from finest crystals.

7, 8 **Sillimanite** Light green (antique 1.78 ct), gray-white, interspersed with quartz (oval 5.95 ct), colorless (hexagonal free form 0.35 ct), also yellowish, brownish, bluish. Violet-brown cat's-eye (no. 8, 1.60 ct). Color of streak white. Transparent to opaque. Mohs' hardness 6½-7½. Density 3.23-3.27. Refractive index 1.655-1.684, double refraction 0.014-0.021. Orthorhombic, $Al_2[O|SiO_4]$. Cleavage perfect.

9 **Monazite** Brown (2.92 and 2.48 ct), also white, pink, brown-red, yellowish. Color of streak white. Translucent to opaque. Mohs' hardness 5-5½. Density 4.98-5.43. Refractive index 1.774-1.849, double refraction 0.049-0.055. Monoclinic, $(Ce,La,Nd,Th)[PO_4]$. Cleavage perfect. If thorium-bearing then radioactive!

10 **Spodumene** Colorless (antique 21.75 ct, octagon 20.26 ct, oval 6.29 ct), also gray-white, yellow. Color of streak white. Transparent, translucent. Mohs' hardness 6½-7. Density 3.15-3.21. Refractive index 1.660-1.681, double refraction 0.014-0.016. Monoclinic, $LiAl[Si_2O_6]$. Cleavage perfect – for the yellow-green variety Hiddenite and the pink-violet variety Kunzite, see p. 130.

11 **Bronzite** Brown-black (1.83 ct), golden-brown (3.71 ct), also colorless, gray, yellow, green. Color of streak white or brownish. Transparent to opaque. Mohs' hardness 5-6. Density 3.30-3.43. Refractive index 1.665-1.703, double refraction 0.015. Orthorhombic, $(Mg,Fe)_2[Si_2O_6]$. Cleavage good.

12 **Goodletite** Metamorphic rock used as gemstone with ruby and/or sapphire as well as pink-gray corundum in a fine basic mass of green tourmaline and translucent chromite glimmer. Can be found on the southern island of New Zealand as rock debris. In New Zealand, the material is also called ruby rock or treasure stone.

13 **Sellaite** Colorless (22.18 ct), also white. Color of streak white. Transparent to translucent. Mohs' hardness 5-5½. Density 3.15. Refractive index 1.378-1.390, double refraction 0.012. Tetragonal, MgF_2. Cut-worthy colorless sellaite crystals were first found in Brazil in 1979.

1
1
2
6
3
4
5
7
10
9
8
11
12
13

1 **Chambersite** Colorless, crimson, also brownish. Color of streak white, pink Transparent. Mohs' hardness 7. Density 3.49. Refractive index 1.732-1.744 double refraction 0.012. Orthorhombic, $Mn_3^{2+}[Cl|BO_3|B_6O_{10}]$.

2 **Scorzalite** Blue-green, violet, also deep blue. Color of streak white. Transparent to opaque. Mohs' hardness 5½-6. Density 3.38. Refractive index 1.627-1.680, double refraction 0.038-0.040. Monoclinic, $(Fe^{2+},Mg)Al_2[OH|PO_4]_2$ Cleavage indistinct.

3 **Rhodizite** Light yellow, also colorless, white, pink, green. Color of streak white. Transparent to translucent. Mohs' hardness 8-8½. Density 3.44. Refractive index 1.690, double refraction none. Cubic, $(K,Cs,Rb)Al_4Be_4[O_4|B_{11}BeO_{24}]$.

4 **Meionite** Yellow, also colorless, white. Color of streak white. Transparent to opaque. Mohs' hardness 5½-6. Density 2.74-2.78. Refractive index 1.566-1.600, double refraction 0.024-0.037. Tetragonal, $Ca_4[CO_3|Al_6Si_6O_{24}]$.

5 **Linarite** Dark azure blue (0.15 ct). Color of streak pale blue. Transparent to translucent. Mohs' hardness 2½. Density 5.30. Refractive index 1.809-1.859, double refraction 0.050. Monoclinic, $PbCu[(OH)_2|SO_4]$.

6 **Durangite** Orange-red (0.15 ct), also pale yellow, dark green. Color of streak yellow. Transparent. Mohs' hardness 5-5½. Density 3.94-4.07. Refractive index 1.634-1.685, double refraction 0.051. Monoclinic, $NaAl[F|AsO_4]$.

7 **Clinoenstatite** Yellowish green (2.17 ct), also colorless, yellow-brown. Color of streak white. Transparent to translucent. Mohs' hardness 5-6. Density 3.19. Refractive index 1.651-1.660, double refraction 0.009. Monoclinic, $Mg_2[Si_2O_6]$. Cleavage good.

8 **Childrenite** Orange-brown (0.27 ct), also dark brown. Color of streak white. Transparent to translucent. Mohs' hardness 5. Density 3.11-3.19. Refractive index 1.630-1.685, double refraction 0.030-0.040. Orthorhombic, $(Fe^{2+}Mn^{2+})Al[(OH)_2|PO_4] \cdot H_2O$. Cleavage indistinct.

9 **Shortite** White (0.17 ct), also colorless, pale yellow. Color of streak white. Transparent. Mohs' hardness 3. Density 2.60. Refractive index 1.531-1.570, double refraction 0.039. Orthorhombic, $Na_2Ca_2[CO_3]_3$. Cleavage good.

10 **Vlasovite** Light brown (cat's-eye 1.56 ct), yellow-brown (octagon 0.97 ct, pentagon 0.33 ct), also colorless. Color of streak white. Transparent. Mohs' hardness 6. Density 2.92-2.97. Refractive index 1.605-1.625, double refraction 0.020. Monoclinic, $Na_2Zr[Si_4O_{11}]$. Cleavage good.

11 **Kaemmererite** (Kämmererite) Red-violet. Color of streak white. Transparent to translucent. Mohs' hardness 2-2½. Density 2.64. Refractive index 1.597-1.600, double refraction 0.003. Monoclinic, $(Mg,Cr)_6[(OH)_8(Al,Si)_3O_{10}]$.

12 **Senarmontite** Colorless (1.72 ct), also white, gray. Color of streak white. Transparent to translucent. Mohs' hardness 2-2½. Density 5.2-5.5. Refractive index 2.087, double refraction none. Cubic, Sb_2O_3.

13 **Actinolite** Green, also colorless, white, gray, brown. Color of streak white. Transparent. Mohs' hardness 5½-6. Density 3.03-3.07. Refractive index 1.614-1.653, double refraction 0.020-0.025. Monoclinic, $Ca_2(Mg,Fe^{2+})_5[OH|Si_4O_{11}]_2$. Cleavage good.

14 **Grandidierite** Blue-green (octagon 1.51 ct, hexagon 3.05 ct, quadrangle 0.73 ct). Color of streak white. Translucent to opaque. Mohs' hardness 7-7½. Density 2.85-3.00. Refractive index 1.590-1.623, double refraction 0.033. Orthorhombic, $(Mg,Fe^{2+})Al_3[O|BO_4|SiO_4]$.

The illustrations are 40 percent larger than the originals.

1
2
3
4
7
8
6
5
9
11
10
12
13
14
14

1 **Goethite** Brown-black (4.70 ct), also yellow-brown. Color of streak brown to yellow. Translucent to opaque. Mohs' hardness 5-5½. Density 3.8-4.3. Refractive index 2.275-2.415, double refraction 0.14. Orthorhombic, $Fe^{3+}O(OH)$. Cleavage perfect.

2 **Schlossmacherite** Light green (on turquoise-blue ceruleite, see p. 234), also gray-green. Color of streak white. Opaque. Mohs' hardness 5. Density 3.00. Refractive index 1.597. Trigonal, $(H_3O,Ca)Al_3[(OH)_6|(SO_4,AsO_4)_2]$. First referred to as a mineral in 1980.

3 **Pumpellyite** Green, bluish green, also white, brown. Color of streak greenish. Aggregate of radial-stellate crystals (6.28 and 7.85 ct). Translucent to opaque. Mohs' hardness 6. Density 3.18-3.33. Refractive index 1.674-1.722, Monoclinic, $Ca_2Fe^{2+}Al_2[(OH)_2|SiO_4|Si_2O_7] \cdot H_2O$.

4 **Inderite** White, colorless, also pink. Color of streak white. Transparent to translucent. Mohs' hardness 2½-3. Density 1.78-1.86. Refractive index 1.486-1.507, double refraction 0.017-0.020. Monoclinic, $Mg[B_3O_3(OH)_5] \cdot 5H_2O$. Cleavage perfect.

5 **Wavellite** Bluish white (12.13 ct), also colorless, yellow, brown, green, blue. Color of streak white. Translucent. Mohs' hardness 3½-4. Density 2.36. Refractive index 1.520-1.561, double refraction 0.025. Orthorhombic, $Al_3[(OH,F)_3|(PO_4)_2] \cdot 5H_2O$. Cleavage perfect.

6 **Powellite** Light yellow (0.71 ct), also gray-white, green-yellow, brown, pale blue. Color of streak gray-white. Transparent. Mohs' hardness 3½-4. Density 4.23. Refractive index 1.967-1.985, double refraction 0.011. Tetragonal, $Ca[MoO_4]$.

7 **Villiaumite** Orange-brown, carmine (together 1.03 ct), also colorless, yellow. Color of streak white. Transparent. Mohs' hardness 2-2½. Density 2.79. Refractive index 1.327, double refraction sometimes anomalous. Cubic, NaF.

8 **Hydroxylapatite** Green-yellow (10.98 ct), also colorless, white, brown, green, black. Color of streak white. Transparent to translucent. Mohs' hardness 5. Density 3.14-3.21. Refractive index 1.642-1.654, double refraction 0.012. Hexagonal, $Ca_5[OH|PO_4)_3]$. Cleavage indistinct.

9 **Manganapatite** Green-black. Gemstone type from the apatite group (compare with p. 210). Physical data similar to Hydroxylapatite (see above no. 8).

10 **Marcasite** Brass yellow, also green-yellow, brownish. Color of streak greenish black. Opaque. Mohs' hardness 6-6½. Density 4.85-4.92. Refractive index indeterminable. Orthorhombic, FeS_2. At times, marcasite can fall apart and release sulfurous acid. Thus, marcasite should only be stored individually. For more on the very similar pyrite, see p. 178.

11 **Clinozoisite** Yellow to yellow-brown, also colorless, greenish, pink. Color of streak white to gray. Transparent to translucent. Mohs' hardness 6-7. Density 3.21-3.38. Refractive index 1.670-1.734, double refraction 0.010. Monoclinic, $Ca_2Al_3[O|OH|SiO_4|Si_2O_7]$. Cleavage perfect.

12 **Labradorite** Green, brown-green, dark red, brown-red. Color of streak white. Such transparent labradorites are rare; usually, labradorite is opaque with play of colors. Further data p. 182.

13 **Bixbite** Black (4.97 ct). Color of streak black. Opaque. Mohs' hardness 6-6½. Density 4.95. Refractive index indeterminable. Cubic, $(Mn^{3+},Fe^{3+})_2O_3$. Cleavage indistinct.

14 **Bustamite** Light pink, also brown-red. Color of streak white. Transparent. Mohs' hardness 5½-6. Density 3.32-3.43. Refractive index 1.622-1.707, double refraction 0.014-0.015. Triclinic, $(Mn^{2+},Ca)_3[Si_3O_9]$.

1
2
3
4
5
8
7
6
9
10
11
12
13
14

1 **Albite** Greenish (12.81 ct), also colorless, white, gray, reddish, yellow. Color of streak white. Transparent to translucent. Mohs' hardness 6-6½. Density 2.57-2.69. Refractive index 1.527-1.538, double refraction 0.011. Triclinic, $Na[AlSi_3O_8]$. Cleavage perfect. See also Jade-albite, p. 170.

2 **Ametrine** Two-colored quartz (15.23 ct) that contains both violet amethyst (see p. 134) and yellow-brown citrine (see p. 136). The only places of economic significance for the finding of ametrine are in Bolivia.

3 **Cacoxenite Included Quartz** Colorless or slightly milky white quartz (see p. 132) with golden-yellow fibrous inclusions of cacoxenite, a phosphate mineral with Mohs' hardness 3-4 and a density of 2.2-2.6.

4 **Hedenbergite** Black-green (19.52 ct), also bluish green, black. Color of streak white, gray. Transparent to opaque. Mohs' hardness 5½-6½. Density 3.50-3.56. Refractive index 1.716-1.751, double refraction 0.025-0.029. Monoclinic, $CaFe^{2+}[Si_2O_6]$. Cleavage good.

5 **Sérandite** Orange-red (0.77 ct), also pale brown. Color of streak white. Transparent. Mohs' hardness 4-5½. Density 3.32. Refractive index 1.660-1.688, double refraction 0.028. Triclinic, $Na(Mn^{2+},Ca)_2[Si_3O_8(OH)]$.

6 **Stolzite** Gray-yellow (2.16 ct), also red, brown, green. Color of streak white. Transparent to translucent. Mohs' hardness 2½-3. Density 7.9-8.34. Refractive index 2.19-2.27, double refraction 0.08. Tetragonal, $Pb[WO_4]$.

7 **Haüynite** (Haüyn) Deep blue, also green-blue, blue-white. Color of streak white to bluish. Transparent to translucent. Mohs' hardness 5½-6. Density 2.4-2.5. Refractive index 1.496-1.510, double refraction none. Cubic, $Na_{5-6}Ca_2[(SO_4,Cl)_2|Al_6Si_6O_{24}]$. Cleavage good.

8 **Hübnerite** Dark red-brown (1.23 and 0.44 ct), also yellow-brown, brown-black. Color of streak reddish brown. Transparent to opaque. Mohs' hardness 4-4½. Density 7.12-7.18. Refractive index 2.17-2.32, double refraction 0.13. Monoclinic, $Mn^{2+}WO_4$. Cleavage perfect.

9 **Pietersite** Trade name for a dark blue-gray breccia aggregate (20.40 ct) made up largely of hawk's-eye (see page 140) and tiger's-eye (see p. 140). Density 2.60. Refractive index 1.544-1.553, double refraction 0.009. Namibia is only place where this aggregate can be found.

10 **Scolecite** Colorless (octagon 2.42 ct, quadrangle 1.68 ct, octagon 0.20 ct), also white. Color of streak white. Transparent to opaque. Mohs' hardness 5-5½. Density 2.21-2.29. Refractive index 1.509-1.525, double refraction 0.007-0.012. Monoclinic, $Ca[Al_2Si_3O_{10}] \cdot 3H_2O$.

11 **Wulfenite** Orange, red. Color of streak white. Transparent to translucent. Mohs' hardness 3. Density 6.50-7.00. Refractive index 2.280-2.400, double refraction 0.120. Tetragonal, $Pb[MoO_4]$.

12 **Fuchsite** A chromian muscovite variety (see page 232). Green-blue (together 37.94 ct), also emerald green. Color of streak light green, white. Translucent. Mohs' hardness 2-3. Density 2.8-2.9. Refractive index 1.552-1.615, double refraction 0.036-0.043. Monoclinic, $K(Al,Cr)_2[(OH,F)_2|AlSi_3O_{10}]$. All stones portrayed are doublets.

13 **Purpurite** Purple colors, also deep pink, brown. Color of streak purple. Translucent. Mohs' hardness 4-4½. Density 3.2-3.4. Refractive index 1.85-1.92, double refraction 0.07. Orthorhombic, $(Mn^{3+},Fe^{3+})[PO_4]$.

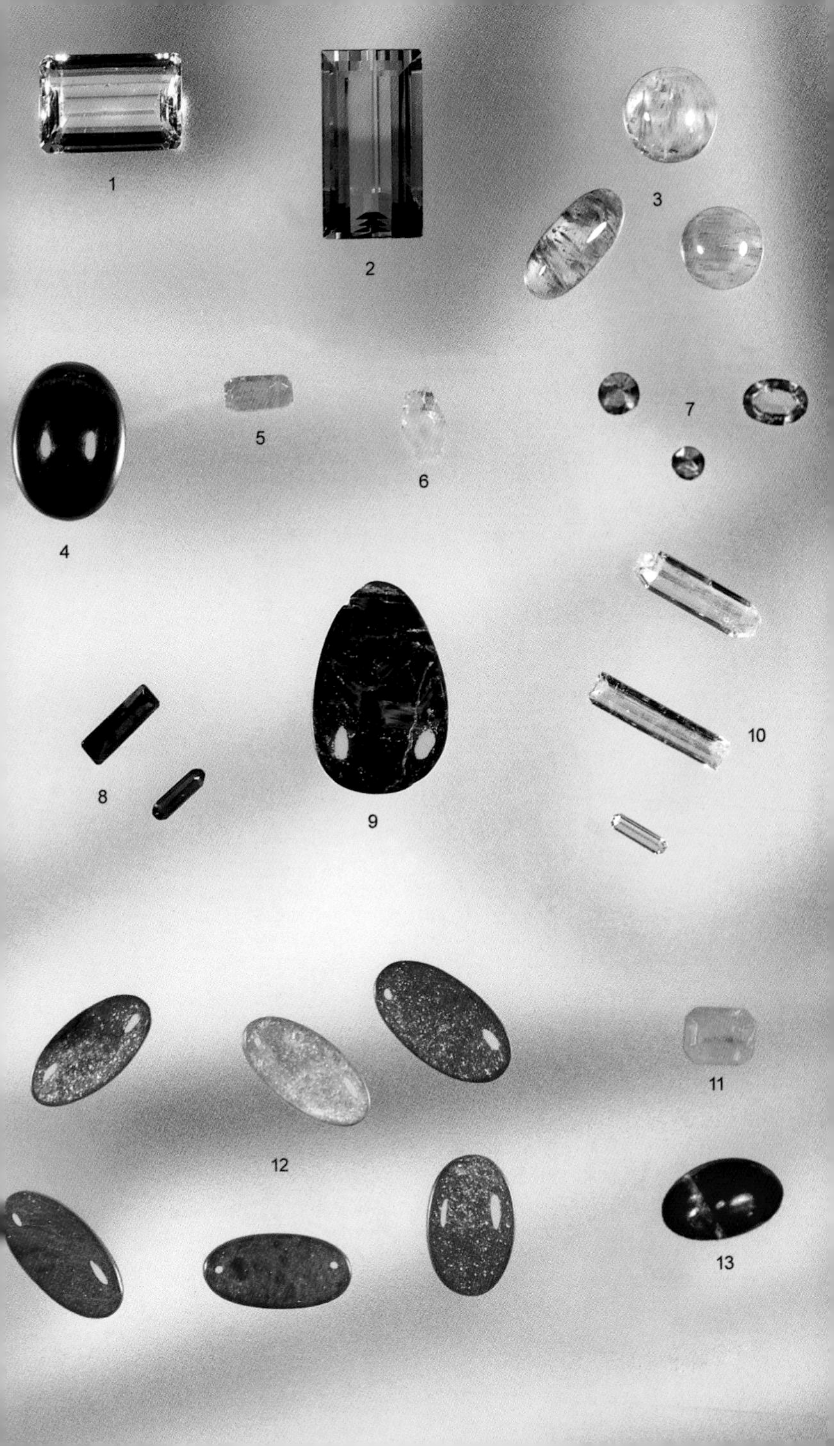

1
2
3
4
5
6
7
8
9
10
11
12
13

1 **Chondrodite** Red-brown (1.78 ct), also red, yellow, gray-green. Color of streak yellow to brown. Transparent to opaque. Mohs' hardness 6-6½. Density 3.16-3.26. Refractive index 1.592-1.646, double refraction 0.028-0.034. Monoclinic, $(Mg,Fe^{2+})_5[(F,OH)_2|SiO_4)_2]$. Cleavage indistinct.

2 **Microlite** Resin yellow, also reddish, brown, green. Color of streak white. Translucent to opaque. Mohs' hardness 5-5½. Density 4.3-5.7. Refractive index 1.98-2.02, double refraction anomalous. Cubic, $(Ca,Na)_2Ta_2O_6(O,OH,F)$.

3 **Anorthite** Colorless, white, yellowish, also reddish gray. Color of streak white. Transparent. Mohs' hardness 6-6½. Density 2.75-2.77. Refractive index 1.577-1.590, double refraction 0.013. Triclinic, $Ca[Al_2Si_2O_8]$.

4 **Hydroxylherderite** Yellowish to brownish (together 5.97 ct), also colorless, blue-green. Color of streak white. Transparent to translucent. Mohs' hardness 5-5½. Density 2.95. Refractive index 1.601-1.631, double refraction 0.030. Monoclinic, $CaBe[OH|PO_4]$. Cleavage indistinct.

5 **Barytocalcite** Yellowish white, also white. Color of streak white. Transparent to translucent. Mohs' hardness 4. Density 3.66. Refractive index 1.525-1.686, double refraction 0.061. Monoclinic, $BaCa[CO_3]_2$.

6 **Hodgkinsonite** Brown-red (0.42 and 0.22 ct), also light pink, orange. Color of streak white. Transparent to translucent. Mohs' hardness 4½-5. Density 3.91-3.99. Refractive index 1.719-1.748, double refraction 0.022-0.026. Monoclinic, $Mn^{2+}Zn_2[(OH)_2|SiO_4]$.

7 **Ilmenite** Black (0.29 ct). Color of streak black. Opaque. Mohs' hardness 5-6. Density 4.5-5.0. Refractive index indeterminable. Trigonal, $Fe^{2+}TiO_3$. Cleavage none.

8 **Vanadinite** Red (0.31 ct), also colorless, yellow, brown. Color of streak yellowish, white. Translucent to opaque. Mohs' hardness 2½-3. Density 6.5-7.1. Refractive index 2.350-2.461, double refraction 0.066. Hexagonal, $Pb_5[Cl|(VO_4)_3]$. Cleavage none.

9 **Shungite** Black glass-like coal (faceted 0.77, cabochon 2.30 ct). Color of streak black. Mohs' hardness 3½-4. Density 1.84-1.98. Consists mainly of non-crystalline carbon. Only region to be found is Karelia/Russia.

10 **Sturmanite** Yellowish, also orange-brown, yellow-green. Color of streak yellowish. Transparent to translucent. Mohs' hardness 2½. Density 1.85. Refractive index 1.500-1.505, double refraction 0.005. Hexagonal, $Ca_6(Fe^{3+},Al,Mn^{2+})_2[(OH)_{12}|B(OH)_4|(SO_4)_2 \cdot 25H_2O$. Cleavage perfect.

11 **Scorodite** Violet-blue (0.84 ct), also yellow-brown, gray-green. Color of streak white. Transparent to translucent. Mohs' hardness 3½-4. Density 3.28-3.29. Refractive index 1.738-1.816, double refraction 0.027-0.031. Orthorhombic, $Fe^{3+}[AsO_4] \cdot 2H_2O$.

12 **Pyroxmangite** Red (octagon 1.66 ct), pink, also brown. Color of streak white. Transparent to opaque. Mohs' hardness 5½-6. Density 3.61-3.80. Refractive index 1.726-1.764, double refraction 0.016-0.020. Triclinic, $(Mn^{2+},Fe^{2+})_7[Si_7O_{21}]$. Cleavage perfect.

13 **Hyalophane** Colorless (1.34 ct), also white, red, yellow. Color of streak white. Transparent to translucent. Mohs' hardness 6-6½. Density 2.58-2.82. Refractive index 1.542-1.546, double refraction 0.004. Monoclinic, $(K,Ba)[Al(Si,Al)_3O_8]$. Cleavage perfect.

The illustrations are 40 percent larger than the originals.

1
2
3
3
4
4
5
6
7
8
9
10
11
12
13

Rocks as Gemstones

Formerly, rocks were used almost exclusively for decorative purposes and for ornamental objects. Nowadays, rocks are becoming more and more important also for personal jewelry, especially for costume jewelry. (Refer also to pages 83 and 84.)

Onyx Marble [1, 2] Also called Marble Onyx

Color: Yellow-green, white, brown, striped	Composition: Calcite or aragonite
Color of streak: According to color of stone	Transparency: Translucent to opaque
Mohs' hardness: 3½–4	Retractive index: 1.486–1.686
Density: 2.72–2.85	Double refraction: –0.156 to –0.172

The rock that is offered in the trade as onyx marble is a limestone formed by the minerals calcite or aragonite (page 224). It must not be confused with chalcedony-onyx (page 158). It is incorrect to name this rock onyx without the addition of the word *marble*. Onyx marble is formed from lime-containing water by layered deposits (therefore always banded) near warm springs or as stalactites or stalagmites in caves. Deposits are found in Egypt, Algeria, Argentina, Morocco, Mexico, and the United States. It is used for ornamental objects such as pendants and brooches. Dyeing produces a wide variety of colors.

Possibilities for Confusion With the mineral serpentine (page 218) as well as with various serpentine rocks, especially with the green-white spotted Connemara from Connaught, Ireland (compare with page 218), or with the banded, greenish ricolite from New Mexico, United States.

Mexican Onyx Misleading trade name for Onyx Marble.

Tufa [5–7] Also called Aragonite Sinter

Color: White, yellow, brown, reddish	Composition: Aragonite
Color of streak: According to color of stone	Transparency: Translucent, opaque
Mohs' hardness: 3½–4	Refractive index: Aragonite 1.530–1.685
Density: 2.95	Double refraction: Aragonite –0.155

Tufa is the calcium carbonate deposit from hot springs in the form of encrustations or stalactites, often with wavy layers. The best-known occurrence is at Karlsbad (Karlovy Vary) in the Czech Republic. Other deposits are found in Argentina, Mexico, New Zealand, Russia, and the United States.

Landscape Marble [8] Also called Ruin Marble

This is a fine-grained limestone, where the layers have been fractured, displaced, and again solidified. Because of the different coloring of the individual layers, images are formed which convey the impression of a landscape. It is used as a decorative stone, also for brooches or pendants, cut en cabochon. The landscape marble shown on the right could make one think of skyscrapers in a metropolis with low, thundery clouds.

1 Onyx marble, bowl
2 Onyx marble, broken piece, partly polished
3 Onyx marble, 2 pendants
4 Onyx marble, figurine
5 Tufa, 2 pieces from Karlsbad, Czech Rep.
6 Tufa brooch and pendant
7 Tufa, New Mexico, United States
8 Landscape marble, Tuscany, Italy

The illustrations are 50 percent smaller than the originals.

1
2
3
4
5
6
7
8

Orbicular Diorite [1, 2] Also called Ball Diorite

Plutonic rock composed of feldspar, hornblende, biotite, and quartz. Formed by rythmic crystallization which separates light and dark materials into spherical shapes. A decorative stone, also cut en cabochon [no. 2].

Kakortokite Trade name for a black-white-red banded nepheline-syenite from Greenland, containing eudialyte.

Obsidian [3–7]

Color: Black, gray, brown, green
Color of streak: White
Mohs' hardness: 5–5½
Density: 2.35–2.60
Cleavage: None
Fracture: Large conchoidal, sharp edged
Composition: Volcanic, amorphous, siliceous glassy rock
Transparency: Transparent to opaque
Refractive Index: 1.45–1.55
Double refraction: None
Dispersion: 0.010
Pleochroism: Absent
Absorption spectrum: Not diagnostic
Fluorescence: None

Obsidian (named after the Roman, Obsius) was used in antiquity for amulets and necklaces. Varieties show golden [Gold o., no. 4] or silver [Silver o., no. 5] sheen, caused by inclusions. Deposits are found in Ecuador, Indonesia, Iceland, Italy, Japan, Mexico, and the United States. See page 282.
Possibilities for Confusion With aegirine-augite (page 232), gadolinite (page 226), jet (page 252), hematite (page 178), pyrolusite (page 232), and wolframite (page 230).

Snowflake Obsidian [6, 7] Trade name for an obsidian with gray-white, ball-shaped inclusions (spherulites). Deposits in Mexico and the United States (Utah, New Mexico).

Moldavite [10–12] Also called Bouteille Stone

Color: Bottle-green to brown-green
Color of streak: White
Mohs' hardness: 5½
Density: 2.32–2.38
Cleavage: None
Fracture: Conchoidal
Crystal system: Amorphous
Chemical composition: $SiO_2(+Al_2O_3)$ silicon dioxide + (aluminum oxide)
Transparency: Transparent to opaque
Refractive index: 1.48–1.54
Double refraction: None
Dispersion: None
Pleochroism: Absent
Absorption spectrum: Not diagnostic
Fluorescence: None

Moldavite (named after Moldau [Vltava], Czech Republic) belongs to the tektite group; possibly formed from condensed rock vapors after being hit by a meteorite. Scarred surfaces; vitreous luster.
Possibilities for Confusion With apatite (page 210), diopside (page 206), precious beryl (page 112), sapphire (page 102), tourmaline (page 126), and with green bottle glass. Other tektites are dark brown to black [8, 9]. Depending on the place of discovery, they have different names, e.g., Australite (Australia), Billitonite (Borneo), Georgiaite (Georgia, United States), Indochinite (Indochina), Javaite (Java), and Philippinice (the Philippines).

1 Orbicular diorite, partly polished, Corsica
2 Orbicular diorite, cabochon, Corsica
3 Obsidian, rough, Mexico
4 Golden obsidian, Mexico
5 Silver obsidian, Mexico
6 Snowflake obsidian, 2 cabochons
7 Snowflake obsidian, partly polished
8 Tektite, rough, Thailand
9 Tektite, 2 faceted stones, Thailand
10 Moldavite, 4 rough pieces, Czech Republic
11 Moldavite, 6 faceted stones, Czech Republic
12 Moldavite, cabochon, Czech Republic

The illustrations are 40 percent smaller than the originals.

1
2
3
5
7
4
6
9
8
11
10
12
10

Alabaster [1, 2]

Color: White, pink, brownish
Color of streak: White
Mohs' hardness: 2
Density: 2.30–2.33
Chemical composition: $Ca[SO_4] \cdot 2H_2O$ hydrous calcium sulfate
Transparency: Opaque, translucent at edges
Refractive index: 1.520–1.530
Double refraction: +0.010
Dispersion: None

Alabaster (from Greek) is the fine-grained variety of gypsum (page 226); in antiquity, the name also referred to microcrystalline limestone. Deposits are found in Germany (Thuringia), England (Derbyshire), France (Parisian Basin), Italy (Tuscany), and the United States (Colorado). Used for ornamental objects, rarely as jewelry. Can easily be dyed because of its porosity.

Possibilities for Confusion With agalmatolite (see below), calcite (page 224), gypsum (page 226), meerschaum (see below), and onyx marble (page 244).

Agalmatolite Also called Picture Stone, Pagoda Stone, Pagodite

Whitish, greenish, or yellowish, dense aggregate of the mineral pyrophyllite; $Al_2Si_4O_{10}(OH)_2$. When heat-treated, the originally soft stone (Mohs' hardness 1–1½) hardens considerably. Deposits are found in Finland, Slovakia, South Africa, and the United States (California). Usage as for alabaster (see above).

Possibilities for Confusion With alabaster (see above), talc (page 226). The greenish variety is used to imitate jade (page 170).

Meerschaum [3–5] Also called Sepiolite

Color: White, also yellowish, gray, reddish
Color of streak: White
Mohs' hardness: 2–2½
Density: 2.0–2.1
Cleavage: Perfect
Fracture: Flat conchoidal, earthy
Crystal system: Orthorhombic; microcrystalline
Chemical composition: $Mg_4[(OH)_2|Si_6O_{15}] \cdot 6H_2O$ hydrous magnesium silicate
Transparency: Opaque
Refractive index: 1.53
Double refraction: None
Dispersion: None
Pleochroism: Absent
Absorption spectrum: Cannot be evaluated
Fluorescence: None

Because of its high porosity, meerschaum can float (German—sea foam). It occurs as a rock as concretion in serpentine. It has a dull greasy luster, feels like soap, and sticks to the tongue. The most important deposit is near Eskischehir, Anatolia (Turkey). It is also found in Greece (Samos), Morocco, Spain, Tanzania, and the United States (Texas). Worked into bowls for pipes and cigarette holders, which, because of the smoke, gradually become golden-yellow [no. 5]; in addition, it is used for costume jewelry. Becomes lustrous when impregnated with grease. Can be confused with alabaster (see above).

Fossils [6–9]

Petrified wood pieces (page 164) and other fossils (petrified animals or parts of animal) are attractive for making jewelry because of their form, structure, color, as well as their old age. For more about their formation, see page 164.

1 Alabaster, 2 pieces dyed red
2 Alabaster ashtray, dyed blue
3 Meerschaum, rough
4 Meerschaum, costume jewelry
5 Meerschaum, cigarette holder
6 Ammonite, shell replaced by pyrite
7 Ammonite, shell replaced by pyrite, partly polished
8 Trilobite, primeval crab in shale
9 Actaeonella, a sea snail, Austria, partly polished

The illustrations are 20 percent smaller than the originals.

1
3
4
5
2
6
7
8
9

Organic Gemstones

A number of gemstones are of organic origin, but have the quality and character of stones. Refer also to the description on page 84.

Coral

Color: Red, pink, white, black, blue	Transparency: Translucent, opaque
Color of streak: White	Refractive index: White and red: 1.486–1.658
Mohs' hardness: 3–4	Double refraction: White and red: –0.160 to –0.172
Density: White and red: 2.60–2.70	Dispersion: None
Cleavage: None	Pleochroism: Absent
Fracture: Irregular, splintery, brittle	Absorption spectrum: Not diagnostic
Crystal system: (Trigonal) microcrystalline	Fluorescence: Weak; violet
Chemical composition: $Ca[CO_3]$ or organic substance	

The coral (name of Greek origin) used as gem material is a branching skeleton-like structure built by small marine animals (coral polyps) related to reef-forming corals. The height of the branches is in the general range 8–10 in (20–40 cm); the branches are up to 2½ in (6 cm) thick.

Deposits are found along the coasts of the western Mediterranean countries, the Red Sea, Bay of Biscay, Canary Islands, Malaysian Archipelago, the Midway Islands, Japan, and Hawaii (United States). Production is increasingly controlled by environmental laws.

The coral is found at depths of 1–1020 ft (3–300 m), and mainly harvested with weighted, wide-meshed nets dredged across the seabed. When harvested by divers, however, not as many corals are damaged. Near Hawaii, minisubmarines have recently been used to collect the coral. Near the Midway Islands, deep-sea coral have been discovered at a depth of 6,562 ft. (2000 m). When it is brought to the surface, the soft parts are rubbed away and the material is sorted as to quality. For more than 200 years, the main trade center has been Torre del Greco, south of Naples, Italy. Three-quarters of the corals harvested all over the world are still processed here. Today, Japan and Taiwan are centers of the international coral trade.

Unworked coral is dull; when polished it has a vitreous luster. It is polished with fine-grained sandstone and emery; finely polished with felt-wheels. It is used for beads for necklaces and bracelets, for cabochons, ornamental objects, and sculptures. Branch-like pieces are pierced transversely and strung as spiky necklaces [no. 13]. Corals are sensitive to heat, acids, and hot solutions. The color can fade when worn.
Possibilities for Confusion With conch pearl (page 258), carnelian (page 142), rhodonite (page 184), and spessartite (page 120). Since the 1970s, imitations are made from glass, horn, rubber (gutta-percha), bone, and plastics.

Noble Coral (Corallium rubrum) Most desired of all coral types. According to place of discovery, it has numerous trade names. The color is uniform: light red to salmon-colored (Momo), medium red (Sardegan), ox-blood red (Moro), tender pink with whitish or light-reddish spots (Angel Skin Coral [no. 11]).

Black Coral consists of an organic horn substance. In the world trade it is, as with blue corals, of no economic importance.

---

1 Noble coral, 3 beads together 23.77ct
2 Noble coral, branch, Sicily
3 Noble coral, 2 figurines, Japan
4 Noble coral, 5 cabochons
5 Noble coral, 2 necklaces
6 Noble coral, figurine, Japan
7 White coral, branches, Japan
8 White coral, 2 engraved beads
9 Noble coral, engraved, Italy
10 Noble coral, branch, Japan
11 Noble coral, 3 cabochons, 14.05ct
12 Noble coral, figurine, Italy
13 White coral, spiky necklace
14 Black coral, branch, Australia

The illustrations are 50 percent smaller than the originals.

1
4
5
2
3
8
7
6
4
9
11
10
12
1
13
1
14

Jet [7, 8]

Color: Deep black, dark brown
Color of streak: Black-brown
Mohs' hardness: 2½–4
Density: 1.19–1.35
Cleavage: None
Fracture: Conchoidal
Chemical composition: Lignite

Transparency: Opaque
Refractive index: 1.640–1.680
Double refraction: None
Dispersion: None
Pleochroism: Absent
Absorption: Cannot be evaluated
Fluorescence: None

Jet (the name comes from a river in Turkey) is a bituminous coal which can be polished. It has a velvety, waxy luster. Deposits are found in England (Whitby), Germany/Wurttemberg, France (Dép. Aude), Poland, Spain (Asturias), and the United States (Colorado, New Mexico, Utah). It is worked on a lathe. Used for mourning jewelry, rosaries, ornamental objects, and cameos.
Possibilities for Confusion With anthracite, asphalt, cannel coal (see below), onyx (page 158), schorl (page 126). Some imitations made with glass, rubber (gutta-percha), plastic.

Cannel Coal [6]

The name is derived from the English "candle," referring to the wax that was extracted from combustible-rich layers in coal seams; formed predominantly from plant spores and pollen. Deposits are found in Germany, England, and Scotland. Due to its homogeneity and density, it can be easily worked on the lathe; a high luster can be achieved through polishing. Serves as substitute for jet.

Ivory [1–5, 9]

Color: White, creamy
Color of streak: White
Mohs' hardness: 2–3
Density: 1.7–2.0
Cleavage: None
Fracture: Fibrous
Chemical composition: Calcium phosphate

Transparency: Translucent, opaque
Refractive index: 1.535–1.570
Double refraction: None
Dispersion: None
Pleochroism: None
Absorption spectrum: Cannot be evaluated
Fluorescence: Various blues

Ivory originally only referred to the material of the elephant's tusk. Today it is also the teeth of hippopotamus, narwhal, walrus, wild boar, and fossilized mammoth. Most comes from Africa, some from Burma (Myanmar), India, and Indonesia (Sumatra). A worldwide ban on any trade in elephant's ivory was enacted in 1989, but it was eased in 2007.

Worked with cutting tools and file, it can be dyed. Used for ornamental objects, pendants, and for costume jewelry. Can be confused with many types of bone [no. 10].

Odontolite

(Also called tooth turquoise) Odontolite (Greek—toothstone) is fossilized tooth or bone substance from extinct prehistoric large animals (such as the mammoth, mastodon, or dinosaurs). Dyed turquoise blue with viviantite (page 224). Deposits are found in Siberia and the south of France. Has become very rare. Can be confused with turquoise (page 186) and ivory that has been dyed blue (see above).

---

1 Ivory, concentric spheres, China
2 Ivory, figurine and snuff bottle
3 Ivory, rough, the Democratic Republic of the Congo
4 Ivory, necklace, China
5 Ivory, bracelet and figurine
6 Cannel coal, rough and partly polished
7 Jet, 3 faceted pieces
8 Jet, 2 cabochons
9 Ivory, brooch, China
10 Bone, dyed, Israel

The illustrations are 50 percent smaller than the originals.

1
2
3
4
5
6
7
8
9
10

Amber Also called Succinite

Color: Yellow, brawn, and also other colors	Transparency: Transparent to opaque
Color of streak: White	Refractive index: 1.539–1.545
Mohs' hardness: 2–2½	Double refraction: None
Density: Mostly 1.05–1.09	Dispersion: None
Cleavage: None	Pleochroism: Absent
Fracture: Conchoidal, brittle	Absorption spectrum: Cannot be evaluated
Crystal system: Amorphous	Fluorescence: Bluish-white to yellow-green Burmite blue
Chemical composition: Approximately $C_{10}H_{16}O$ mixture of various resins	

Amber is the fossilized, hardened resin of the pine tree, *Pinus succinifera,* formed mainly in the Eocene epoch of the Tertiary period, about 50 million years ago; found mostly in the Baltic, although younger ambers are known from the Dominican Republic. Mostly amber is drop or nodular shaped with a homogeneous structure, or has a shell-like formation, often with a weathered crust. Pieces weighing over 22 lb (10 kg) have been found. It is often turbid because of numerous bubbles [no. 8], fine hair lines, or tension fractures. It is possible to clear air bubbles and enclosed liquids from the material by boiling in rape-seed oil. Yellow and brown are predominant colors. There are occasional inclusions of insects or parts of plants (no. 7, and illustration on page 60) and of pyrites.

Amber is sensitive to acids, caustic solutions, and gasoline, as well as alcohol and perfume. Can be ignited by a match, smelling like incense. When rubbed with a cloth, amber becomes electrically charged and can attract small particles. It has a vitreous luster; when it is polished, a resinous luster.

Deposits The largest deposit in the world is west of Kaliningrad, Russia. Under 100 ft (30 m) of sand is a 30 ft (9 m) layer of amber-containing clay, the so-called blue earth. It is surface-mined with dredging chain buckets. First the amber is washed out, then picked by hand. Only 15 percent is suitable for jewelry. The remainder is used for pressed amber (see Ambroid below) or used for technical purposes.

There are large reserves on the seabed of the Baltic. After heavy storms, amber is found on the beaches and in shallow waters of bordering countries. This sea amber is especially solid and used to be regularly fished for by fishermen. Further deposits are found in Sicily/Italy (called Simetite), Rumania (Rumanite), Burma (Myanmar—Bunnite), China, the Dominican Republic, Japan, Canada, Mexico, the United States (Alaska, New Jersey).

Uses Amber has been used since prehistoric times for jewelry and religious objects, accessories for smokers, also as amulets and mascots. The Baltic amber, the "gold of the North," is among the earliest-used gem materials. Used today for ornamental objects, ring stones, pendants, brooches, necklaces, and bracelets.

Possibilities for Confusion With citrine (page 136), fluorite (page 214), meerschaum (page 248), onyx marble (page 244), sphalerite (page 216), ambroid (see below). Imitated by newly created resins (copal), synthetic resins, and yellow glass.

Ambroid Natural-looking pressed amber made from smaller pieces and the remains of the genuine amber. These bits are welded at 284–482 degrees F (140–250 degrees C) and 3000 atmospheres pressure into a substance that can be easily mistaken for natural amber.

1 Amber, rough
2 Amber, partly polished
3 Amber, 3 cabochons
4 Amber, 2 bead necklaces
5 Amber, 2 baroque necklaces
6 Amber, various colors
7 Amber, with insect inclusion
8 Amber, with bubble inclusion

1
2
3
4
5
6
7
8

Pearl

Color: White, pink, silver-, cream-, golden-colored, green, blue, black
Color of streak: White
Mohs' hardness: 2½–4½
Density: 2.60–2.85
Cleavage: None
Fracture: Uneven
Crystal system: (Orthorhombic) microcrystalline
Chemical composition: Calcium carbonate + organic substances + water
Transparency: Translucent to opaque
Refractive index: 1.52–1.66 Black: 1.53–1.69
Double refraction: -0.156
Dispersion: None
Pleochroism: Absent
Absorption spectrum: Not diagnostic
Fluorescence: Weak, cannot be evaluated
Genuine black p.: Red to reddish River-p.: Strong: pale green

Most pearls are products of bivalve mollusks mainly of the oyster type (family *Ostreidae*). They are built up of mother-of-pearl (nacre), which is mainly calcium carbonate (in the form of aragonite), and an organic horn substance (conchiolin) that binds the microcrystals concentrically around an irritant.

Although the Mohs' hardness is only 2½–4½, pearls are extraordinarily compact, and it is very difficult to crush them.

The derivation of the name pearl is uncertain, but may be from a type of shell (Latin—*perna*) or from its spherical shape (Latin—*sphaerula*).

The size of pearls varies between a pin head and a pigeon's egg. One of the largest fine pearls ever found (called the Hope Pearl after a former owner) is 2 in (5 cm) long and weighs 454ct (1814 grains = 90.8 grams); it is in the South Kensington Museum in London.

The typical pearly luster is produced by the overlapping platelets of aragonite and film of conchiolin nearer to the pearl surface. This formation also causes the interference of light and the resulting iridescent colors (called orient) that can be observed on the pearl surface. The color of pearl varies with the type of mullusk and the water, and is dependent on the color of the upper conchiolin layer. If the conchiolin is irregularly distributed, the pearl becomes spotty.

Formation Pearls are formed by saltwater oysters (genus *Pinctada*), some freshwater mussels *(Unio)*, and more rarely by other shellfish. They are formed as a result of an irritant that has intruded between the shell of the mollusk and the mantle or into the interior of the mantle. The outer skin of this mantle—the epithelium—normally forms the shell by secretion of mother-of-pearl, and also encrusts all foreign bodies within its reach. And such an encrustation will develop into a pearl. If a pearl is formed as a wart-like growth on the inside of the shell, it must be separated from the shell when it is collected. Therefore, its shape is always semispherical. It is called blister or shell pearl. In the trade they sometimes cement these to mother-of-pearl backings to form mabé pearls.

1 Mother-of-pearl shell with cultured pearls
2 Pearl necklace, Biwa Lake cultured, baroque
3 Pearl necklace, 4 chokers, silver white
4 Pearl necklace, baroque
5 Pearl necklace, Biwa Lake cultured, graduated
6 4 baroque pearls
7 Pearl, Mabé cultured, silver-white, 20mm
8 Pearl, Mabé cultured, gray, oval
9 6 baroque pearls, 35.71ct
10 10 baroque pearls, Biwa Lake cultured
11 Mother-of-pearl, 2 cut pieces
12 Pearl choker, gray
13 3 pearls, Biwa Lake cultured, 29.67ct
14 4 cultured pearls, 16.16ct
15 Pearl necklace, choker, gray
16 6 black baroque pearls
17 Pearl necklace, black baroque
18 6 gray pearls, 17.28ct
The illustrations are 60 percent smaller than the originals.

1
8
6
3
7
4
5
9
12
10
11
3
11
13
14
16
18
17
15
3
3

When a foreign body enters the inner part of the mantle—the connective tissue—the mollusk forms a nonattached, rounded pearl as a type of immunity defense. The epithelial tissue, which has been drawn into the connective tissue together with the foreign body, forms a pearl sac around the intruder and isolates it by secretion of nacre. As we now know, the nacre can also produce a pearl without any foreign body. It is sufficient that a part of the epithelium may for any reason (for instance, an injury from the outside) be drawn into the connective tissue of the mantle.

Natural Pearls

Natural pearls are those pearls that come into being without intervention by human beings, in the ocean as well as in freshwater.

Sea Pearls Pearl-producing sea mollusks live along long stretches of coast at a depth of about 50-65 ft (15-20 m). The various species range in size from about 2½ to 12 in (6 to 30 cm); their life span is about 13 years.

Their habitat is the warmer regions on both sides of the equator. The most important occurrences yielding the best qualities (rose and creamy white) have long been those in the vicinity of the Persian Gulf. Because of this occurrence, all natural seawater pearls, wherever they come from, have been called "oriental pearls" in the trade. US FTC guidelines now restrict this term to those "of the distinctive type and appearance of pearls obtained from mollusks inhabiting the Persian Gulf."

Also, in the Gulf of Manaar (between India and Sri Lanka) there are ancient beds (pearls of pink-red and soft yellow color), but the pearls are mostly small (so-called seed pearls). Other important occurrences are along the coasts of Madagascar, Burma (Myanmar), the Philippines, many islands in the South Pacific, northern Australia, and the coastal lines of Central America and northern South America. In Japan, the most important country of cultured pearl production, there are only some small beds with natural pearls. Pearls are harvested by divers. Formerly the work was done without any special tools; today the most modem diving gear is sometimes used. Only every 30th or 40th oyster contains a pearl. Near Sri Lanka in 1958, dragnets were experimentally used; the result was catastrophic, as the next growth was almost completely destroyed.

The giant conch (*Strombus gigas*) is a sea snail that produces pearls. Its product (called conch or pink pearl) has a silky sheen that resembles porcelain. Commercially it is not important, and most gemologists do not consider it a true pearl because it lacks nacre.

Cross section of pearl formation in the mantle area of a pearl mollusk.

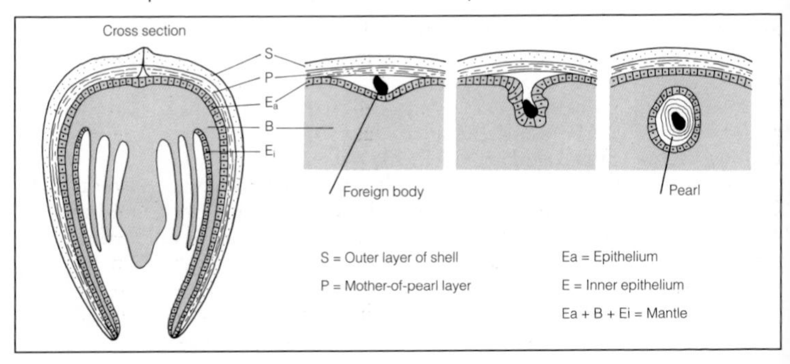

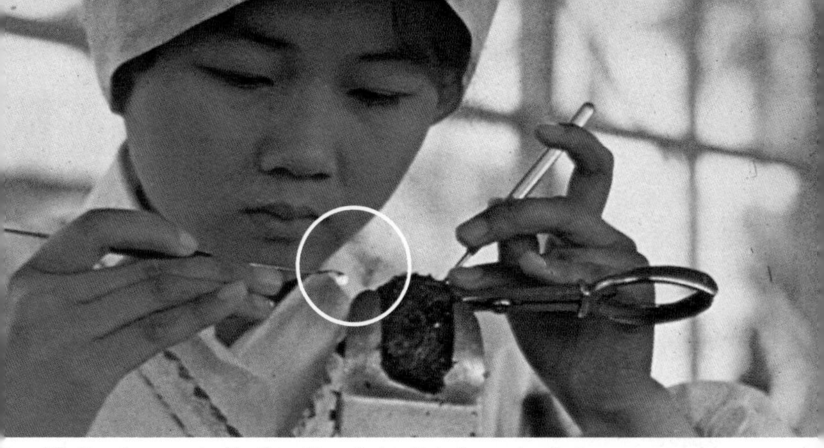

'ery skilled hands are required for the setting of a bead into the pearl mollusk.

River Pearls Fishing for natural pearls in freshwater, the river pearls, is today of no great importance commercially; they are rarely of good quality. During the Middle Ages and just after, fishing for pearls in the rivers, which are low in lime and rich in oxygen, of Central Europe as well as in rivers of similar habitats in Asia and North America was of some importance.

In Europe, fishing for pearls was the absolute privilege of princes; fished pearls had to be delivered to the ruler personally. Because of pollution of the water, pearl mussels have not flourished or become extinct. Even though the supply in some rivers has partially regenerated due to the improvement of water quality, their existence continues to be threatened because of elevated nitrate levels in the water. In the Scandinavian countries and in Central Europe, pearl mussels are environmentally protected; *pearl fishing is forbidden.* A limited number of natural freshwater pearls is still obtained from rivers in the United States.

Cultured Pearls

In the 20th century depletion and pollution drastically reduced the supply of natural pearls. The increased demand for pearls has led to their cultivation in large quantities. Such cultured pearls are not an imitation, but a product which has been produced with human assistance. Today cultured pearls amount to 90 percent of the total pearl trade. There are cultured-pearl farms in the ocean as well as in freshwater rivers.

Saltwater Cultured Pearls The principle behind pearl culturing is simple. Humans cause the mollusk to produce a pearl by insertion of a foreign body (compare formation of the pearl, page 256). In China as early as the 13th century, small objects were fixed to the inner wall of a mollusk shell so that they would be covered with pearl material. Round pearls were first produced, it is thought, by the Swedish naturalist Carl von Linné (Linnaeus) in 1761 in river mussels.

Modern cultivation of round pearls is based on the experimental work of the German zoologist F. Alverdes, as well as the Japanese, T. Nishikawa, O. Kuwabara, T. Mise, and K. Mikimoto, in the late 1800s and early decades of the twentieth century. To stimulate the mollusks to produce pearls, rounded mother-of-pearl beads from the shell of the North American freshwater mussel are at first wrapped with a piece of tissue from the mantle of a pearl mollusk (*Pinctada martensii*), and are then inserted into the cell lining of the mantle of another pearl mollusk.

The inserted tissue continues to grow and has the effect of a pearl sac in which pearl material is secreted. The most important element in the production of a pearl is the tissue, not the foreign body. It has been proven that one can do without the bead, but then the process will not remain economical since the culture of a large pearl takes too much time.

Only one layer of nacre is necessary for the bead to acquire the characteristic pearly luster. Since 1976, coreless pearls from the Pacific have been on the market; they are called *Keshi pearls* in the trade. These grow spontaneously (without nucleation) in mollusks previously used for pearl-culturing.

The insertion of the bead into the mollusk requires agile, skilled hands; commonly, this job is filled by women. They operate on 300 to 1000 oysters a day. The normal size of a bead (0.24–0.27 in/6–7 mm) requires a three-year-old mollusk; for smaller beads, younger mollusks can be used. When the bead is greater than 0.35 in (9 mm), the mortality rate of the oysters rises to 80 percent.

Prepared mollusks are kept underwater in the bay in wire cages or, preferably nowadays, plastic cages suspended at a depth of 6½–20 ft (2–6 m) and hung from bamboo floats or ropes which are fixed to buoys. Several times a year the mollusks and their cages must be cleaned and freed from seaweed and other deposits. Their natural enemies are fishes, crabs, polyps, and various parasites—mainly a zooplankton which appears in large quantities (a "red tide") and endangers a whole cultivation farm because it consumes large quantities of oxygen.

The temperature of the water has a great influence on the growth of the mollusks; the Japanese variety dies at 51 degrees F (11 degrees C). If the temperature

Mollusk cages must always be supervised and several times a year cleaned of seaweed and other unwanted deposits.

Rafts of cultivated pearl farms in an ocean bay in southern Japan.

suddenly falls before the winter, the floats with the submerged burden must be dragged from northern farms to warmer waters. The yearly growth rate of the pearl layer surrounding the bead in Japan used to be 0.09 mm; it has now reached 0.3 mm and is said to be 1.5 mm in southern seas.

Some cultivation farms have been transferred from bays to the open sea, as it was thought that the water current would activate the mollusks to produce faster, with better shapes. At the same time, the bays with their innumerable floats would be less crowded and conditions improved for other pearl mollusks.

The mollusks remain in the water about two to three years; by then, the nacre layers around the bead are about 0.5–1.0 mm. If they remain longer in the water, there is a danger that they will become ill, die, or mar the shape of the pearl. No mother-of-pearl is secreted after the 7th year.

A saltwater mollusk can usually be used only once. After the pearl has been removed, most of them die. Accordingly, it is important to make sure that there is sufficient aftergrowth. Cultured pearls with a very thin nacre shell are considered to be of inferior quality.

The best times for harvesting in Japan are the dry winter months, November to January, as the secretion of the mother-of-pearl is halted and an especially good luster is present. The pearls are taken from the mollusks, washed, dried, and sorted according to color, size, and quality. The whole production yields roughly 10 percent for good-quality jewelry; 60 percent are of minor quality, and 15–20 percent are rejected.

In order to improve and/or change the colors, cultured pearls are treated in various ways, such as bleaching, dyeing, or with radiation. Occasionally, the color is also influenced by insertion of colored cores. Colors that are achieved through radiation are not always permanent.

The first Japanese farms were founded in 1913 in southern Honshu; today there are also some on Shikoku and Kyushu. Since 1956, pearls of large size and good quality have been grown in coastal waters of northern and western Australia, as well as cultured blister pearls with a diameter of 0.6–1 in (15–22 mm), called Mabé pearls in the trade. Today, numerous farms exist in South-Southeast Asia, among others in southern Myanmar, Malaysia, and Indonesia, in several Pacific Island states, as well as in the Gulf of California/Mexico.

Cultured Freshwater Pearls Since the 1950s there has been a freshwater pearl farm in the Biwa Lake (Japanese—*biwako*), north of Kyoto on Honshu, Japan. Pieces of tissue measuring 4 × 4 mm, usually without solid beads, are inserted into the freshwater mussels (*Hyriopsis schlegeli*). As these mussels are very large (8 × 4.3 in/20 × 11 cm), ten insertions can be made into each half and, sometimes, an additional one with a mother-of-pearl bead. For each insertion a sac with pearl is formed. After one to two years the pearls are already 0.24–0.28 in (6–8 mm) large, but rarely round. Therefore, they are taken out of the mussel, covered with new tissue, and inserted into the same or another mussel to improve the shape. Biwa cultured pearls reach a diameter of 0.5 in (12 mm), but rarely have a perfect round shape. "Rice" and button shapes are most common.

The life of a freshwater mussel is 13 years, but after the operation it produces mother-of-pearl only for three more years. Many mussels can be harvested three times.

Cultivation methods are the same as for the sea mollusks. Cages are hung on a bamboo frame at about 3–6½ ft (1–2 m) depth. The success rate is about 60 percent, which is clearly higher than in seawater, probably because there are fewer dangers in Biwa Lake.

Since the beginning of the 1970s, freshwater pearls are also cultivated in China. They are brought onto the market now in large quantities. But their quality is not as good as that of the Japanese cultured pearls. Since the early 1990s, round bead-nucleated Chinese freshwater cultured pearls have been increasingly available.

A skilled eye is necessary for the sorting and quality evaluation of pearls.

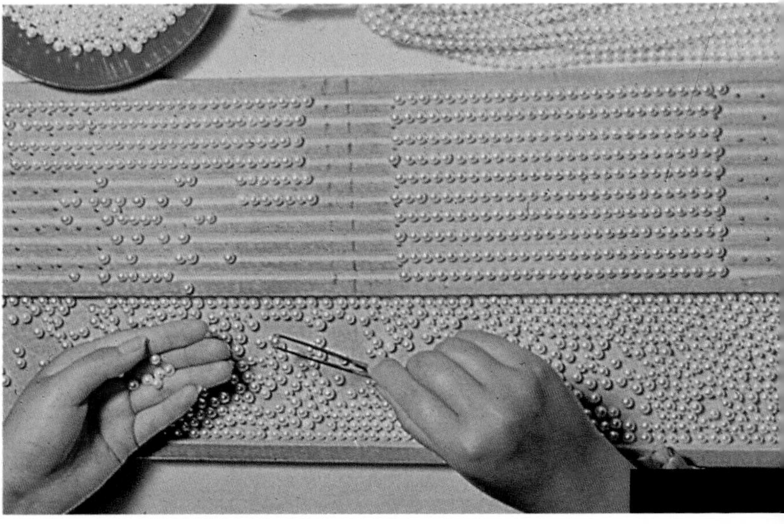

Pearls are drilled in such a way that damaged or less perfect spots "disappear" at the same time.

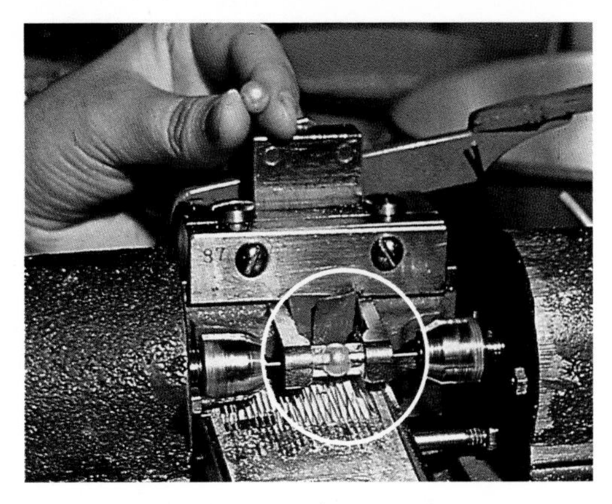

Use and Valuation of Pearls

Pearls have been regarded as one of the most valuable gem materials. They have been used for adornment for 6,000 years. In 2500 B.C., there was a substantial pearl trade in China. Pearls are also popular because they do not require any processing; in their natural state they show their full gloss, the desired luster.

As much as 70 percent of all pearls are strung and worn as necklaces. In the United States, the most popular length, known as the "Princess," is about 17 to 19 in (43 to 48 cm). If pearls of the same size are used, one speaks of a uniform necklace; if they are graduated in size with the biggest in the middle and the smallest at the ends, one speaks of the graduated necklace. The combination, i.e., the careful selection of the pearls for necklaces or collars, is done by eye.

Uses A point of the pearl, which either has a mark or is less perfect, is chosen for drilling a hole, thus eliminating the mark. The diameter of the drill hole, according to international agreement, should be 0.3 mm. To fix the pin for earrings, brooches, pins, and rings, a drill hole to the depth of two-thirds or three-quarters of the pearl's diameter should suffice. Blue pearls should never be drilled, as they may change color when air reaches the drill hole.

Spotted or damaged pearls can be peeled, i.e., the outer layer can be removed. Badly damaged parts can be cut away; the remaining part is traded as half or three-quarter pearl (not to be confused with blister pearls). They are used mainly for earrings and brooches.

For decades, the United States has been the largest buyer of cultured pearls.

Valuation The pearl is valued according to shape, color, size, surface condition, and luster. The most valuable is the spherical shape. Those flattened on one side or half-spherical pearls are called bouton (French—button) or button pearls; irregular pearls are baroque pearls. Pearls long worn in a necklace take on the shape of a little barrel; one speaks of barrel pearls.

Fair-skinned women in Europe and the United States prefer white or rose color; dark-haired ladies prefer cream-colored pearls.

Natural pearls are weighed in grains (1 grain 0.05 g = 0.24ct or 1/4 carat), and today increasingly in carats. The Japanese weight of *momme* (= 3.75 g = 18.75ct) is becoming rarer in European trade circles. Traditionally, natural pearl prices were calculated according to weight. Except in large wholesale transactions, cultured pearl prices are normally quoted with regard to size.

The word *pearl* without addition should be used only for natural pearls. Cultured pearls must be designated as such.

Taking Care of Pearls

Since conchiolin is an organic substance, it is prone to change, especially to drying out. This can lead to an "aging" of the pearls, limiting their useful life. At first they became dull, then fissures occur, and finally the beads spall. There is no guarantee possible for the life span of a pearl; on average, one estimates 100 to 150 years. But there are pearls which are some hundreds of years old and still look their best. Proper care can certainly preserve and extend the life of pearls. Extreme dryness is damaging; pearls are also sensitive to acids, perspiration, cosmetics, and hair spray.

Regular examination and maintenance by a firm specializing in pearls can also prolong their life.

Since pearls have a low hardness, they can be easily scratched. Therefore, wear and store them in such a way that the pearl surface is never in contact with metal or other gemstones.

Discerning Pearls from Imitations

As is the case for all gemstones, there are numerous imitations of pearls on the market. To recognize these is just as important as the differentiation between natural and cultured pearls, because there is a huge disparity in respective prices.

Differentiating between Natural and Cultured Pearls There is little or no difference in the appearance of natural and cultured pearls. To differentiate between them is difficult. Their density can help, as it is greater than 2.73 in the case of most (but not all) cultured pearls, whereas the density of natural pearls is usually lower.

Sometimes an examination with certain radiation can clarify. Under ultraviolet light, for instance, cultured pearls have a yellow luminescence, and under X rays a green one, but these reactions are not completely dependable. One reliable method of differentiating between cultured and natural pearls is by examining their inner structure. Natural pearls have a concentrically layered structure, whereas the inner structure of the cultured pearl varies according to the type of bead. Typically, an expert uses an instrument like an endoscope to explore the structure of the pearl inside the drilling hole.

Radiography methods with X rays, such as the X ray diffraction (lauegram) method and the X ray silhouette (shadowgraph) procedure, are effective. They can be applied to both drilled and undrilled pearls. In cultured pearls, they detect the bead structure or the thickness of the "natural" pearl layer around the foreign body.

Imitations A fine imitation is the so-called fish-scale pearl. It consists of glass or enamel coated with *essence d'orient,* which is produced from the scales of certain fish. In other imitations, part of a sea snail (antilles pearls), mussels (takara pearls from Japan), or teeth (of the sea cow-dugong pearl) are used. There are also plastic products on the market. Under FTC guidelines, all of these must be clearly identified as imitations.

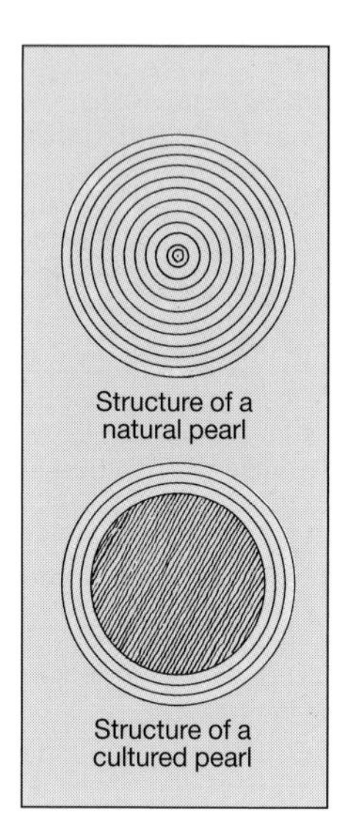

Natural and cultured pearls distinguish themselves through their inner structure.

The mabé pearl (sometimes called Japanese pearl in the trade) may also be included among the imitations, because this is not a cultured pearl in the trade sense. It consists of a thin mother-of-pearl layer and some other artificial part. A clay or resin bead is fixed to the inner part of a shell and is then covered with a thin pearl layer. After the harvest the bead is removed and replaced by a mother-of-pearl half-bead. The nacre shell is then filled with epoxy and cemented to a mother-of-pearl backing.

Operculum

The operculum (misleadingly called Chinese cat's-eye) has a structure similar to a half-pearl, with a porcelain-type coloring, but is actually the slightly arched lid of a sea snail found in the Australasian islands area, where it is used for adornment. It is not well known in Europe or the United States.

Mother-of-Pearl

The inner nacreous layer of a mollusk shell, or sometimes of a snail shell which has an iridescent play of color, is called mother-of-pearl ("mother of the pearl").

Mother-of-Pearl of the Pearl Mollusk The mother-of-pearl of the pearl mollusk is most often used. Accordingly, the main suppliers are the pearl farms. The basic color is usually white; it is naturally dark in the mother-of-pearl from Tahiti.

For structure, formation, and distribution, see the discussion of pearls starting on page 256. Mother-of-pearl is favorably used for ornamental purposes, for clock faces, buttons, costume jewelry, and for inlaid work (such as the handles of knives and pistols).

Mother-of-Pearl of the Paua (abalone) The mother-of-pearl of the Paua (*Haliotis australis*) from New Zealand, which has a blue-green iridescent color play, has been used by the indigenous Maori people for centuries for inlays in mystical carvings. Similar animals found along the U.S. coasts of Florida and California are called abalone.

The mother-of-pearl of this mollusk has now been used for some time in the Western world, especially for costume jewelry (see photo below). It is called sea opal, because of a resemblance to the color effects seen in opal.

Mother-of-pearl of the Paua with striking iridescence.

Imitation and Synthetic Gemstones

There is nothing against the law in producing imitation or synthetic gemstones, as long as no one is harmed or defrauded by it. These products are indeed an important element of the gemstone trade. Those who do not want the security risk of, or cannot afford, genuine gemstones can use these for their adornment. But when imitations or syntheses are passed off as more valuable true gemstones at inflated prices, then that is fraud. Imitations and syntheses must always be designated as such; they have to be named correctly. A distinction is also made between imitations, which only look similar to the gems, and syntheses, which are artificial versions of natural gemstones (compare the definitions on pages 10 and 11).

It is almost impossible for amateurs to tell the difference between real gemstones or decorative stones and imitations. The sophistication of gemstone imitations is becoming more and more refined. Often, costly examinations are necessary in order to determine the authenticity of the stone. Experts need an extensive number of instruments which exceed the possibilities of an average gemological laboratory. Only large institutions have the necessary know-how and the appropriate equipment. They work with x-rays, scanning electron microscopes, lasers, and the most diverse spectroscopic analyses.

Imitations

Ancient Egyptians were the first who feigned gemstones with glass and glaze, because genuine were too expensive and/or too rare.

In 1758, a Viennese, Joseph Strasser, developed a type of glass which served for a long time as a substitute for diamond. It could be cut and was, in fact, very similar to diamond in appearance, due to its high refractive index. Even though production and sale was prohibited by the Empress Maria Theresa, this diamond imitation, called *strass,* reached the European trade via Paris.

Until 1945, Gablonz and Turnau in Czechoslovakia were important centers for the glass-jewelry industry. Then this tradition was taken over by Neugablonz in the Allgäu, Bavaria. For costume jewelry, cheap glass was used. For more valuable gemstone imitations, lead or flint glass with a high refractive index is used. Porcelain, enamel, and resins as well as other plastics also serve as gemstone imitations. Most imitations have only a color similar to that of the gemstone; other properties, such as hardness or fire, could not be satisfactorily imitated.

A selection of gemstone doublets and gemstone triplets

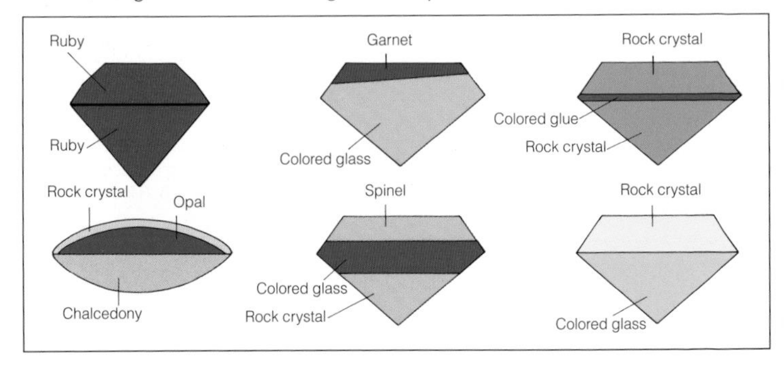

Combined Stones

A popular type of gemstone imitations are fabrications called combined or assembled stones (see the illustration on opposite page), in which one part may be a genuine gem, another part glass, foil, or plastic. There are many combinations; sometimes two natural gemstone pieces are made even more beautiful with a colorful glue layer to make a larger stone. Stones with two parts are called doublets; stones with three parts, triplets (see the illustration on page 266). Stones which are carefully constructed are difficult to recognize, especially when the seams are covered by the setting.

Synthetic Gemstones

The dream of mankind to produce artificial stones that are really the same as the natural gemstones was realized at the end of the 19th century. The French chemist A. V. Verneuil succeeded around 1888 in synthesizing rubies at a commercial price. In fact, 50 years earlier the first gemstones had been produced synthetically, but they were only of scientific interest.

The flame fusion process (see page 268) developed by Verneuil is still largely used today. The method is as follows: Powdered raw material is melted out at about 2000 degrees Celsius. Aluminum oxide with dyeing additives is dropped through a high-temperature oxyhydrogen flame that melts the powder. The molten drops fall on a cradle, where they crystallize and form a pear-shaped "boule" (see page 269). Although this boule does not have any recognizable crystal faces, the inner structure is the same as that of a natural crystal. The boules grow to about 3 in (8 cm) thick and several inches high. The growth time is several hours. Nowadays the boules usually have cylindrical shapes.

Verneuil first produced rubies, followed in 1910 by synthetic sapphires; later on, colorless, yellow, green, and alexandrite-colored corundums were produced. By adding rutile components to the smelting, the cultivation of synthetic star rubies and star sapphires succeeded in the United States in 1947, following Verneuil's method.

Following the Verneuil process, synthetic spinels have also been produced since 1910. Their composition, though, is somewhat different from that of natural spinels. With the addition of heavy metals, very good color tints of other gemstones can be achieved, for instance, those of aquamarine and of tourmaline.

Synthetic emeralds of usable gem size have existed only since the 1940s, although experiments had been conducted for over a hundred years.

Large synthetic crystals of highest purity can also be cultivated with the drawing procedure that was developed in 1918 by the German chemist I. Czochralski. The cultivation product is drawn out of the nutrient melt, after a crystal nucleus has initiated the growth of the boule (see the illustration on page 268).

In 1953/1954, diamond synthesis succeeded in Sweden and the United States. Initially not gem quality, these industrial-grade synthetic diamonds have become very important as abrasives. Gem-quality synthetic diamonds first appeared in 1970, and have also found essential applications in science and industry. In recent years gem synthetic diamonds have begun to appear in the jewelry market. Many experts believe they will become increasingly available in the future.

Today, there are hardly any gemstones left for which a synthetic version has not or could not be produced. But whether each possible synthesis can succeed on the market is more a question of the relationship between cost of production and market value. Over a dozen different synthesis procedures are known. Certainly there are others because companies generally regard these procedures as proprietary information and so keep their advances and details secret.

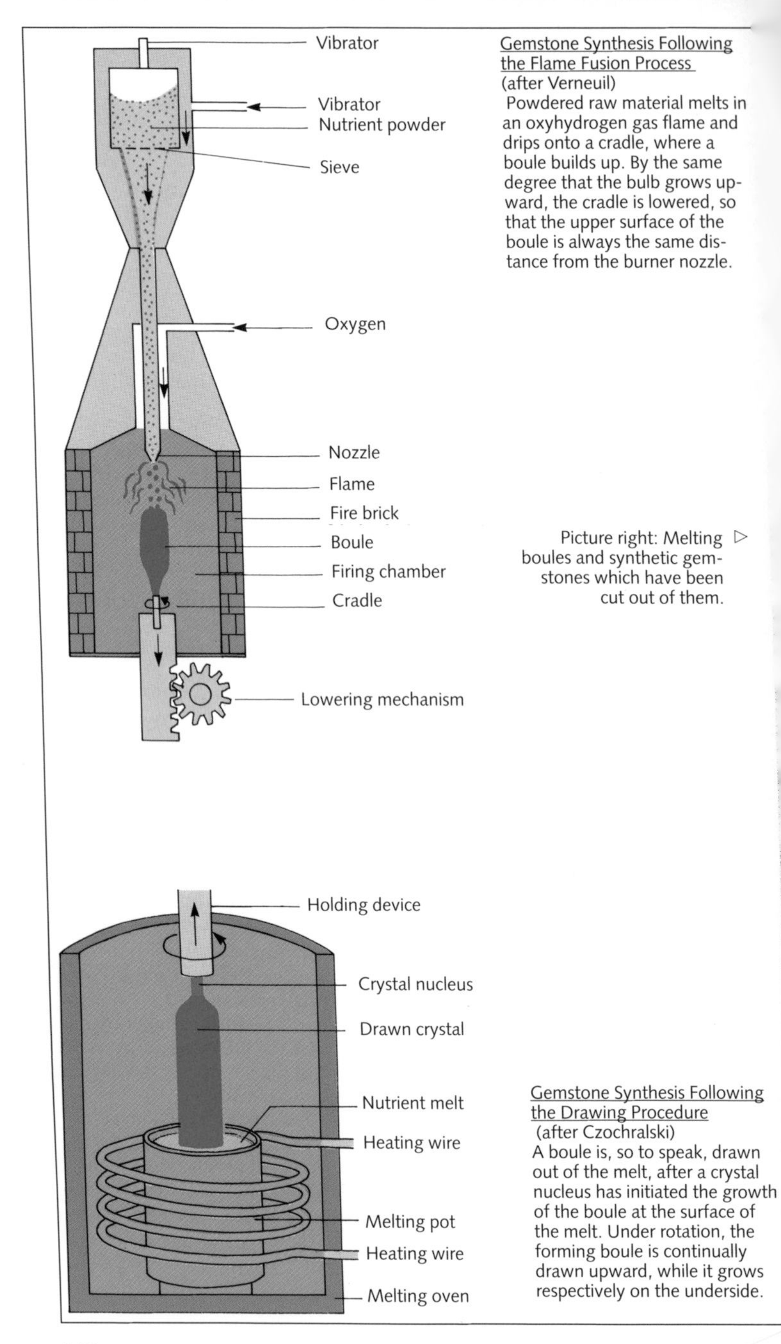

Gemstone Synthesis Following the Flame Fusion Process
(after Verneuil)
Powdered raw material melts in an oxyhydrogen gas flame and drips onto a cradle, where a boule builds up. By the same degree that the bulb grows upward, the cradle is lowered, so that the upper surface of the boule is always the same distance from the burner nozzle.

Picture right: Melting ▷ boules and synthetic gemstones which have been cut out of them.

Gemstone Synthesis Following the Drawing Procedure
(after Czochralski)
A boule is, so to speak, drawn out of the melt, after a crystal nucleus has initiated the growth of the boule at the surface of the melt. Under rotation, the forming boule is continually drawn upward, while it grows respectively on the underside.

Synthetic Gemstones without a Natural Model (Selection)

A special kind of imitation includes the artifically produced, synthetic stone that does not have a counterpart in nature, but rivals natural gemstones in physical characteristics and has optical effects. These synthetic stones are ranked among gemstones. The idea behind these synthetic stones is to combine as many desirable gemstone properties as possible and in a cost-effective way. In nature, the perfect or almost perfect gemstone is very rare.

As the provisions of international agreements state, names of synthetic gemstones "need to be named unnmistakenly and unambiguously with the correct name of the material," and to this end, "trademarks, labels or imaginative names are prohibited to reveal a similarity to the name of a gemstone."

The reality is, however, quite different. In the trade you may very well find designations which feign a relation to known gemstones, such as Zirconia or Diamonair.

On the other hand, some synthetic stones are known merely as formulas, which have the danger of becoming known only as imitations of real, natural stone (particularly diamond) instead of true independent gemstones.

The gemstone syntheses stated in the following are usually transparent and colorless. While they can also be produced in various colors, they serve primarily as diamond imitators.

Fabulite (also called Diagem) Strontium Titanate, $SrTiO_3$. It was first produced in the USA in 1953, and has been on the market in gemstone quality since 1969.

Gadolinium Gallium Garnet (also called GGG or Galliant) $Gd_3Ga_5O_{12}$. On the market since the beginning of the 1970s.

Linobate Lithium Niobite, $LiNbO_3$. Produced since 1967 in the USA.

YAG (also called Diamonair) Yttrium Aluminum Garnet, $Y_3Al_5O_{12}$. See ill. below. Developed at the beginning of the 1960s, has been on the market in gemstone quality since 1969.

Zirconia (also called Fianite, Phianite Yttrium zirconium oxide $ZrO_2|Y_2O_3$ or calcium zirconium oxide, ZrO_2/CaO. First commercial use in 1973. It was long considered the best diamond imitation until synthetic moissanite and GE POL diamond (see page 275) came on the market. It is offered in all colors.

The synthetically made gemstone YAG is a good diamond imitation. This photograph has been enlarged to ten times its normal size.

Treatment of Jewels and Gemstones

Many jewels and gemstones are changed in their natural characteristics through technical intervention in order to gain better quality in color, transparency, and firmness, to prevent the aging of the stones, and to be able to use them as imitations. Experts call such a targeted manipulation of natural stones treatment.

Depending on the kind of treatment, we differentiate between dyeing and bleaching, impregnating, heat treatment, irradiation, and laser coating. These work processes often overlap and complement each other.

Coloring

For the past 2,000 years, stones have been beautified through dyeing. With the agate, a small-scale industry has developed in this way. Whereas people used natural colors in earlier times, today aniline colors are used. They allow for a greater depth of color and more resistance. See agate coloring, page 152.

Besides agate, the following stones are also color treated: amber, chalcedony, chrysocolla, chrysoprase, cordierite, jade, jasper, coral, labradorite, lapis lazuli, opal, mother-of-pearl, ruby, sapphire, serpentine, sodalite, emerald, turquoise, cultured pearls.

Colored soapstone is used as an imitation for jade, lapis lazuli, and turquoise; marble (a calcitic rock), which can be easily dyed, is used as an imitation for hemimorphite, jade, smithsonite, and turquoise, colored sillimanite as a ruby substitute.

With superficial dyeing, manipulation can sometimes be detected by dabbing the gem with acetone. With color-drenched gemstones, patchy color substances might be detected under the microscope. Unnatural colors can be identified by either method.

Today, all gemstones must be declared as "color modified" or "treated" according to CIBJO guidelines (see page 12). This also holds true for agates, for which coloring did not have to be declared until 2007.

While coloring has been done for centuries by adding colored substances either as liquids or dissolved in liquids, stones (e.g., quartz, spinel, topaz, tourmaline) have been completely altered in their original color by steaming open the thinnest metal layers ever since the 1980s. This type of color change has to be declared.

Bleaching

The bleaching of gemstone materials is not only used for making too-dark stones lighter (e.g., with ivory and tiger's-eye) but also for removing blemishes, such as unattractive spots or unnatural foreign substances in pores and fissures.

Pearls, for instance, obtain a uniform appearance by being bleached with hydrogen peroxide, which then allows for a much greater effect of the pearls when used in a necklace. With jade, the primary goal is to remove disturbing foreign substances in the pores of the stone.

Many decolorized gemstones can be recognized as treated with the help of absorption spectra.

Impregnating

With so-called impregnating, gemstones are treated with substances in order to fill fissures and pores. By doing so, the color can be deepened and the shine enhanced, but also the firmness of the stone can be strengthened. The Romans beautified gems that contained fissures (e.g., turquoise) with oil. Later on, bee wax and natural resins served as filling material. Today, above all, synthetic resins are employed. Since the 1980s, lead-containing glass in conjunction with high temperatures of up to 1832° F (1000° C) are pressed into the gemstone. Through the thermal influence of the viscosity of the filling material, the impregnation can be made easier and can be accelerated.

The idea behind impregnation is to make color-altering fissures no longer visible to the eye. Therefore, only such colorless filling materials are used that have roughly the same refractive index as the respective stone.

Some impregnated gemstones include: aquamarine, azurite, charoite, jadeite, coral, lapis lazuli, malachite, opal, quartz, rhodochrosite, rhodonite, ruby, sapphire, serpentine, emerald, sodalite, topaz, turquoise, tourmaline, zircon, cultured pearls.

The amateur can check stones that are treated with wax or resin with a hot needle. Minor melting processes can be spotted with a magnifying glass. Sometimes also a typical smell of wax or resin arises. A person familiar with a microscope can detect the beautification at the different structures and at prominent inclusions; the professional will know by looking at the physical data, in particular the density and refractive index.

The beautification with colorless, natural substances does not have to be declared; however, the buyer must be informed of any glass impregnation.

So-called reconstructed or recrystallized stones can be counted among the group of impregnated gemstones. These are jewels or decorative stones that are melted together, sintered, pressed together, or glued together with the help of a binding agent from small splitters or from the powder of real gemstones into larger pieces. Particularly with amber, hematite, corals, lapis lazuli, malachite, and turquoise, these types of formed stones exist. The treatment of the gemstone fragments serves primarily to stabilize the material in order to process it like a gemstone. This type of gemstone treatment must be declared.

Heating

With many gemstones (such as amethyst, aquamarine, quartz, ruby, sapphire, spinel, tiger's eye, topaz, tourmaline, zircon), less appealing models can be improved in color and transparency through heat treatment and thus can be moved into a new price range. Stones that undergo heat treatment show colors that even Nature itself could never create.

Although some gemstones had been beautified through heating already in ancient times, the commercial use of heat treatment began around 1900. Today, the heat treatment of gemstones has become a significant branch of the gemstone trade.

The heating of gemstones often happens in the countries in which they were found, and even with primitive equipment, the results are usually quite good. In consumer countries, the heat treatment of the gemstones is carried out in many places industrially. Gemstones and decorative stones that have been given a permanent and irreversible color change through heat treatment do not need to be explicitly declared as such in the trade according to the CIBJO guidelines.

This holds also true for amber, the natural appearance of which is often clouded through the inclusion of small air bubbles. When it is slowly heated, amber loses the air bubbles and appears clear (see also page 254).

Most of the citrine on the market is heat-treated amethyst. Even moonstone can often be traced back to amethyst. Inferior garnets receive other colors or are turning lighter in color through heat treatment.

The duration of heat treatment lies between a couple of minutes and several hours, depending on the type of gemstone, the color effect desired, and the necessary temperatures. The latter in general fluctuates between 482-1112° F (250-600° C), with heat-treated corundum between 2192-3452° F (1200-1900° C). About 2912° F (1600° C) is necessary to change the color of diamond. The technical procedure is done in gas-tight, strong-walled devices called autoclaves.

A microscope and changed-absorption spectra (see graphic below) can frequently be used as proof for heat treatment; however, sometimes a costly analysis procedure is necessary.

A special type of heat treatment is the so-called diffusion treatment with corundum that has been practiced ever since the mid-1970s. Corundum is embedded in special substances and is heat treated by adding color-giving elements. In this way, corundum obtains an extremely thin metal layer that optimizes the color or changes it altogether, similar to adhered foil.

Irradiating

Through radiation with electromagnetic rays rich in energy or through bombardment with elementary particles, the color of gemstones can be improved or altered entirely. However, color obtained this way does not always last. Some types of stones fade in sunlight relatively quickly. Nonetheless, since about 1970, this radiation treatment has been increasingly used with gemstones.

Proof of radiation treatment is possible by means of altered absorption spectra. Sometimes only an expert can identify the radiation treatment with ultra-modern instruments. All artificially irradiated gemstones and decorative stones have to be identified as "radiated" or "treated" without exception. The following gemstones are frequently irradiated: beryl, chrysoberyl cat's-eye, coelestine, diamond, garnet, kunzite, mother-of-pearl, quartz, ruby, sapphire, topaz, tourmaline, zircon, and cultured pearls. Rock crystal turns smoky quartz- or amethyst-colored when irradiated; colorless topaz turns blue, yellow, or green; brown zircon colorless or blue.

Lasering

Ever since the end of the 1980s, lasers are used with diamonds in order to take care of impurities and to optimize color and transparency.

Traces of the procedure can be recognized under the microscope by identifying the created hollow channels, fillings, and fissures.

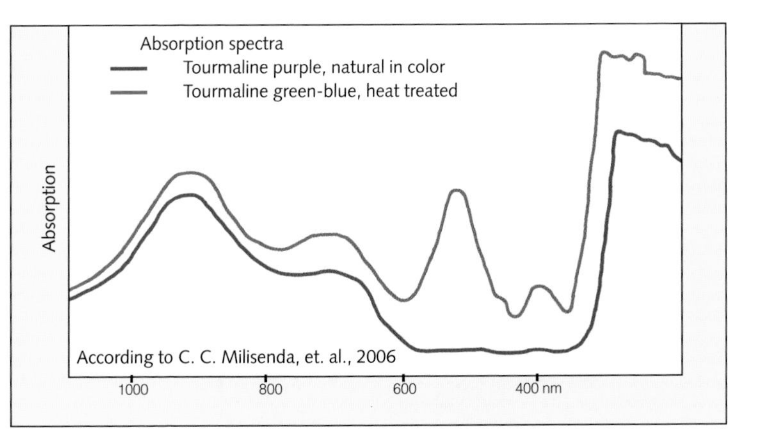

Genuineness Test for Diamond

It is often very difficult to distinguish genuine gemstones from synthetics, imitations, or some other fabrications. The refinement and delicacy of execution of gemstone imitations are becoming more and more perfect. Often, a test for genuineness is indicated only after very close examination.

A testing device that is easy to handle, and that in many cases results quickly in a reliable answer, does exist for diamonds. The tester (a heat resistance tester) uses the different abilities of diamond and stones that are diamond substitutes to conduct heat. A special advantage of the device is that it can also be used for stones that are already set. Even the smallest pieces can be tested with its fine metal tip.

Diamond imitations prior to 1996 are recognized beyond doubt with such a device. However, the device does not identify the synthetic moissanite (on the market since 1997) since the stone possesses a similar thermal resistance as diamond. This has led to a great feeling of insecurity in the diamond sector. Many synthetic moissanite imitations were sold as real diamonds and even given certificates of authenticity. Now, however, there are also moissanite testing devices (ill. page 275).

In order to exclude all kinds of syntheses through diamond identification, two operations must be carried out: first, one has to check with a heat resistance tester whether the device spectrum indicates synthesis or real diamond. As soon as the device indicates diamond, one has to perform further tests with a moissanite testing device. Moissanite testers examine the different electric conductance of the stones and can, thus, comfirm the authenticity of a diamond, or, if applicable, refute it.

Synthetic moissanite has not always been recognized by the gemstone trade. For too long, one merely relied on a single testing device without carrying out further checks. With a little care and effort, it is clearly possible to recognize the synthetic moissanite by comparing fixed data of both diamonds and syntheses (see table on page 275).

The devices that were considered reliable in determining the genuineness of diamonds have been rendered ineffective since about 2003. These measuring instruments continue to recognize many diamond imitations; however, there is no guarantee for the genuineness of the diamonds examined. Companies like Gemesis in Florida and Apollo Diamonds in Boston have introduced synthetic decorative diamonds into the market that has caused tremendous confusion in the international diamond trade. Even in well-equipped laboratories, a great number of instruments are needed in order to filter out synthetic diamonds, like the ones mentioned here.

Synthetic diamonds of jewelry quality, as developed by Gemesis and Apollo, mean a direct attack on the monopoly of De Beers because of lower prices (see page 90).

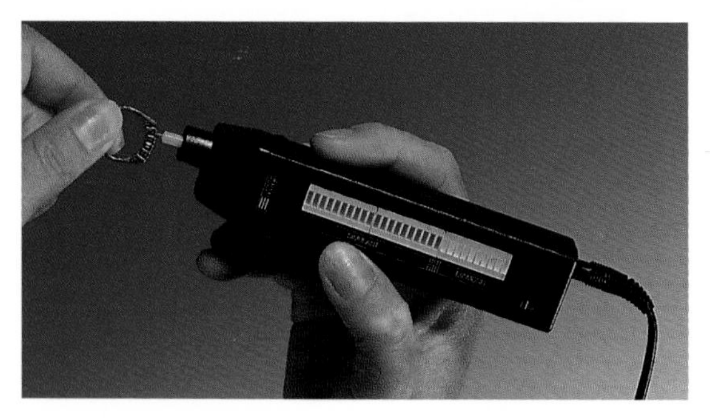

Heat resistance tester to identify diamond imitations except moissanite.

Diamond and Its Imitations (Selection)

Product	Mohs' hardness	Density	Refractive index	Double refraction	Dispersion BG	Dispersion CF
Diamond	10	3.50–3.53	2.417–2.419	anomalous	0.044	0.025
Synth. Moissanite	$9^1/_2$	3.10–3.22	2.648–2.691	0.043	0.104	
Sapphire	9	3.95–4.03	1.762–1.778	0.008	0.018	0.011
YAG	$8^1/_2$	4.55–4.65	1.833–1.835	none	0.028	0.015
Zirconia	$8^1/_2$	5.50–6.00	2.150–2.180	none	0.060	0.035
Spinel	8	3.54–3.63	1.712–1.762	none	0.020	0.011
Synth. Spinel	8	3.63–3.65	1.720–1.740	anomalous	0.020	0.010
Topaz	8	3.49–3.57	1.609–1.643	0.008–0.016	0.014	0.008
Beryl	$7^1/_2$–8	2.66–2.87	1.562–1.602	0.004–0.010	0.014	0.011
Zircon	$6^1/_2$–$7^1/_2$	3.93–4.73	1.810–2.024	0.002–0.059	0.039	0.022
Synth. Rutile	$6^1/_2$–7	4.24–4.28	2.62–2.97	0.287	0.330	0.190
GGG	$6^1/_2$	7.00–7.09	1.970–2.020	none	0.038	0.022
Fabulite	$5^1/_2$–6	5.11–5.15	2.409	none	0.190	0.109
Linobate	$5^1/_2$	4.64–4.66	2.21–2.30	0.090	0.13	0.075
Glass	5–$6^1/_2$	2.0–4.5	1.44–1.90	none		to 0.098
Strass	5	3.15–4.20	1.57–1.69	none	0.041	

GE POL Diamonds Whereas the almost perfect diamond imitation, the "synthetic moissanite," can be identified when carefully examined, since 1999 there have been diamonds on the market that cannot be identified as "manipulated" despite their near-flawless color, brilliance, and purity. These are diamonds that have been treated by General Electric (GE) in the USA and are sold by the Antwerp company Pegasus Overseas Limited (POL) as Pegasus diamonds or as Bellataire.

Starting materials are far less valuable brown or brownish natural diamonds. The beautification process (probably high pressure and high temperature treatment) is highly protected.

With the treated diamonds, no specific particularity is known that could be grasped by the usual gemological instruments. It is possible that further developed spectroscopic examinations could provide identification. In order to respond to the accusation of doubtful marketing, General Electric recently started to mark the treated gemstones with a laser inscription on its band. Allegedly, though, it is possible to remove this inscription through polishing.

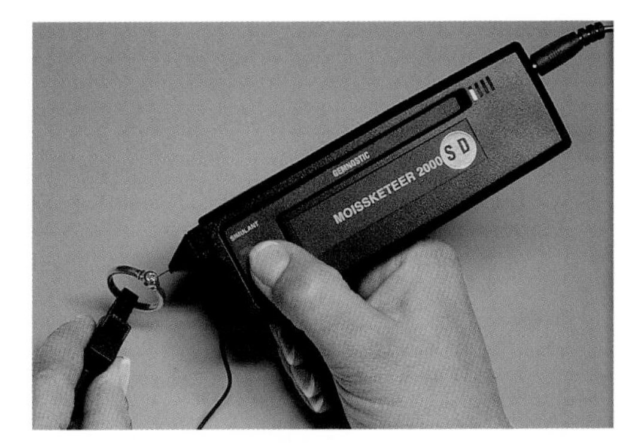

Moissanite testing device; works on the basis of electric conductance.

New on the Market

"New on the market" gemstones have not always been recently found. Some may have been known for some time, but their importance for the gemstone trade has remained minor or insignificant due to their rarity of occurrence or inadequate marketing.

1 **Verdite** Light to dark green, often spotty serpentine rock. Named for its green color. Translucent to opaque. Mohs' hardness about 3. Density 2.8-3.0. Used for sculptures and costume jewelry.

2 **Charoite** Lilac-colored to violet. Through admixture of accompanying minerals, it becomes decoratively white or black spotted or rather flamed. Color of streak white. Translucent to opaque. Mohs' hardness 4½-5. Density 2.54-2.78. Refractive index 1.550-1.561, double refraction 0.004-0.009. Monoclinic, $(K,Na)_4Ca_8[(OH,F)_2|Si_6O_{15}|Si_{12}O_{30}] \cdot 3H_2O$. Cleavage good.

3 **Eclogite** Opaque metamorphic rock with red pyrope and almandine, green diopside, blue glaucophane and light gray zoisite. Density 3.2-3.6. Used en cabochon.

4 **Lepidolite** Rose-colored, red violet. Color of streak white. Transparent to translucent. Mohs' hardness 2½-3. Density 2.80-2.90. Refractive index 1.525-1.587, double refraction 0.018-0.038. Monoclinic, $K(Li,Al)_3[(F,OH)_2|(Si,Al)Si_3O_{10}]$.

5 **Bikitaite** Colorless, white. Color of streak white. Transparent to translucent. Mohs' hardness 6. Density 2.28. Refractive index 1.510-1.523, double refraction 0.013. Monoclinic, $Li[AlSi_2O_6] \cdot H_2O$. Cleavage perfect.

6 **Gneiss** Opaque gray, greenish, brownish or reddish metamorphic rock with slaty structure. The chief constituents are feldspars and quartz. Density 2.65-3.05. Used for costume jewelry.

7 **Unakite** Opaque granitic rock with the chief constituents being quartz and feldspar as well as greenish epidote. Density 2.85. Cut en cabochon and barrel-shaped for costume jewelry.

8 **Nuummite** Opaque metamorphic rock with almost black basic color with chief constituents being gedrite and anthophyllite; accessory constituents are pyrite, pyrrhotine, chalcopyrite. Because of a lamellar, fibrous structure, there is an iridescent play of colors. Mohs' hardness 5½-6. Density about 3. It is worked flat or en cabochon.

9, 10 **Ammolite** (also known as Korite) Fossilized shell of ammonites which in turn are composed primarily of aragonite. Because of a lamellar structure, it reveals an iridescent opal-like play of color. Mohs' hardness 4. Density 2.75-2.80. Refractive index 1.52-1.68, double refraction 0.155. Can be found in Alberta, Canada/USA. On the market since 1969. Frequently worked into doublets and triplets.

11 **Ussingite** pink, violet red. Color of streak white. Transparent to translucent. Mohs' hardness 6-6½. Density 2.46-2.50. Refractive index 1.504-1.543, double refraction 0.039. Triclinic, $Na_2[OH|AlSi_3O_8]$.

12 **Carletonite** Deep blue, also pale blue. Color of streak white. Transparent. Mohs' hardness 4-4½. Density 2.45. Refractive index 1.517-1.521, double refraction 0.004. Tetragonal, $KNa_4Ca_4[(OH,F)|(CO_3)_4|(Si_4O_9)_2] \cdot H_2O$. Cleavage perfect.

13 **Catapleiite** Colorless. Color of streak white. Transparent. Mohs' hardness 5-6. Density 2.72. Refractive index 1.590-1.629, double refraction 0.039. Hexagonal, $Na_2Zr[Si_3O_9] \cdot 2H_2O$. Cleavage perfect.

14 **Larimar** Blue variety of pectolite (p. 228 and p. 288). On the market since the 1970s. Only found in the Dominican Republic. Easy to polish.

15 **Gaspeite** Light green. Color of streak yellow-green. Opaque. Mohs' hardness 4½-5. Density 3.71. Refractive index 1.61-1.81, double refraction 0.22. Trigonal, $(Ni,Mg,Fe^{2+})[CO_3]$. Cleavage good. Discovered in 1977.

16 **Sugilite** Violet. Color of streak white. Transparent to translucent. Mohs' hardness 6-6½. Density 2.76-2.80. Refractive index 1.607-1.611, double refraction 0.001-0.004. Hexagonal, $Na_3KLi_2(Fe^{3+},Mn^{3+},Al)_2[Si_{12}O_{30}]$. Cleavage indistinct.

The illustrations are 10 percent larger than the originals.

1
2
3
4
5
6
7
8
9
10
11
12
13
16
15
14

1 **Poudretteite** Violet, also colorless, pale pink. Color of streak white. Transparent. Mohs' hardness 5. Density 2.51-2.53. Refractive index 1.511-1.532, double refraction 0.021. Hexagonal, $KNa_2B_3[Si_{12}O_{30}]$. Cleavage none. Known in gemstone quality since 2001.

2 **Parrot Wing** Named after its multi-colored appearance. Mixture of jasper (p. 162) with copper minerals, particularly with chrysocolla (p. 216).

3 to 6 **Blue and Green Quartz** Dull blue opaque quartz aggregate [3].Transparent, faceted blue quartz has been on the market ever since 2005 which is called saphyr in Brazil. Blue color through treatment with gamma rays followed by heating from green quartz [5]. The latter receives its green color through the heating of amethyst. The blue and green quartz stones treated in this way are heat and light resistant. The only locality of its basic material is Minas Gerais/Brazil. Less light resistant are green quartz gemstones [6] from colorless quartz that comes from Uruguay and from the Rio Grande, as well as from Minas Gerais/Brazil. Color through gamma radiation. All green quartz gemstones are called prasiolite in the trade.

7 **Bytownite** Belongs to the group of plagioclase feldspar (p. 180). Colorless, gray, greenish. Color of streak white. Transparent. Mohs' hardness 6-6½. Density 2.72-2.74. Refractive index 1.565-1.574, double refraction 0.009. Triclinic, $(Ca,Na)\ [(Al,Si)_2Si_2O_8]$. Cleavage perfect.

8 **Synthetic Fresnoite** Brownish to orange. Color of streak white. Transparent. Mohs' hardness 3-4. Density 4.45. Refractive index 1.765-1.775, double refraction 0.010. Tetragonal, $Ba_2Ti[O|S_2O_7]$. Cleavage good. On the market since 2000. Natural fresnoite not suitable for jewelry.

9 **Shomiokite** Pale pink. Color of streak white. Transparent. Mohs' hardness 2-3. Density 2.64. Refractive index 1.530-1.539, double refraction 0.009. Orthorhombic, $Na_3(Y,Dy)[CO_3]_3 \cdot 3H_2O$. Cleavage perfect. Known in gemstone quality since 1999. Locality Kola Peninsula in Russia.

10 to 15 **Zoisite Varieties** Apart from tanzanite and thulite (p. 176), more recently also colorless and tinted zoisite stones [13-15] are known. Color of streak white. Transparent. Mohs' hardness 6-6½. Density 3.15-3.36. Refractive index 1.690-1.707, double refraction 0.008-0.009. Orthorhombic, $Ca_2\ Al_3[O|OH|SiO_4|Si_2O_7]$. Transparent zoisite [10-12] colored green through chromium on the market since 1990. According to IMA, it is supposed to be called green tanzanite.

16 to 18 **Mandarin Garnet** Trade name for vivid orange-colored spessartite (see p. 120). Color-giving substance is manganese. Most noble species come from northern Namibia.

19 **Tsavolite** Green to emerald green variety of the grossular garnet (see p. 122). Color-giving substances are chromium and vanadium.

20 **Rainbow Garnet** Trade name for iridescent andradite (see p. 122).

21 **Enstatite** Colorless variety, faceted. See also page 208.

22 **Californite** Greenish variety, cabochon. See also page 202.

23 to 27 **Moonstone Varieties** Belong to the feldspar group (p. 180). The body color can range from milky to slightly colored sheen. Different types on the market. So-called Ceylon moonstone colorless to white with bluish sheen (ill. p. 181). The Indian moonstone often has bold body colors with cat's-eye effect [24], sporadically also asterism. With rainbow moonstone [23] due to adularescence together with labradorescence, milky to bluish sheen with bluish and golden-yellow play of colors.

1
2
3
4
5
6
7
8
9
10
11
12
13
14
15
16
17
18
19
20
21
22
23
24
25
26
27

1 to 3 **Pezzotaite** Discovered first in Madagascar in 2002. Dark pink to rose-red. Color of streak white. Transparent to translucent. Mohs' hardness 8. Density 2.9-3.1. Refractive index 1.608-1.615, double refraction 0.007. Trigonal, $CsLiBe_2Al_2[Si_6O_{18}]$. Cleavage imperfect.

4 **Paraiba Tourmaline** Trade name for vivid colored blue, green, and blue-green elbaite tourmalines 5 (compare with p. 126). Caused by manganese and copper trace elements. First on the market in 1987. Locality originally only in the state of Paraiba (name), Brazil.

6 **Thorite** Brown to orange-yellow. Color of streak orange to brown. Transparent to opaque. Mohs' hardness 4½-5. Density 6.59. Refractive index 1.78-1.840, double refraction 0.06. Tetragonal, $(Th, U)[SiO_4]$. Radioactive! Cleavage good. First found in gemstone quality in 2001.

7 **Kyanite** Light green variety, octagon (compare also with p. 212).

8 **Heterosite** Brown to purple. Color of streak brown-red to purple. Opaque. Mohs' hardness 4-4½ . Density 3.2-3.4. Refractive index 1.86-1.91, double refraction 0.05. Orthorhombic, $(Fe^{3+}, Mn^{3+})[PO_4]$. On the market since 2006.

9 **Mellite** Honey to wax-yellow, also white. Color of streak white. Transparent. Mohs' hardness 2-2½. Density 1.58-1.60. Refractive index 1.509-1.541, double refraction 0.030. Tetragonal, $Al_2 [C_6 (COO)_6] \cdot 16H_2O$.

10 **Katoite** Belongs to garnet group (p. 120). Pink, green, also colorless, white. Color of streak white. Transparent to opaque. Mohs' hardness 6-7. Density 3.25-3.50. Cubic, $Ca_3Al_2 [(SiO_4)_{1\text{-}1.5} (OH)_{8\text{-}6}]$.

11 **Sphalerite** Dark green variety, hexagon (compare also p. 216).

12 **Richterite** Blue, gray purple. Color of streak white. Opaque. Mohs' hardness 5-6. Density 2.97-3.45. Refractive index 1.604-1.664, double refraction 0.016-0.020. Monoclinic, $(Na,K) (Ca,Na)_2 (Mg,Mn^{2+},Fe^{2+})_5[(OH,F)|Si_4O_{11}]_2$. Gemstone quality since the 1990s. Ill. 12 is light blue richterite in paragenesis with sugilite (p. 276).

13 **Chrysoberyl** Light blue-green variety. Color-giving substance is vanadium. Locality Tanzania. Compare also with p. 114.

14 **Lazulite Varieties** Many transitions from blue color (p. 208) to greenish shades. Grass-green lazulite 15 [15] first discovered in 1998.

16 **Synthetic Quartz** Green with induced fissures. Emerald substitute.

17 **Covelline** Indigo blue. Color of streak gray to black. Opaque. Mohs' hardness 1½-2. Density 4.60-4.76. Hexagonal, CuS.

18 **Bastnasite** Orange, brown. Color of streak white. Transparent. Mohs' hardness 4-4½. Density 4.78-5.20. Refractive index 1.72-1.82, double refraction 0.100. Hexagonal, $(La, Ce) [(F, OH)|CO_3]$. Cleavage perfect. On the gemstone market since 1998.

19 **Painite** Red. Color of streak white. Transparent. Mohs' hardness 7 ½-8. Density 4.01. Refractive index 1.787-1.816, double refraction 0.029. Hexagonal, $CaZrAl_9 [O_{15}|BO_3]$. Cleavage none. Gemstone quality since 2001.

20 **Danburite** Yellow variety, octagon. Compare also with page 198.

21 to 26 **Opal Varieties** Since the 1970s, opals have been found in Peru without the play of colors. They are called Andean Opals in the trade and belong to the group of common opals (page 168). Opaque blue, green-blue or pink [25]. Frequently dendritic with fern-like inclusions [26]. Green-blue stones can become milky. By letting them soak in water, they will re-gain their color. Cat's-eye is found with Andean Opals of strong color [22]. In many places, new opal deposits are discovered. Transparent opals are faceted [21, 24], translucent ones are used en cabochon [23].

1
2
3
4
5
5
6
7
8
9
10
11
12
13
14
15
16
17
18
19
20
21
22
23
24
25
26

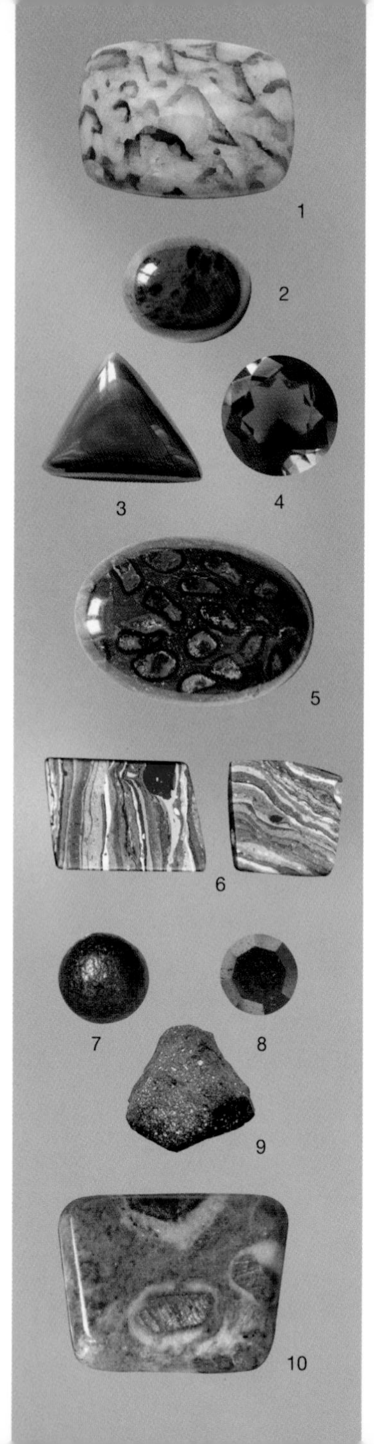

1 **Graphic Granite** Magmatic rock with inclusions of potash feldspar and rod-shaped quartz. The darker quartz pieces rest in light feldspar fields like oriental characters or Germanic runes. It is the darkness of the hollows that shines through the transparent quartz.

2 to 4 **Obsidian Varieties** Volcanic rock (compare with page 246) with numerous varieties suitable for decorative purposes. Rainbow obsidian [3] shimmers in several colors of the rainbow depending on one's point of view. Transparent, smoky obsidian is called marekanite in the trade. Apache Tears [4] is the name for a rounded, often transparent obsidian from Arizona (see also page 291).

5 **Star Stone/Petrified Wood** Name for wood found in petrified forest of the city of Chemnitz/Saxony (see page 164). Used for jewelry ornaments for the Saxon court in the second half of the 18th century.

6 **Rainbow Calsilica** Trade name for a multi-colored veined artificial stone. Consists of dyed sand that has been firmed with artificial resin. An imitation often turned to for various decorative stones.

7 to 9 **Carbonado** Granular aggregate of small diamond crystals, brown-black to black [9]. Natural diameter 0.31 in. (8 mm). Color of streak white. Opaque. Mohs' hardness 10. Density 3.07-3.45. Round formations [7] are called bort. Natural diameter 0.24 in. (6 mm). Origin unknown. Due to its toughness and unstructured mixture of the diamond crystals, a mechanical treatment is not possible. Most recently, they can be faceted with laser technique.

10 **Corundum-Fuchsite** Metamorphic rock with red ruby (see page 98), grey-blue kyanite (see page 212) in light green fuchsite matrix (see page 240). On the market ever since 2003.

Symbolic and Beneficial Stones

Gemstones represent something special due to their color, luster and form, but also due to the rarity of them. They have therefore always been surrounded by a touch of the mysterious. It is assumed that they possess powers protecting against injury from the outside, or for gaining inner strength, but especially also for medicinal purposes.

Cosmic-Astral Symbolic Stones

Gemstones are awarded symbolic meaning in several respects. Some states, for example, identify themselves with a precious stone mined in their country. Sometimes, precious stones are a symbol for power, status, and wealth. Frequently they are an amalgamation of a wishful imagination of supernatural powers and effective magic.

In relation to mystical notions about connections between man, earth, and cosmos, precious stones become a symbol for fascination and magic, and also become amulet and talisman.

Gemstones as Amulets and Talismans

All primitive races practice in some way warding off evil forces of nature and court the good spirits to be well disposed to them. With increasing cultural progression, precious stones are gaining a growing importance as amulets and talismans.

Amulets are worn on the body. They are to stave off harm, to repel the "evil eye." Sometimes, they are formed in a defensive posture with a finger that is straight or in any other way conspicuously pointed. Besides metal and ceramic, primarily coral serves as the material.

Raw and cut gemstones and pearls are also made into such amulets that are worn as protection, for example, on a neck chain, the arm, or the ear.

Devotionals and rosaries, in the last century often artfully produced from cut precious stones, are amulets blessed by the Church.

Talismans are not worn on the body, but kept in front of the house, in the apartment, or taken along in the car. They are to guarantee health, bring good luck, and be helpful in other ways.

Many gemstones, especially the larger ones such as agate aggregates, mineraloids, drusies, and geodes, are used as talismans. The scarabaeus, once regarded as holy in Egypt, is also nowadays cut as a talisman out of various gemstones.

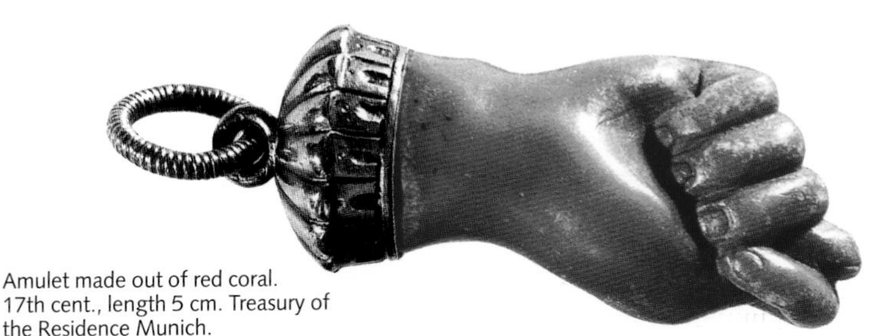

Amulet made out of red coral. 17th cent., length 5 cm. Treasury of the Residence Munich.

Planet Stones

In Antiquity, and also in the Middle Ages, the formation of gemstones was seen as an efflux of the stars. There are many colorful precious stones that glisten and sparkle in the same way as the stars emit light and colors. The cosmos reflects, so to speak, in the gemstones.

To every planet a gemstone is assigned magic qualities. No assignment system can be detected. Each historian of Antiquity reports different planet stones.

At the present, the planet stones are undergoing a renaissance, with numerous additions in the list of offerings.

Assignment of Gemstones to the Planets in Modern Literature

	Uyldert, 1983	Raphaell, 1987	Richardson/Huett, 1989	Ahlborn, 1996
Mercury	Citrine yellow Sapphire Topaz	Chrysocolla Turquoise	Garnet	Chrysolite, Heliotrope, Nephrite Tiger's-eye
Venus	Nephrite Rose quartz blue Sapphire Emerald	Kunzite, rose Tourmaline	Chrysoberyl Malachite, Moonstone, Pearl, Sapphire, Topaz	Malachite, Topaz Turquoise
Mars	Garnet, Ruby, Silex	Bloodstone, Carnelian	Bloodstone, Jasper, Onyx, Sardonix	Heliotrope, Carnelian, Rhodochrosite, Rhodonite, Tiger's-eye
Jupiter	Hyacinth, orange Carnelian	Azurite, Lapis lazuli	Jade, Turquoise	Citrine, Cross Stone Sardonyx
Saturn	Gagate, Onyx, Spinel	Malachite, Peridot, green Tourmaline	Quartz, Tiger's-eye	Chalcedony
Uranus	Amazonite, Malachite, Turquoise	Aquamarine, Chrysoberyl, Coelestine		
Neptune	Amethyst, Opal	Amethyst, Fluorite	Aquamarine, Azurite, Diamond, Coral, Moonstone, Opal Quartz, Spinel, Tourmaline	
Pluto	Almandine, Bloodstone, Pyrope	Garnet, Obsidian, Smoky quartz, Ruby	Amethyst, Jade, Kunzite, Spinel, Zircon	

Zodiac Stones

Similarly to the planet stones, there existed in Antiquity belief in the interrelation of certain constellations, the so-called zodiac signs, to man and gemstones.

These ideas were further nourished in the Middle Ages, and with many people they are nowadays equally topical. Jewelers and dealers in precious stones add to these notions in order to advance their business.

Here too, there is no recognizable coordination system of the gemstones, nor did one ever exist. The gemstones of the zodiac signs listed on the following page were taken from contemporary literature.

Zodiac Stones

Zodiac signs	Illustration	Additional Zodiac Stones
Aries 3/21–4/20	Red Jasper, Carnelian	Bloodstone, Chalcedony, Chrysoprase, Ruby, Silex
Taurus 4/21–5/20	Carnelian, Rose quartz	Golden Topaz, Coral, Lapis lazuli, Quartz, Sapphire, Sard, Emerald
Gemini 5/21–6/21	Citrine, Tiger's-eye	Agate, Aquamarine, Rock crystal Chrysocolla, Jasper, Onyx, Topaz, Turquoise
Cancer 6/22–7/22	Chrysoprase, Aventurine	White Chalcedony, Chrysolite, Diamond, Carnelian, Moonstone, Rhodochrosite, Emerald
Leo 7/23–8/23	Rock crystal Goldquartz	Almandine, Amber, Chrysolite, Citrine, Diamond, Carnelian, Onyx, Ruby, Sulfur
Virgo 8/24–9/23	Citrine, yellow Agate	Amazonite, Beryl, Jasper, Carnelian, Sardonyx, Turquoise, Zircon
Libra 9/24–10/23	Orange Citrine, Smoky quartz	Aventurine, Beryl, Diamond, Jade, Cunzite, Nephrite, Opal, Peridot, Sardonyx, Emerald, Topaz, Rose Tourmaline
Scorpio 10/24–11/22	Deep red Carnelian	Agate, Aquamarine, Chalcedony, Chrysoprase, Garnet, Obsidian, Smoky quartz, Ruby, Topaz
Sagittarius 11/23–12/21	Sapphire, Chalcedony	Amethyst, Rock crystal, Beryl, Garnet, Pyrope, Sapphire quartz, Sodalite, Spinel, Topaz
Capricorn 12/22–1/20	Onyx, Quartz–Cat's-Eye	Amethyst, Beryl, Gagate, Malachite, Obsidian, Peridot, Smoky quartz, Rose quartz, Ruby, green Tourmaline
Aquarius 1/21–2/19	Turquoise, Hawk's-eye	Amazonite, Amethyst, Aquamarine, Chrysocolla, Coelestine, Garnet, Jasper, Malachite, Obsidian, blue Sapphire
Pisces 2/20–3/20	Amethyst Amethyst Quartz	Aquamarine, Blue-quartz, Diamond, Jade, Moonstone, Opal, Sapphire, Sugilite

Gemstones of the Months

January
Garnet
Rose quartz

February
Amethyst
Onyx

March
Tourmaline
Blood Jasper

April
Sapphire
Diamond
Rock crystal

May
Emerald
Chrysoprase

June
Pearl
Moonstone

July
Ruby
Carnelian

August
Onyx
Sardonyx

September
Peridot

October
Aquamarine
Opal

November
Topaz
Tiger's-eye

December
Zircon
Turquoise

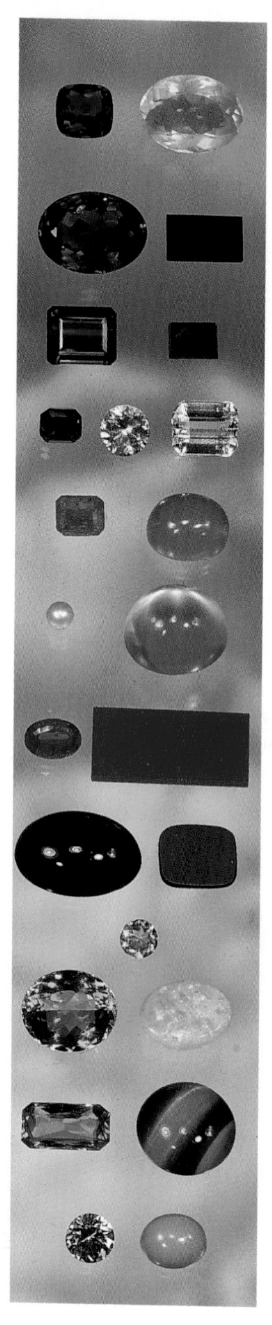

Gemstones of the Months

Originally, the respective zodiac signs were regarded also as the birthstones. They were either amulets or talismans. Only in more recent times, gemstones of the months are according to the calendar also propagated as birthstones, aside from the zodiac gemstones. Here again, arbitrarily assembled gemstone groups are supposed to unfold magic effects. Whenever people feel called upon to convey their own mythological-astrological interpretations, various gemstones are always named as indicators of magic effect. The German poet Theodore Koerner (1791–1813) names in his poem "The Stones of the Months" the following assignments of gemstones to the months:

January	Hyacinth
February	Amethyst
March	Heliotrope
April	Sapphire
May	Emerald
June	Chalcedony
July	Carnelian
August	Onyx
September	Chrysolite
October	Aquamarine
November	Topaz
December	Chrysoprase

Currently, gemstones are also offered for the times of the year, for spring, summer, fall, and winter, and even for every day of the week.

In a publication of 1985, the following recommendation for gemstones for the days of the week can be found:

Sunday	Amber, Gold Topaz
Monday	Moonstone, Pearl
Tuesday	Ruby, Garnet
Wednesday	Turquoise, Sapphire, Lapis lazuli
Thursday	Amethyst
Friday	Emerald, Malachite
Saturday	Diamond

Medicinal Stones

Just as the gemstones are regarded as symbols for a supernatural link of man to the sun, the moon, and the stars, they also symbolize magic and curative science.

Powers attributed to the gemstones can prevent or heal illnesses and mitigate or eliminate other infirmities. No authentication or scientific proof exists for the curative powers of gemstones.

Historical Survey

Written accounts about healing or the prevention of illnesses with gemstones have been extant since Antiquity. Eminent writers of the old Age, such as Aristotle, Gajus Plinius Secundus, Dioscurides, and later Marbod, Albertus Magnus, and Konrad von Megenberg, wrote about it.

Under the influence of the Church, the curative effects of the gemstone described in the Antiquity were in part given new meanings to eliminate influences of paganism, and to lead the faithful to an ethical mode of life.

Over centuries, the healing prescriptions by the Abbess Hildegard von Bingen (1098-1179), written down in her book *Physica*, found great acceptance. This gemstone medicinal science (lithotherapy) is experiencing a resurrection. Whole bookstore shelves are filled with discourses about the "Hildegard medicine."

Since according to Hildegard's views, gemstones form through the combined actions of fire and water, they possess powers corresponding to those natural phenomena. In addition, there is God's influence, which wants the gemstones to be seen as a blessing for that which is honorable and useful.

Hildegard discusses at length 24 gemstones and their medicinal effects; some other stones are mentioned in passing. An example about the sapphire follows:

> Who is dull and would like to be clever, should, in a sober state, frequently lick with the tongue on a sapphire, because the gemstone's warmth and power, combined with the saliva's moisture, will expel the harmful juices that affect the intellect. Thus, the man will attain a good intellect.

Hildegard von Bingen (1098–1179). Statue in the middle-shrine of the Hildegard-Altar, in the Rochus Chapel, Bingen (Germany).

Healing Stones Nowadays

If healing gemstones are mentioned these days, it does not mean that illnesses are cured with gemstones, but that—according to the beliefs of the users—all kinds of negative effects on man can be influenced positively.

Powers and Effects of the Gemstones

All gemstones, a few other stones, and even solidified organogenic (developed from organic materials) products, are used.

The energetic powers ascribed to them by their users stem from Mother Earth during the formation of the stones, or from the sun's warming rays. Therefore, it is recommended to put from time to time healing stones into the sun, or expose them to the (not really warming) moonlight, so that they can replenish the energy that has meanwhile been used. Example: "It is most effective if the stones are put on the terrace or the window sill two nights before full moon."

Another manner of energy-replenishment can be achieved by "burying [the stone] overnight in the ground and afterwards rinsing it for a short time." Synthetic stones may not be used as healing stones. "Only the genuine gemstones grown in nature in the course of millions of years have stored the powers of nature. The artificially produced gemstones lack the soul that has mastery of all."

The much-praised powers of the healing stones have not yet been scientifically proved. All comments such as "proved manifold," "carefully tested," "researched over years," and "scientifically documented" are sheer assertions.

The question as to whether any side elements of the gemstones and the trace elements can influence the human body and the psyche is completely open at the present time. Therefore, traditional medicine rejects a therapy with gemstones.

Without any doubt, there are certain healing results with lithotherapy (treatment with stones). But that probably is not due to the supposed medicinal powers of the gemstones; it is a so-called placebo effect, an imaginary effect, a success by suggestion. The lithotherapy obviously is in general nothing else but a "psychotherapy."

Drum Stones, i.e., gemstones rounded on all sides. Above, from the left: Amethyst quartz, Rose quartz, Aventurine. Below, from the left: Rhodonite, Larimar, Agate.

Preparation and Usage of the Gemstones

In order to maximize the effect of the healing stones, at times mystical and symbolic signs are engraved on the stone, as was practiced in Antiquity and the Middle Ages. The shape of the gemstones may also be useful in the curative treatment, according to therapists. As a pendulum, for example, rock crystals are used that are formed as symmetrical as possible, but also quartz and other gemstones that have been shaped. A pronounced tip is supposed to increase the success of an individual search for effective stones, and in diagnosing sicknesses of the body.

Very popular are baroque or drum stones rounded on all sides—the so-called fondling-stones that are held in the hand, or carried in a pocket. Yet any other natural or cut form of a gemstone can supposedly be just as helpful.

The use of the gemstones is direct, by skin contact, ground in the form of powder or as a pill, as a means for meditation, or indirect as an essence for curative drinks, and in gemstone elixir for poultices. Stomachaches are helped by ruby-water, and for protection and illumination one is to take diamond- and rock crystal-water (according to an author in 1985).

According to a recent publication, for beneficial results the following is required: "Discharging of the stone before the treatment (by rinsing with cold water, or burying for some days), especially with newly bought, gifted, or inherited stones."

Sometimes, modern literature about beneficial stones is a reflection of the darkest Middle Ages. The borderline between magic, sorcery, fantasy, and therapeutic healing is not recognizable in lithotherapy.

Pendulums with a pronounced tip used in healing with gemstones. From the left: Rock crystal (smoothed), Citrine (untreated), Sugulite (cut), Smoky topaz (smoothed).

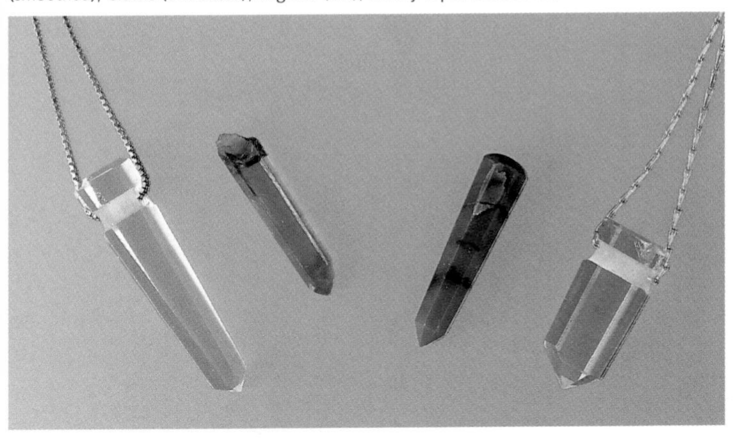

Healing with Gemstones

Examples of illnesses that allegedly can be healed. Assembled from current literature (selection).

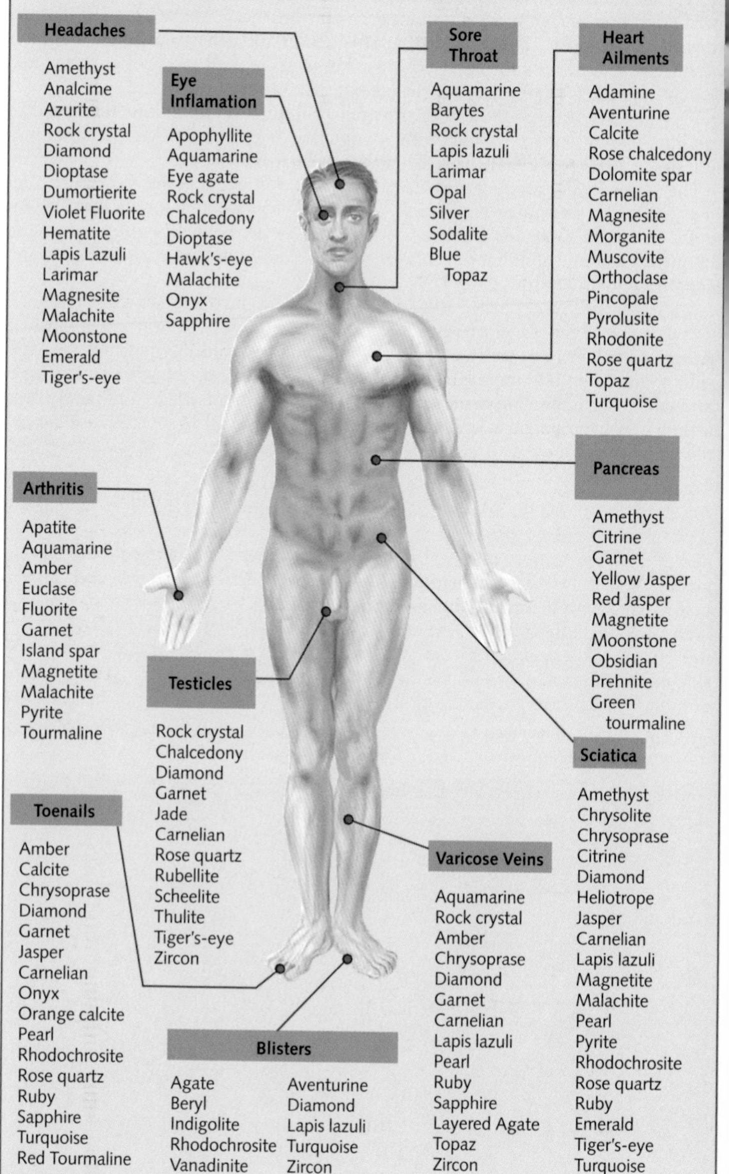

Apache tears, Arizona/USA. Natural size.

Apache Tears Trade name for rounded or drop-like shapes for the rock glass obsidian (see page 246). In prehistoric times it is believed to have been used as an amulet. It was a healing stone among Native Americans. Today, it is a common healing stone. "Apache Tears help with indigestion as well as stomach and intestinal problems and even cure stomach ulcers," as contemporary literature of 1997 states.

Amulet Stones (see ill. below) Contemporary literature describes agate as particularly effective against the evil eye. Sugilite allegedly has a harmonizing effect on nerves and the brain. Jasper protects against hyper-functioning of the thyroid gland. Women between the ages of 30 and 40 in particular are advised to carry beige jasper with them. Sodalite gives the body more strength and strengthens immunity against infectious diseases and inflammations. Snowflake obsidian protects through its earthy property from cold feet. Fluorite stimulates the regeneration of skin and mucous membranes and strengthens bones and teeth. The various shapes are supposed to enhance the effect of the stones.

Top from left: Agate, Sugilite, Jasper. Bottom from left: Sodalite, Snowflake Obsidian, Fluorite.

Moqui-Marbles
Trade name for a globular limonite aggregate from Arizona and Utah. These marbles are intensified in their effect through touch, closeness of body, tenderness, affection, and light. They allegedly loosen congestions and blockages and create in people a greater feeling of brotherly love and togetherness. It has its greatest effect as a stone pair with the more globular female stone shapes (left) and the more angular male ones. Slightly scaled down.

Sky Stone
Trade name for a fine-grained mesentery layered light-colored limestone (not gypsum, as stated in literature!). Serves as a healing stone against various types of indisposition: as a pyramid put under the bed, it is especially effective against sleep deprivation; when pulverized, it is also effective against joint pains. Solely mined in the Salzburg, Austria, area. 50 percent scaled down.

Boji Stones
Trade name for a bulbous aggregate from pyrite and limonite. Only place of discovery is Kansas, USA. These stones are said to have energy centers that can be "activated through stroking." They develop their power best as pairs, that is, with the so-called male (more angular, left) and the female (smoother, arched, right) stones. Natural size.

Pyrite Sun
Disc-shaped, radial pyrite aggregate (see page 178). Serves as costume jewelry and as a healing stone. In an esoteric book of 1997, among other things, one can read the following: "The pyrite sun cures stomach disease and indigestion. When worn in silver around the neck, it strengthens the immune system immensely." Its only occurence is in the coal mines of Sparta, Illinois, USA.
Diameter of the original, 8 cm.

Bibliography

Anderson, B.W. 1976. *Gemstones for Everyman*. Van Nostrand Reinhold Co., New York.
Anderson, B.W. 1990. *Gem Testing*. Butterworths, London.
Anthony, J.W., Bideaux, R.A., Bladh, K.W. and M.C. Nichols 1990/2003. *Handbook of Mineralogy*. 5 vol. Mineral Data Publ., Tucson.
Arem, J.E. 1987. *Color Encyclopedia of Gemstones*. Van Nostrand Reinhold Co., New York.
Bancroft, P. 1984. *Gem and Crystal Treasures*. Western Enterprises, Tucson.
Brocardo, G. 1982. *Minerals and Gemstones: An Identification Guide*. Hippocrene Books INC., New York.
Elwell, D. 1979. *Man-Made Gemstones*. Ellis Horwood, Chichester.
Federman, D. 1990. *Consumer Guide to Colored Gemstones*. Van Nostrand Reinhold, New York.
Gill, J.O. 1978. *Gill's Index to Journals, Articles and Books relating to Gems and Jewelry*. Gemological Institute of America, Santa Monica.
Hughes, R.W. 1997. *Ruby and Sapphire*. RWH Publishing, Boulder.
Hurlbut, C.S. 1991. *Gemology*. John Wiley and Sons. New York, USA.
Kievlenko, E.Y. 2003. *Geology of Gems*. Ocean Pictured Ltd.
Liddicoat, R.T. 1993. *Handbook of Gem Identification*. Gemological Institute of America, Santa Monica.
Lieber, W. 1994. *Amethyst*. Christian Weise Verlag, Munich.
Mandarino, J.A. and M.E. Back 2004. *Fleischer's Glossary of Mineral Species*. The Mineral Record Inc., Tucson.
Miller, A.M. and J. Sinkankas 1994. *Standard Catalog of Gem Values*. Geo-science Press, Tucson.
Nassau K. 1980. *Gems Made By Man*. Chilton Book Co., Radnor.
Nassau K. 1996. *Gemstone Enhancement*. Butterworth Heinemann, Oxford.
Palache, C., Berman, H. and C. Frondel 1966. *Dana's System of Mineralogy*. 3 Vol. Wiley, New York.
Read, P.G. 1978. *Gemmological Instruments*. Newnes-Butterworths, London.
Read, P.G. 1994. *Dictionary of Gemmology*. Butterworth-H, Oxford.
Read, P.G. 1997. *Gemmology*. London.
Roberts, W.L., Campbell, T.J. and G.R. Rapp 1990. *Encyclopedia of Minerals*. Van Nostrand Reinhold Company, New York.
Rose, J.D. 1986. *Garnet*. Butterworths, London.
Sevdermish, M. and A. Mashiah 1996. *The Dealer's Book of Gems and Diamonds*. Kal Printing House, Israel.
Shipley, R.M. 1974. *Dictionary of Gems and Gemology*. Gemological Institute of America, Santa Monica.
Shirai, S. 1970. *The Story of Pearls*. Japan Publications, Inc. Tokyo.
Sinkankas, J. 1984. *Gem Cutting*. Chapman and Hall, New York.
Sinkankas J. 1989. *Emerald and Other Beryls*. Chilton Book, Radnor.
Webster, R. 1976. *Practical Gemmology*. N.A.G. Press Ltd., London.
Webster, R. 1998. *The Gemmologists' Compendium*. N.A.G., London.
Webster, R. 1997. *Gems*. Butterworths, London.
Zeitner, J.C. 1996. *Gem and Lapidary Materials for Cutters, Collectors and Jewellers*. Geoscience Press, Tucson.

Journals

The American Mineralogist. Mineralogical Society of America, Washington.
The Australian Gemmologist. Gemmological Association of Australia, Sidney.
The Australian Mineralogist. Mineralogical Society of Australia, Sidney.
The Canadian Mineralogist. Mineralogical Association of Canada, Ottawa.
Gems and Gemology. Gemological Institute of America, Los Angeles.
The Journal of Gemmology. Gemmological Association of Great Britain.
Lapidary Journal. San Diego.
The Mineralogical Magazine. Mineralogical Society of Great Britain, London.
The Mineralogical Record. Tucson.
Rocks and Minerals. Washington.

Table of Constants

How to Use the Table

This table has been compiled to aid both the professional and the amateur to identify individual gemstones. It is designed to be used in conjunction with the standard gemological tests. It is not totally comprehensive, but it should enable the tester to eliminate some possibilities and perhaps suggest some, if he or she is faced with one of the more unusual examples.

Suppose that you have an unknown yellow gemstone to identify. Using the Table of Constants, this is how you go about identifying it:

1 Turn to pages 302-303, which cover yellow, orange, and brown stones
2 Test for specific gravity (pages 25-27); this gives you a density of 3.65.
3 Test the refractive index (pages 36-37); this gives a reading of 1.738.
4 Turn to page 303 and run down the 1.700-1.799 column and across the 3.50-3.99 line until they meet. This gives you several possibilities.
5 Test for double refraction (page 40). If there is none, the field is narrowed to periclase or two members of the garnet group (pyrope and grossularite).
6 Turning to the garnet section in the main text, the tester will observe that pyrope and grossularite have distinctive absorption spectra. Using the spectroscope (pages 44-47), you can identify the gem if it shows one of these spectra.

Gem Colors: White + Colorless + Gray

Density \ Refr. index	1.400–1.499	1.500–1.599	1.600–1.699
1.00-1.99	Ulexite 2-2½ I 0.029 Gaylussite 2½-3 I 0.080 Kurnakovite 3 I 0.036 Opal 5½-6½ I -	Ulexite 2-2½ I 0.029 Amber 2-2½ I - Ivory 2-3 I - Gaylussite 2½ - 3 I 0.080	
2.00-2.49	Yugawaralite 4½ I 0.012 Analcime 5-5½ I - Natrolite 5-5½ I 0.013 Obsidian 5-5½ I - Cancrinite 5-6 I 0.026 Haüyn 5½-6 I - Sodalite 5½-6 I - Opal 5½-6½ I -	Sepiolite 2-2½ I - Ivory 2-3 I - Colemanite 4½ I 0.029 Apophyllite 4½-5 I 0.002 Haüyn 5½-6 I - Leucite 5½-6 I 0.001 Petalite 6-6½ I 0.014 Hambergite 7½ I 0.072	Whewellite 2½-3 I 0.161 Howlite 3-3½ I 0.019 Colemanite 4½ I 0.029 Hambergite 7½ I 0.072
2.50-2.99	Calcite 3 I 0.172 Coral 3-4 I 0.166 Creedite 3½-4 I 0.024 Onyx Marble 3½-4 I 0.163 Obsidian 5-5½ I - Cancrinite 5-6 I 0.026 Haüyn 5½-6 I - Opal 5½-6½ I -	Vivianite 1½-2 I 0.062 Pearl 2½-4½ I 0.156 Calcite 3 I 0.172 Howlite 3-3½ I 0.019 Coral 3-4 I 0.166 Anhydrite 3½ I 0.044 Aragonite 3½-4 I 0.155 Dolomite 3½-4 I 0.185 Augelite 4½-5 I 0.017 Beryllonite 5½-6 I 0.009 Leucite 5½-6 I 0.001 Scapolite 5½-6 I 0.021 Sanidine 6-6½ I 0.008 Labradorite 6-6½ I 0.009 Moonstone 6-6½ I 0.008 Rock Crystal 7 I 0.009 Smoky Quartz 7 I 0.009 Precious Beryl 7½-8 I 0.007	Vivianite 1½-2 I 0.062 Pearl 2½-4½ I 0.156 Calcite 3 I 0.172 Howlite 3-3½ I 0.019 Coral 3-4 I 0.166 Anhydrite 3½ I 0.044 Aragonite 3½-4 I 0.155 Dolomite 3½-4 I 0.185 Onyx Marble 3½-4 I 0.163 Magnesite 3½-4½ I 0.208 Datolite 5-5½ I 0.045 Tremolite 5-6 I 0.022 Meionite 5½-6 I 0.030 Nephrite 6-6½ I 0.027 Danburite 7-7½ I 0.007 Tourmaline 7-7½ I 0.023 Precious Beryl 7½-8 I 0.007 Phenakite 7½-8 I 0.016
3.00-3.49	Fluorite 4 I -	Phosphophyllite 3-3½ I 0.027 Magnesite 3½-4½ I 0.208 Herderite 5-5½ I 0.027 Meliphanite 5-5½ I 0.019 Tremolite 5-6 I 0.022 Montebrasite 5½-6 I 0.22 Amblygonite 6 I 0.027	Magnesite 3½-4½ I 0.208 Apatite 5 I 0.004 Hemimorphite 5 I 0.022 Datolite 5-5½ I 0.045 Diopside 5-6 I 0.027 Enstatite 5½ I 0.010 Amblygonite 6 I 0.027 Nephrite 6-6½ I 0.027 Jadeite 6½-7 I 0.020 Danburite 7-7½ I 0.007 Tourmaline 7-7½ I 0.023 Dumortierite 7-8½ I 0.026 Euclase 7½ I 0.022
3.50-3.99		Barytocalcite 4 I 0.061	Celestine 3-3½ I 0.011 Hemimorphite 5 I 0.022 Willemite 5½ I 0.030 Topaz 8 I 0.012
4.00-4.99		Witherite 3-3½ I 0.148	Baryte 3-3½ I 0.012 Celestine 3-3½ I 0.011 Witherite 3-3½ I 0.148 Willemite 5½ I 0.030

The numbers following the gemstone refer to Mohs' hardness I double refraction

<table>
<tr><th>Refr. index
Density</th><th>1.700-1.799</th><th>1.800-1.899</th><th>1.900 and higher</th></tr>
<tr><td>1.00-1.99</td><td></td><td></td><td></td></tr>
<tr><td>2.00-2.49</td><td></td><td></td><td></td></tr>
<tr><td>2.50-2.99</td><td>Magnesite 31/2-41/2 | 0.208</td><td></td><td></td></tr>
<tr><td>3.00-3.49</td><td>Magnesite 3½-4½ | 0.208
Bronzite 5-6 | 0.015
Diopside 5-6 | 0.027
Zoisite 6-6 ½ | .008
Katoite 6-7|-
Clinozoisite 6-7 | 0.010
Diaspore 6½-7 | 0.048
Chambersite 7 | 0.012
Sapphirine 7½ | 0.005</td><td></td><td></td></tr>
<tr><td>3.50-3.99</td><td>Kyanite 4-7 | 0.024
Legrandite 4½-5 | 0.060
Willemite 5½ | 0.030
Periclase 5½-6 | -
Benitoite 6-6½ | 0.047
Grossular 6½-7½ | -
Sapphirine 7½ | 0.005
Taaffeite 8-8½ | 0.006
Sapphire 9 | 0.008</td><td>Benitoite 6-6½ | 0.047
Low Zircon 6½-7½ | -
Zircon 6½-7½ | 0.030</td><td>Sphalerite 3½-4 | -
Anatase 5½-6 | 0.056
Low Zircon 6½-7½ | -
Zircon 6½-7½ | 0.030
Diamond 10 | -</td></tr>
<tr><td>4.00-4.99</td><td>Adamite 3½ | 0.049
Legrandite 4½-5 | 0.060
Monazite 5-5½ | 0.052
Willemite 5½ | 0.030
Sapphire 9 | 0.008</td><td>Monazite 5-5½ | 0.052
Low Zircon 6½-7½ | -
Zircon 6½-7½ | 0.030
YAG 8½ | -</td><td>Powellite 3½-4 | 0.011
Sphalerite 3½-4 | -
Linobate 5½ | 0.090
Low Zircon 6½-7½ | -
Zircon 6½-7½ | 0.030</td></tr>
<tr><td>5.00-5.99</td><td>Monazite 5-5½ | 0.052</td><td>Monazite 5-5½ | 0.052</td><td>Senarmontite 2-2½ | -
Scheelite 4½-5 | 0.014
Hematite 5½-6½ | 0.287
Fabulite 5½-6 | -
Simpsonite 7-7½ | 0.058 |
Zirconia 8½ | -</td></tr>
<tr><td>6.00-6.99</td><td></td><td>Anglesite 3-3½ | 0.017
Cerussite 3-3½ | 0.274</td><td>Phosgenite 2-3 | 0.028
Vanadinite 2½-3 | 0.066
Cerussite 3-3½ | 0.274
Scheelite 4½-5 | 0.014
Cassiterite 6-7 | 0.097
Zirconia 8½ | -</td></tr>
<tr><td>7.00 and higher</td><td></td><td></td><td>Cinnabar 2-2½ | 0.351
Stolzite 2½-3 | 0.08
Vanadinite 2½-3 | 0.066
Mimetesite 3½-4 | 0.015
Cassiterite 6-7 | 0.097
GGG 6½ | -</td></tr>
</table>

Gem Colors: Red + Pink + Orange

<table>
<tr><th>Refr. index
Density</th><th>1.400–1.499</th><th>1.500–1.599</th><th>1.600–1.699</th></tr>
<tr><td>1.00-1.99</td><td>Inderite 2½-3 | 0.018
Kurnakovite 3 | 0.036
Opal 5½-6½ | -</td><td>Amber 2-2½ | -
Kurnakovite 3 | 0.036
Opal 5½-6½ | -</td><td></td></tr>
<tr><td>2.00-2.49</td><td>Analcime 5-5½ | -
Natrolite 5-5½ | 0.013
Cancrinite 5-6 | 0.026
Tugtupite 5½-6 | 0.006
Opal 5½-6½ | -</td><td>Stichtite 1½-2½ | 0.026
Gypsum 2 | 0.009
Sepiolite 2-2½ | -
Apophyllite 4½-5 | 0.002
Thomsonite 5-5½ | 0.015
Cancrinite 5-6 | 0.026
Tugtupite 5½-6 | 0.006
Opal 5½-6½ | -
Petalite 6-6½ | 0.014</td><td></td></tr>
<tr><td>2.50-2.99</td><td>Calcite 3 | 0.172
Coral 3-4 | 0.166
Cancrinite 5-6 | 0.026
Tugtupite 5½-6 | 0.006 |
Opal 5½-6½ | -</td><td>Pearl 2½-4½ | 0.156
Calcite 3 | 0.172
Coral 3-4 | 0.166
Anhydrite 3½ | 0.044
Aragonite 3½-4 | 0.155
Dolomite 3½-4 | 0.185
Apophyllite 4½-5 | 0.002
Cancrinite 5-6 | 0.026
Scapolite 5½-6 | 0.021
Tugtupite 5½-6 | 0.006
Opal 5½-6½ | -
Petrified Wood 5½-7 | -
Orthoclase 6-6½ | 0.008
Sunstone 6-6½ | 0.010
Jasper 6½-7 | -
Amethyst 7 | 0.009
Aventurine 7 | 0.009
Rose Quartz 7 | 0.009
Precious Beryl 7½-8 | 0.007</td><td>Kämmererite 2-2½ | 0.003
Muscovite 2-3 | 0.039
Pearl 2½-4½ | 0.156
Calcite 3 | 0.172
Coral 3-4 | 0.166
Anhydrite 3½ | 0.044
Aragonite 3½-4 | 0.155
Dolomite 3½-4 | 0.185
Eudialyte 5-5½ | 0.006
Tremolite 5-6 | 0.022
Nephrite 6-6½ | 0.027
Danburite 7-7½ | 0.007
Tourmaline 7-7½ | 0.023
Precious Beryl 7½-8 | 0.007
Phenakite 7½-8 | 0.016
Pezzottaite 8 | 0.007</td></tr>
<tr><td>3.00-3.49</td><td>Fluorite 4 | -</td><td>Meliphanite 5-5½ | 0.019
Tremolite 5-6 | 0.022
Chondrodite 6-6½ | 0.031</td><td>Rhodochrosite 4 | 0.214
Apatite 5 | 0.004
Nephrite 6-6½ | 0.027
Kunzite 6½-7 | 0.015
Jadeite 6½-7 | 0.020
Danburite 7-7½ | 0.007
Tourmaline 7-7½ | 0.023
Dumortierite 7-8½ | 0.026
Andalusite 7½ | 0.010
Topaz 8 | 0.012
Rhodizite 8-8½ | -</td></tr>
<tr><td>3.50-3.99</td><td></td><td>Strontianite 3½ | 0.150</td><td>Celestine 3-3½ | 0.011
Siderite 3½-4½ | 0.242
Rhodochrosite 4 | 0.214
Willemite 5½ | 0.030
Topaz 8 | 0.012</td></tr>
<tr><td>4.00-4.99</td><td></td><td></td><td>Baryte 3-3½ | 0.012
Celestine 3-3½ | 0.011
Smithsonite 5 | 0.228
Willemite 5½ | 0.030</td></tr>
</table>

The numbers following the gemstone refer to Mohs' hardness | double refraction

Refr. index. / Density	1.700–1.799	1.800–1.899	1.900 and higher
1.00-1.99			
2.00-2.49			
2.50-2.99			
3.00-3.49	Rhodochrosite 4 | 0.214 Bustamite 5½-6 | 0.014 Rhodonite 5½-6½ | 0.012 Clinozoisite 6-7 | 0.010	Rhodochrosite 4 | 0.214 Purpurite 4-4½ | 0.07	Purpurite 4-4½ | 0.07
3.50-3.99	Siderite 3½-4½ | 0.242 Rhodochrosite 4 | 0.214 Hodgkinsonite 4½-5 | 0.024 Willemite 5½ | 0.030 Rhodonite 5½-6½ | 0.012 Benitoite 6-6½ | 0.047 Almandine 6½-7½ | - Hessonite 6½-7½ | - Pyrope 6½-7½ | - Rhodolite 6½-7½ | - Spinel 8 | - Taaffeite 8-8½ | 0.006 Alexandrite 8½ | 0.009 Chrysoberyl 8½ | 0.009 Ruby 9 | 0.008 Sapphire 9 | 0.008	Siderite 3½-4½ | 0.242 Rhodochrosite 4 | 0.214 Titanite 5-5½ | 0.146 Benitoite 6-6½ | 0.047 Almandine 6½-7½ | - Low Zircon 6½-7½ | - Zircon 6½-7½ | 0.030	Sphalerite 3½-4 | - Titanite 5-5½ | 0.146 Anatase 5½-6 | 0.056 Low Zircon 6½-7½ | - Zircon 6½-7½ | 0.030 Diamond 10 | -
4.00-4.99	Smithsonite 5 | 0.228 Monazite 5-5½ | 0.052 Willemite 5½ | 0.030 Almandine 6½-7½ | - Spessartine 6½-7½ | - Gahnite 7½-8 | - Painite 7½-8 | 0.029 Ruby 9 | 0.008 Sapphire 9 | 0.008	Smithsonite 5 | 0.228 Monazite 5-5½ | 0.052 Almandine 6½-7½ | - Spessartine 6½-7½ | - Low Zircon 6½-7½ | - Zircon 6½-7½ | 0.030 Gahnite 7½-8 | - Painite 7½-8 | 0.029 Sphaerocobaltite 4 | 0.254	Greenockite 3-3½ | 0.023 Sphalerite 3½-4 | - Microlite 5-5½ | - Rutile 6-6½ | 0.287 Low Zircon 6½-7½ | - Zircon 6½-7½ | 0.030
5.00-5.99	Bastnasite 4-4½ | .100 Monazite 5-5½ | 0.052	Bastnasite 4-4½ | .100 Monazite 5-5½ | 0.052	Proustite 2½ | 0.203 Crocoite 2½-3 | 0.270 Cuprite 3½-4 | - Zincite 4-5 | 0.016 Scheelite 4½-5 | 0.014 Hematite 5½-6½ | 0.287 Tantalite 6-6½ | 0.160
6.00-6.99			Crocoite 2½-3 | 0.270 Wulfenite 3 | 0.120 Cuprite 3½-4 | - Scheelite 4½-5 | 0.014 Tantalite 6-6½ | 0.160
7.00 and higher			Cinnabar 2-2½ | 0.351 Stolzite 2½-3 | 0.08 Vanadinite 2½-3 | 0.066 Wulfenite 3 | 0.120 Hübnerite 4-4½ | 0.13 Tantalite 6-6½ | 0.160

Gem Colors: Yellow + Orange + Brown

<table>
<tr><th>Refr. index
Density</th><th>1.400–1.499</th><th>1.500–1.599</th><th>1.600–1.699</th></tr>
<tr><td>1.00-1.99</td><td>Gaylussite 2½-3 | 0.080
Opal 5½-6½ | -</td><td>Amber 2-2½ | -
Ivory 2-3 | -</td><td>Jet 2½-4 | -</td></tr>
<tr><td>2.00-2.49</td><td>Whewellite 2½-3 | 0.161
Natrolite 5-5½ | 0.013
Obsidian 5-5½ | -
Cancrinite 5-6 | 0.026
Moldavite 5½ | -
Opal 5½-6½ | -</td><td>Sepiolite 2-2½ | -
Serpentine 2½-5½ | 0.011
Apophyllite 4½-5 | 0.002
Obsidian 5-5½ | -
Leucite 5½-6 | 0.001
Petalite 6-6½ | 0.014
Hambergite 7½ | 0.072</td><td>Whewellite 2½-3 | 0.161
Hambergite 7½ | 0.072</td></tr>
<tr><td>2.50-2.99</td><td>Calcite 3 | 0.172
Onyx Marble 3½-4 | 0.163
Obsidian 5-5½ | -
Cancrinite 5-6 | 0.026
Opal 5½-6½ | -</td><td>Pearl 2½-4½ | 0.156
Serpentine 2½-5½ | 0.011
Calcite 3 | 0.172
Aragonite 3½-4 | 0.155
Obsidian 5-5½ | -
Beryllonite 5½-6 | 0.009
Scapolite 5½-6 | 0.021
Moonstone 6-6½ | 0.008
Orthoclase 6-6½ | 0.008
Sanidine 6-6½ | 0.008
Sunstone 6-6½ | 0.010
Tiger's-Eye 6½-7 | -
Aventurine 7 | 0.009
Citrine 7 | 0.009
Smoky Quartz 7 | 0.009
Cordierite 7-7½ | 0.010
Precious Beryl 7½-8 | 0.007</td><td>Muscovite 2-3 | 0.039
Pearl 2½-4½ | 0.156
Calcite 3 | 0.172
Aragonite 3½-4 | 0.155
Dolomite 3½-4 | 0.185
Onyx Marble 3½-4 | 0.163
Datolite 5-5½ | 0.045
Tremolite 5-6 | 0.022
Brazilianite 5½ | 0.020
Meionite 5½-6 | 0.030
Vlasovite 6 | 0.020
Nephrite 6-6½ | 0.027
Prehnite 6-6½ | 0.030
Danburite 7-7½ | 0.007
Tourmaline 7-7½ | 0.023
Precious Beryl 7½-8 | 0.007
Phenakite 7½-8 | 0.016</td></tr>
<tr><td>3.00-3.49</td><td>Fluorite 4 | -</td><td>Magnesite 3½-4½ | 0208
Leukophane 4 | 0.025
Ekanite 4½-6½ | 0.001
Herderite 5-5½ | 0.027
Meliphanite 5-5½ | 0.019
Tremolite 5-6 | 0.022
Montebrasite 5½-6 | 0.22
Amblygonite 6 | 0.027
Chondrodite 6-6½ | 0.031</td><td>Rhodochrosite 4 | 0.214
Apatite 5 | 0.004
Diopside 5-6 | 0.027
Hypersthene 5-6 | 0.013
Enstatite 5½ | 0.010
Actinolite 5½-6 | 0.022
Amblygonite 6 | 0.027
Nephrite 6-6½ | 0.027
Axinite 6½-7 | 0.011
Hiddenite 6½-7 | 0.015
Jadeite 6½-7 | 0.020
Cornerupine 6½-7 | 0.014
Peridote 6½-7 | 0.037
Sinhalite 6½-7 | 0.039
Danburite 7-7½ | 0.007
Tourmaline 7-7½ | 0.023
Dumortierite 7-8½ | 0.026
Andalusite 7½ | 0.010
Topaz 8 | 0.012</td></tr>
<tr><td>3.50-3.99</td><td></td><td>Strontianite 3½ | 0.150
Barytocalcite 4 | 0.061</td><td>Siderite 3½-4½ | 0.242
Rhodochrosite 4 | 0.214
Willemite 5½ | 0.030
Topaz 8 | 0.012</td></tr>
<tr><td>4.00-4.99</td><td></td><td>Witherite 3-3½ | 0.148</td><td>Baryte 3-3½ | 0.012
Witherite 3-3½ | 0.148</td></tr>
</table>

The numbers following the gemstone refer to Mohs' hardness | double refraction

<table>
<tr><th>Refr. index
Density</th><th>1.700–1.799</th><th>1.800–1.899</th><th>1.900 and higher</th></tr>
<tr><td>1.00-1.99</td><td></td><td></td><td></td></tr>
<tr><td>2.00-2.49</td><td></td><td></td><td>Sulfur 1½-2½ | 0.291</td></tr>
<tr><td>2.50-2.99</td><td>Magnesite 3½-4½ | 0.208</td><td></td><td></td></tr>
<tr><td>3.00-3.49</td><td>Rhodochrosite 4 | 0.214
Hypersthene 5-6 | 0.013
Epidote 6-7 | 0.032
Clinozoisite 6-7 | 0.010
Vesuvianite 6½ | 0.007
Axinite 6½-7 | 0.011
Peridote 6½-7 | 0.037
Sinhalite 6½-7 | 0.039</td><td>Scorodite 3½-4 | 0.029
Rhodochrosite 4 | 0.214
Heterosite 4-4 ½ | 0.05
Purpurite 4-4½ | 0.007</td><td>Heterosite 4-4½ | 0.05
Purpurite 4-4½ | 0.07</td></tr>
<tr><td>3.50-3.99</td><td>Siderite 3½-4½ | 0.242
Rhodochrosite 4 | 0.214
Kyanite 4-7 | 0.024
Hypersthene 5-6 | 0.013
Willemite 5½ | 0.030
Periclase 5½-6 | -
Epidote 6-7 | 0.032
Sinhalite 6½-7 | 0.039
Grossular 6½-7½ | -
Hessonite 6½-7½ | -
Pyrope 6½-7½ | -
Staurolite 7-7½ | 0.012
Spinel 8 | -
Chrysoberyl 8½ | 0.009
Sapphire 9 | 0.008</td><td>Siderite 3½-4½ | 0.242
Rhodochrosite 4 | 0.214
Titanite 5-5½ | 0.146
Andradite 6½-7½ | -
Low Zircon 6½-7½ | -
Zircon 6½-7½ | 0.030</td><td>Sphalerite 3½-4 | -
Goethite 5-5½ | 0.14
Titanite 5-5½ | 0.146
Anatase 5½-6 | 0.056
Andradite 6½-7½ | -
Low Zircon 6½-7½ | -
Zircon 6½-7½ | 0.030
Diamond 10 | -</td></tr>
<tr><td>4.00-4.99</td><td>Monazite 5-5½ | 0.052
Willemite 5½ | 0.030
Spessartine 6½-7½ | -
Painite 7½-8 | 0.029
Sapphire 9 | 0.008</td><td>Monazite 5-5½ | 0.052
Andradite 6½-7½ | -
Spessartine 6½-7½ | -
Low Zircon 6½-7½ | -
Zircon 6½-7½ | 0.030</td><td>Sphalerite 3½-4 | -
Rutile 6-6½ | 0.287
Andradite 6½-7½ | -
Low Zircon 6½-7½ | -
Zircon 6½-7½ | 0.030</td></tr>
<tr><td>5.00-5.99</td><td>Bastnasite 4-4 ½ | 0.100
Monazite 5-5½ | 0.052</td><td>Bastnasite 4-4 ½ | 0.100
Monazite 5-5½ | 0.052</td><td>Crocoite 2½-3 | 0.270
Zinkite 4-5 | 0.016
Scheelite 4½-5 | 0.014
Hematite 5½-6½ | 0.287
Tantalite 6-6½ | 0.160
Zirconia 8½ | -</td></tr>
<tr><td>6.00-6.99</td><td></td><td>Anglesite 3-3½ | 0.017
Cerussite 3-3½ | 0.274
Thorite 4 ½ -5 | 0.6</td><td>Phosgenite 2-3 | 0.028
Crocoite 2½-3 | 0.270
Wulfenite 3 | 0.120
Cerussite 3-3½ | 0.274
Tantalite 6-6½ | 0.160
Cassiterite 6-7 | 0.097</td></tr>
<tr><td>7.00 and higher</td><td></td><td></td><td>Stolzite 2½-3 | 0.08
Vanadinite 2½-3 | 0.066
Wulfenite 3 | 0.120
Hübnerite 4-4½ | 0.13
Wolframite 5-5½ | 0.14
Tantalite 6-6½ | 0.160
Cassiterite 6-7 | 0.097</td></tr>
</table>

Gem Colors: Green + Yellow-Green + Blue-Green

Density / Refr. index	1.400–1.499	1.500–1.599	1.600–1.699
1.00-1.99	Opal 5½-6½ I -	Amber 2-2½ I -	
2.00-2.49	Chrysocolla 2-4 I 0.031 Analcime 5-5½ I - Obsidian 5-5½ I - Moldavite 5½ I - Haüyne 5½-6 I - Opal 5½-6½ I -	Chrysocolla 2-4 I 0.031 Serpentine 2½-5½ I 0.011 Variscite 4-5 I 0.031 Apophyllite 4½-5 I 0.002 Obsidian 5-5½ I - Moldavdite 5½ I - Haüyne 5½-6 I -	Turquoise 5-6 I 0.040
2.50-2.99	Calcite 3 I 0.172 Onyx Marble 3½-4 I 0.163 Obsidian 5-5½ I - Haüyne 5½-6 I - Opal 5½-6½ I -	Vivianite 1½-2 I 0.062 Pearl 2½-4½ I 0.156 Serpentine 2½-5½ I 0.011 Calcite 3 I 0.172 Variscite 4-5 I 0.031 Apophyllite 4½-5 I 0.002 Wardite 4½-5 I 0.009 Obsidian 5-5½ I - Amazonite 6-6½ I 0.008 Chrysoprase 6½-7 I 0.006 Aventurine 7 I 0.009 Prasiolite 7 I 0.009 Aquamarine 7½-8 I 0.004 Precious Beryl 7½-8 I 0.007 Emerald 7½-8 I 0.006	Vivianite 1½-2 I 0.062 Pearl 2½-4½ I 0.156 Calcite 3 I 0.172 Onyx Marble 3½-4 I 0.163 Datolite 5-5½ I 0.045 Tremolite 5-6 I 0.022 Turquoise 5-6 I 0.040 Brazilianite 5½ I 0.020 Montebrasite 5½-6 I 0.22 Nephrite 6-6½ I 0.027 Prehnite 6-6½ I 0.030 Grandidierite 7-7½ I 0.033 Tourmaline 7-7½ I 0.023 Precious Beryl 7½-8 I 0.007 Emerald 7½-8 I 0.006
3.00-3.49	Fluorite 4 I -	Phosphophyllite 3-3½ I 0.027 Leukophane 4 I 0.025 Ekanite 4½-6½ I 0.001 Herderite 5-5½ I 0.027 Tremolite 5-6 I 0.022 Montebrasite 5½-6 I 0.22 Chondrodite 6-6½ I 0.031 Grandidierite 7-7½ I 0.033	Malachite 3½-4 I 0.254 Apatite 5 I 0.004 Dioptase 5 I 0.052 Hemimorphite 5 I 0.022 Datolite 5-5½ I 0.045 Diopside 5-6 I 0.027 Hypersthene 5-6 I 0.013 Tremolite 5-6 I 0.022 Enstatite 5½ I 0.010 Actinolite 5½-6 I 0.022 Nephrite 6-6½ I 0.027 Smaragdite 6-6½ I 0.022 Hiddenite 6½-7 I 0.015 Jadeite 6½-7 I 0.020 Cornerupine 6½-7 I 0.014 Peridote 6½-7 I 0.037 Sinhalite 6½-7 I 0.039 Tourmaline 7-7½ I 0.023 Andalusite 7½ I 0.010 Euclase 7½ I 0.022
3.50-3.99			Celestine 3-3½ I 0.011 Malachite 3½-4 I 0.254 Hemimorphite 5 I 0.022 Hypersthene 5-6 I 0.013 Willemite 5½ I 0.030 Topaz 8 I 0.012
4.00-4.99			Baryte 3-3½ I 0.012 Smithsonite 5 I 0.228

The numbers following the gemstone refer to Mohs' hardness I double refraction

Refr. index / Density	1.700–1.799	1.800–1.899	1.900 and higher
1.00-1.99			
2.00-2.49			Sulfur 1½-2½ I 0.291
2.50-2.99			
3.00-3.49	Malachite 3½-4 I 0.254 Scorodite 3½-4 I 0.029 Triphylite 4-5 I 0.007 Dioptase 5 I 0.052 Bronzite 5-6 I 0.015 Diopside 5-6 I 0.027 Hypersthene 5-6 I 0.013 Aegirine-Augite 6 I 0.040 Zoisite 6-6 ½ I 0.008-0.009 Katoite 6-7 I - Epidote 6-7 I 0.032 Clinozoisite 6-7 I 0.010 Vesuvianite 6½ I 0.007 Peridot 6½-7 I 0.037 Serendibite 6½-7 I 0.005 Sinhalite 6½-7 I 0.039 Diaspore 6½-7½ I 0.048 Sapphirine 7½ I 0.05	Malachite 3½-4 I 0.254 Scorodite 3½-4 I 0.029 Aegirine-Augite 6 I 0.040 Uvarovite 6½-7½ I -	Malachite 3½-4 I 0.254
3.50-3.99	Malachite 3½-4 I 0.254 Kyanite 4-7 I 0.024 Hypersthene 5-6 I 0.013 Willemite 5½ I 0.030 Periclase 5½-6 I - Hedenbergite 5½-6½ I 0.027 Aegirine-Augite 6 I 0.040 Epidote 6-7 I 0.032 Serendibite 6½-7 I 0.005 Sinhalite 6½-7 I 0.039 Grossular 6½-7½ I - Sapphirine 7½ I 0.05 Spinel 8 I - Taaffeite 8-8½ I 0.006 Alexandrite 8½ I 0.009 Chrysoberyl 8½ I 0.009 Sapphire 9 I 0.008	Malachite 3½-4 I 0.254 Titanite 5-5½ I 0.146 Aegirine-Augite 6 I 0.040 Andradite 6½-7½ I - Demantoid 6½-7½ I - Low Zircon 6½-7½ I - Uvarovite 6½-7½ I - Zircon 6½-7½ I 0.030	Malachite 3½-4 I 0.254 Sphalerite 3½-4 I - Titanite 5-5½ I 0.146 Andradite 6½-7½ I - Low Zircon 6½-7½ I - Zircon 6½-7½ I 0.030 Diamond 10 I -
4.00-4.99	Adamite 3½ I 0.049 Malachite 3½-4 I 0.254 Smithsonite 5 I 0.228 Willemite 5½ I 0.030 Gahnite 7½-8 I - Sapphire 9 I 0.008	Malachite 3½-4 I 0.254 Smithsonite 5 I 0.228 Gadolinite 6½-7 I 0.02 Andradite 6½-7½ I - Zircon 6½-7½ I 0.030 Gahnite 7½-8 I -	Malachite 3½-4 I 0.254 Powellite 3½-4 I 0.011 Sphalerite 3½-4 I - Bayldonite 4½ I 0.040 Microlite 5-5½ I - Andradite 6½-7½ I - Zircon 6½-7½ I 0.030
5.00-5.99			Microlite 5-5½ I - Zirconia 8½ I -
6.00-6.99		Anglesite 3-3½ I 0.017	Phosgenite 2-3 I 0.028
7.00 and higher			Stolzite 2½-3 I 0.008

Gem Colors: Blue + Blue-Green + Blue-Red

<table>
<tr><th>Refr. index.
Density</th><th>1.400–1.499</th><th>1.500–1.599</th><th>1.600–1.699</th></tr>
<tr><td>1.00-1.99</td><td>Opal 5½-6½ | -</td><td>Amber 2-2½ | -
Opal 5½-6½ | -</td><td></td></tr>
<tr><td>2.00-2.49</td><td>Chrysocolla 2-4 | 0.031
Cancrinite 5-6 | 0.026
Haüyn 5½-6 | -
Sodalite 5½-6 | -
Opal 5½-6½ | -</td><td>Gypsum 2 | 0.009
Chrysocolla 2-4 | 0.031
Variscite 4-5 | 0.031
Apophyllite 4½-5 | 0.002
Cancrinite 5-6 | 0.026
Haüyn 5½-6 | -
Opal 5½-6½ | -</td><td>Turquoise 5-6 | 0.040</td></tr>
<tr><td>2.50-2.99</td><td>Calcite 3 | 0.172
Cancrinite 5-6 | 0.026
Lapis Lazuli 5-6 | -
Haüyn 5½-6 | -
Opal 5½-6½ | -</td><td>Vivianite 1½-2 | 0.062
Pearl 2½-4½ | 0.156
Calcite 3 | 0.172
Coral 3-4 | 0.172
Anhydrite 3½ | 0.044
Variscite 4-5 | 0.031
Apophyllite 4½-5 | 0.002
Wardite 5 | 0.009
Cancrinite 5-6 | 0.026
Lapis Lazuli 5-6 | -
Haüyn 5½-6 | -
Opal 5½-6½ | -
Amazonite 6-6½ | 0.008
Chalcedony 6½-7 | 0.006
Chrysoprase 6½-7 | 0.006
Jasper 6½-7 | -
Aventurine 7 | 0.009
Prasiolite 7 | 0.009
Cordierite 7-7½ | 0.010
Grandidierite 7-7½ | 0.033
Aquamarine 7½-8 | 0.004
Emerald 7½-8 | 0.006</td><td>Vivianite 1½-2 | 0.062
Pearl 2½-4½ | 0.156
Calcite 3 | 0.172
Anhydrite 3½ | 0.044
Aragonite 3½-4 | 0.155
Dolomite 3½-4 | 0.185
Pectolite 4½-5 | 0.038
Ceruleite 5-6 |-
Richterite 5-6 | 0.16-0.020
Herderite 5-5½ | 0.027
Turquoise 5-6 | 0.040
Montebrasite 5½-6 | 0.22
Nephrite 6-6½ | 0.027
Boracite 7-7½ | 0.010
Grandidierite 7-7½ | 0.033
Tourmaline 7-7½ | 0.023
Emerald 7½-8 | 0.006</td></tr>
<tr><td>3.00-3.49</td><td>Fluorite 4 | -
Lapis Lazuli 5-6 | -</td><td>Phosphophyllite 3-3½ | 0.027
Herderite 5-5½ | 0.027
Lapis Lazuli 5-6 | -
Montebrasite 5½-6 | 0.22
Grandidierite 7-7½ | 0.033</td><td>Apatite 5 | 0.004
Dioptase 5 | 0.052
Hemimorphite 5 | 0.022
Lazulite 5-6 | 0.033
Nephrite 6-6½ | 0.027
Clinozoisite 6-7 | 0.010
Axinite 6½-7 | 0.011
Jadeite 6½-7 | 0.020
Tanzanite 6½-7 | 0.009
Sillimanite 6½-7½ | 0.017
Tourmaline 7-7½ | 0.023
Dumortierite 7-8½ | 0.026
Euclase 7½ | 0.022
Topaz 8 | 0.012</td></tr>
<tr><td>3.50-3.99</td><td></td><td></td><td>Celestine 3-3½ | 0.011
Hemimorphite 5 | 0.022
Topaz 8 | 0.012</td></tr>
<tr><td>4.00-4.99</td><td></td><td></td><td>Baryte 3-3½ | 0.012
Celestine 3-3½ | 0.011
Smithsonite 5 | 0.228</td></tr>
</table>

The numbers following the gemstone refer to Mohs' hardness | double refraction

Refr. index / Density	1.700–1.799	1.800–1.899	1.900 and higher
1.00-1.99			
2.00-2.49			
2.50-2.99			
3.00-3.49	Lithiophilite 4-5 | 0.01 Triphylite 4-5 | 0.007 Dioptase 5 | 0.052 Clinozoisite 6-7 | 0.010 Vesuvianite 6½ | 0.007 Axinite 6½-7 | 0.011 Diaspore 6½-7 | 0.048 Serendibite 6½-7 | 0.005 Tanzanite 6½-7 | 0.009 Sapphirine 7½ | 0.005	Purpurite 4-4½ | 0.07	Purpurite 4-4½ | 0.07
3.50-3.99	Azurite 3½-4 | 0.109 Lithiophilite 4-5 | 0.01 Kyanite 4½-7 | 0.024 Hedenbergite 5½-6½ | 0.027 Benitoite 6-6½ | 0.047 Serendibite 6½-7 | 0.005 Sapphirine 7½ | 0.005 Spinel 8 | - Taaffeite 8-8½ | 0.006 Ruby 9 | 0.008 Sapphire 9 | 0.008	Azurite 3½-4 | 0.109 Benitoite 6-6½ | 0.047 Low Zircon 6½-7½ | - Zircon 6½-7½ | 0.030	Anatase 5½-6 | 0.056 Low Zircon 6½-7½ | - Zircon 6½-7½ | 0.030 Diamond 10 | -
4.00-4.99	Smithsonite 5 | 0.228 Gahnite 7½-8 | - Ruby 9 | 0.008 Sapphire [MG2]0.008	Smithsonite 5 | 0.228 Low Zircon 6½-7½ | - Zircon 6½-7½ | 0.030 Gahnite 7½-8 | -	Powellite 3½-4 | 0.011 Low Zircon 6½-7½ | - Zircon 6½-7½ | 0.030
5.00-5.99		Linarite 2½ | 0.050	Boleite 3-3½ | 0.020 Zirconia 8½ | -
6.00-6.99			
7.00 and higher			

Gem Colors: Violet + Blue-Red

Refr. index. Density	1.400–1.499	1.500–1.599	1.600–1.699
1.00-1.99	Opal 5½-6½ ǀ -	Amber 2-2½ ǀ - Opal 5½-6½ ǀ -	
2.00-2.49	Tugtupite 5½-6 ǀ 0.006 Opal 5½-6½ ǀ -	Stichtite 1½-2½ ǀ 0.026 Tugtupite 5½-6 ǀ 0.006 Opal 5½-6½ ǀ - Ussingite 6-6½ ǀ 0.039	
2.50-2.99	Calcite 3 ǀ 0.172 Coral 3-4 ǀ 0.166 Creedite 3½-4 ǀ 0.024 Lapis Lazuli 5-6 ǀ - Tugtupite 5½-6 ǀ 0.006 Opal 5½-6½ ǀ -	Kämmererite 2-2½ ǀ 0.003 Calcite 3 ǀ 0.172 Coral 3-4 ǀ 0.166 Anhydrite 3½ ǀ 0.044 Charoite 4½-5 ǀ 0.006 Poudretteite 5 ǀ 0.021 Lapis Lazuli 5-6 ǀ - Scapolite 5½-6 ǀ 0.021 Tugtupite 5½-6 ǀ 0.006 Opal 5½-6½ ǀ - Petrified Wood 5½-7 ǀ - Ussingite 6-6½ ǀ 0.039 Jasper 6½-7 ǀ - Pollucite 6½-7 ǀ - Amethyst 7 ǀ 0.009 Amethyst Quartz 7 ǀ 0.009 Rose Quartz 7 ǀ 0.009 Cordierite 7-7½ ǀ 0.010	Kämmererite 2-2½ ǀ 0.003 Calcite 3 ǀ 0.172 Coral 3-4 ǀ 0.166 Anhydrite 3½ ǀ 0.044 Nephrite 6-6½ ǀ 0.027 Sugilite 6-6½ ǀ 0.002 Sogdianite 6-7 ǀ 0.002 Tourmaline 7-7½ ǀ 0.023
3.00-3.49	Fluorite 4 ǀ - Lapis Lazuli 5-6 ǀ -	Lapis Lazuli 5-6 ǀ - Amblygonite 6 ǀ 0.027	Apatite 5 ǀ 0.004 Spurrite 5 ǀ 0.039 Richterite 5-6 ǀ 0.016-0.020 Scorzalite 5½-6 ǀ 0.039 Amblygonite 6 ǀ 0.027 Nephrite 6-6½ ǀ 0.027 Axinite 6½-7 ǀ 0.011 Jadeite 6½-7 ǀ 0.020 Kunzite 6½-7 ǀ 0.015 Tanzanite 6½-7 ǀ 0.009 Sillimanite 6½-7½ ǀ 0.017 Tourmaline 7-7½ ǀ 0.023 Dumortierite 7-8½ ǀ 0.026 Topaz 8 ǀ 0.012
3.50-3.99			Topaz 8 ǀ 0.012
4.00-4.99			Smithsonite 5 ǀ 0.228
5.00-5.99			
6.00-6.99			
7.00 and higher			

The numbers following the gemstone refer to Mohs' hardness ǀ double refraction

Density / Refr. index	1.700–1.799	1.800–1.899	1.900 and higher
1.00-1.99			
2.00-2.49			
2.50-2.99			
3.00-3.49	Scorodite 3½-4 | 0.029 Axinite 6½-7 | 0.011 Tanzanite 6½-7 | 0.009 Chambersite 7 | 0.012 Sapphirine 7½ | 0.005 |	Scorodite 3½-4 | 0.029 Heterosite 4-4½ | 0.05 Purpurite 4-4½ | 0.07	Heterosite 4-4½ | 0.05 Purpurite 4-4½ | 0.07
3.50-3.99	Benitoite 6-6½ | 0.047 Almandine 6½-7½ | - Sapphirine 7½ | 0.005 Spinel 8 | - Taaffeite 8-8½ | 0.006 Ruby 9 | 0.008 Sapphire 9 | 0.008	Benitoite 6-6½ | 0.047 Almandine 6½-7½ | - Low Zircon 6½-7½ | - Zircon 6½-7½ | 0.030	Low Zircon 6½-7½ | - Zircon 6½-7½ | 0.030
4.00-4.99	Adamite 3½ | 0.049 Smithsonite 5 | 0.228 Almandine 6½-7½ | - Gahnite 7½-8 | - Ruby 9 | 0.008 Sapphire 9 | 0.008	Smithsonite 5 | 0.228 Almandine 6½-7½ | - Low Zircon 6½-7½ | - Zircon 6½-7½ | 0.030 Gahnite 7½-8 | -	Low Zircon 6½-7½ | - Zircon 6½-7½ | 0.030
5.00-5.99			Proustite 2½ | 0.203 Cuprite 3½-4 | - Zirconia 8½ | -
6.00-6.99			Cuprite 3½-4 | -
7.00 and higher			

Gem Colors: Black + Gray

<table>
<tr><th>Refr. index
Density</th><th>1.400–1.499</th><th>1.500–1.599</th><th>1.600–1.699</th></tr>
<tr><td>1.00-1.99</td><td>Opal 5½-6½ | -</td><td>Amber 2-2½ | -
Coral 3-4 | 0.16
Opal 5½-6½ | -</td><td>Jet 2½-4 | -</td></tr>
<tr><td>2.00-2.49</td><td>Obsidian 5-5½ | -
Sodalite 5½-6 | -
Opal 5½-6½ | -</td><td>Sepiolite 2-2½ | -
Obsidian 5-5½ | -
Opal 5½-6½ | -
Hambergite 7½ | 0.072</td><td>Hambergite 7½ | 0.072</td></tr>
<tr><td>2.50-2.99</td><td>Calcite 3 | 0.172
Obsidian 5-5½ | -
Opal 5½-6½ | -</td><td>Pearl 2½-4½ | 0.156
Calcite 3 | 0.172
Aragonite 3½-4 | 0.155
Obsidian 5-5½ | -
Opal 5½-6½ | -
Petrified Wood 5½-7 | -
Bytownite 6-6½ | 0.009
Labradorite 6-6½ | 0.009
Chalcedony 6½-7 | 0.006
Jasper 6½-7 | -
Smoky Quartz 7 | 0.009</td><td>Pearl 2½-4½ | 0.156
Calcite 3 | 0.172
Aragonite 3½-4 | 0.155
Hornblende 5-6 | 0.02
Nephrite 6-6½ | 0.027
Tourmaline 7-7½ | 0.023</td></tr>
<tr><td>3.00-3.49</td><td>Fluorite 4 | -</td><td></td><td>Hydroxylapatite 5 | 0.012
Bronzite 5-6 | 0.015
Hornblende 5-6 | 0.02
Hypersthene 5-6 | 0.013
Neptunite 5-6 | 0.037
Nephrite 6-6½ | 0.027
Jadeite 6½-7 | 0.020
Tourmaline 7-7½ | 0.023</td></tr>
<tr><td>3.50-3.99</td><td></td><td></td><td>Hypersthene 5-6 | 0.013</td></tr>
<tr><td>4.00-4.99</td><td></td><td></td><td></td></tr>
<tr><td>5.00-5.99</td><td></td><td></td><td></td></tr>
<tr><td>6.00-6.99</td><td></td><td></td><td></td></tr>
<tr><td>7.00 and higher</td><td></td><td></td><td></td></tr>
</table>

The numbers following the gemstone refer to Mohs' hardness | double refraction

Density / Refr. index	1.700–1.799	1.800–1.899	1.900 and higher
1.00-1.99			
2.00-2.49			
2.50-2.99			
3.00-3.49	Bronzite 5-6 | 0.015 Hypersthene 5-6 | 0.013 Neptunite 5-6 | 0.037 Aegirine-Augite 6 | 0.040 Epidote 6-7 | 0.032	Aegirine-Augite 6 | 0.040	
3.50-3.99	Hypersthene 5-6 | 0.013 Hedenbergite 5½-6½ | 0.027 Aegirine-Augite 6 | 0.040 Epidote 6-7 | 0.032 Staurolite 7-7½ | 0.012 Spinel 8 | - Sapphire 9 | 0.008	Aegirine-Augite 6 | 0.040 Andradite 6½-7½ | -	Goethite 5-5½ | 0.14 Anatase 5½-6 | 0.056 Andradite 6½-7½ | - Diamond 10 | -
4.00-4.99	Gadolinite 6½-7 | 0.02 Gahnite 7½-8 | - Sapphire 9 | 0.008	Gadolinite 6½-7 | 0.02 Andradite 6½-7½ | - Gahnite 7½-8 | -	Goethite 5-5½ | 0.14 Davidite 5-6 | - Chromite 5½ | - Brookite 5½-6 | 0.117 Rutile 6-6½ | 0.287 Andradite 6½-7½ | -
5.00-5.99			Descloizite 3-3½ | 0.165 Hematite 5½-6½ | 0.287 Magnetite 5½-6½ | - Tantalite 6-6½ | 0.160
6.00-6.99		Cerussite 3-3½ | 0.274	Cerussite 3-3½ | 0.274 Descloizite 3-3½ | 0.165 Tantalite 6-6½ | 0.160
7.00 and higher			Cinnabar 2-2½ | 0.351 Hübnerite 4-4½ | 0.13 Wolframite 5-5½ | 0.14 Tantalite 6-6½ | 0.160

Gem Colors: Multicolored + Phenomenal

Density / Refr. index	1.400–1.499	1.500–1.599	1.600–1.699
1.00-1.99	Opal 5½-6½ | -	Amber 2-2½ | - Opal 5½-6½ | -	
2.00-2.49	Tugtupite 5½-6 | 0.006 Opal 5½-6½ | -	Serpentine 2½-5½ | 0.011 Howlite 3-3½ | 0.019 Tugtupite 5½-6 | 0.006 Opal 5½-6½ | -	Howlite 3-3½ | 0.019 Turquoise 5-6 | 0.040
2.50-2.99	Onyx Marble 3½-4 | 0.163 Lapis Lazuli 5-6 | - Tugtupite 5½-6 | 0.006 Opal 5½-6½ | -	Serpentine 2½-5½ | 0.011 Howlite 3-3½ | 0.019 Onyx Marble 3½-4 | 0.163 Aragonite 3½-4 | 0.155 Ammonite 4 | 0.155 Charoite 4½-5 | 0.006 Lapis Lazuli 5-6 | - Tugtupite 5½-6 | 0.006 Opal 5½-6½ | - Petrified Wood 5½-7 | - Labradorite 6-6½ | 0.009 Moonstone 6-6½ | 0.008 Peristerite 6-6½ | 0.009 Sunstone 6-6½ | 0.010 Agate 6½-7 | 0.006 Chalcedony 6½-7 | 0.006 Jasper 6½-7 | - Moss Agate 6½-7 | 0.006 Tiger's-Eye 6½-7 | - Amethyst Quartz 7 | 0.009 Aventurine 7 | 0.009	Howlite 3-3½ | 0.019 Onyx Marble 3½-4 | 0.163 Aragonite 3½-4 | 0.155 Ammonite 4 | 0.155 Turquoise 5-6 | 0.040 Nephrite 6-6½ | 0.027 Tourmaline 7-7½ | 0.023
3.00-3.49	Fluorite 4 | - Lapis Lazuli 5-6 | -	Lapis Lazuli 5-6 | -	Malachite 3½-4 | 0.254 Rhodochrosite 4 | 0.214 Nephrite 6-6½ | 0.027 Jadeite 6½-7 | 0.020 Tourmaline 7-7½ | 0.023
3.50-3.99			Malachite 3½-4 | 0.254 Rhodochrosite 4 | 0.214
4.00-4.99			Malachite 3½-4 | 0.254
5.00-5.99			
6.00-6.99			
7.00 and higher			

The numbers following the gemstone refer to Mohs' hardness | double refraction

Density / Refr. index	1.700–1.799	1.800–1.899	1.900 and higher
1.00-1.99			
2.00-2.49			
2.50-2.99			
3.00-3.49	Malachite 3 1/2-4 | 0.254 Rhodochrosite 4 | 0.214 Rhodonite 5 1/2-6 1/2 | 0.012	Malachite 3 1/2-4 | 0.254 Rhodochrosite 4 | 0.214	Malachite 3 1/2-4 | 0.254
3.50-3.99	Malachite 3 1/2-4 | 0.254 Rhodochrosite 4 | 0.214 Rhodonite 5 1/2-6 1/2 | 0.012 Alexandrite 8 1/2 | 0.009	Malachite 3 1/2-4 | 0.254 Rhodochrosite 4 | 0.214	Malachite 3 1/2-4 | 0.254
4.00-4.99	Malachite 3 1/2-4 | 0.254	Malachite 3 1/2-4 | 0.254	Malachite 3 1/2-4 | 0.254
5.00-5.99			
6.00-6.99			
7.00 and higher			

Index

Note: Page references in **bold** indicate primary references, and *italics* indicate photographs of gemstones

An Imprint of Sterling Publishing Co., Inc.
1166 Avenue of the Americas
New York, NY 10036

Originally published by BLV Buchverlag GmbH & Co. KG under the title *Edelsteine and Schmucksteine*
This 2013 edition published by Sterling Publishing Co., Inc.

Translated by Daniel Shea and Nicole Shea

Acknowledgements of Photographs
Archive for Art and History, Berlin: 287
H. Bank, Idar-Oberstein: 227/4 top; 239 top; 277/9; 279/7; 12, 13; 281/4, 9, 14, 21
Bavarian Administration of Stately Castles, Gardens and Lakes, Munich: 283
G. Becker, Idar-Oberstein: 279/11
De Beers, London: 61, 66, 67 bottom, 77 left, 77 right, 78 left, 78 right, 79, 89
De Beers Consolidated Mines Ltd., Johannesburg, South Africa: 91
E.A. Bunzel, Idar-Oberstein: 69 left, 70, 76 top
Chudoba-Gübelin, *Edelsteinkundliches Handbuch*, Wilhelm Stollfuß Verlag, Bonn: 45
Department of Mines and Energy, Parkside/ Australia: 651
D.J. Edelmann GmbH (GPRA), Rund um die Perle, Frankfurt: 259, 261, 262, 263
System Eickhorst, Hamburg: 16, 36, 49, 274, 275
H. Eisenbeiss, Bad Kohlgrub: 14 bottom left, 14 bottom right, 40 bottom left, 109, 265, 288, 289, 291 top, 291 bottom, 292 (all)
W. Eisenreich/W. Schumann: 24, 55
Groh + Ripp, Idar-Oberstein: 279/16, 18, 19, 23, 25, 27; 281/5
E. Gübelin, Luzern: 52 bottom, 59 top, 59 center, 59 bottom, 62, 63, 64, 65 right, 71, 72 bottom, 73 bottom, 101, 105, 108
H. A. Hänni, Basel: 179/9; 279/1, 2, 10, 14, 17, 20; 281/1, 2, 13, 16, 17, 19; 282/1, 6, 7, 8
K. Hartmann, Sobernheim: 52 top, 53, 54, 58, 60, gemstone tableaus p. 87-225, p. 245-257, p. 269
K. Hartmann/Diaverleih T. Sachs, Sobernheim: 1, 6, 15, 31, 285, 286
U. Henn, Idar-Oberstein: 279/8; 281/25
HMSO, Norwich: 9 (Crown copyright is reproduced with the permission of the Controller of HMSO)
R. Hochleitner, Archiv LAPIS, Munich: 270
G. Holzey, Erfurt: 279/15; 281/3, 10, 12, 23; 282/2, 3, 4, 5
Homberg + Brusius, Kirschweiler: 73 top
Foto Hosser, Idar-Oberstein: 74 bottom, 227, 229, 231, 233, 235, 237, 239, 241, 243, 267
Jain Zuchtperlen, Seeheim/Bergstraße: 260
Krantz, Bonn: 76 bottom
A. Krüss GmbH, Hamburg: 44
Lichtblick-Fotodesign, Schwollen: 179/6; 229/8 top; 277/4, 5, 11; 279/3, 21, 22, 24, 26; 281/6, 7, 8, 11, 20, 22, 26; 282/10
U. Medenbach, Witten: 2-3, 4-5, 33
Museum Association of the Prehistoric State Collection, Munich/M. Eberlein: 68
E. Pauly, Veitsrodt: 160
A. Ruppenthal KG, Idar-Oberstein: 154, 156
R. Schultz-Güttler, Sao Paulo: 279/4, 5, 6
W. Schumann, Munich: 67 top, 69 right
State Gemstone Institute, Vienna: 279/9; 281/15, 18; 282/9
Welsch, Hottenbach: 75 (all)

Graphics
Daniela Farnhammer: front endpaper, 282
Marlene Gemke: 37 top, 258, 268, 270
Cartography Huber: back endpaper
Manuela Hutschenreiter: 21 right, 27 bottom, 40 bottom right, 81 bottom
Computer graphics Jörg Mair: 72 top, 74, 88
Studio Pachlhofer: Crystals in the margins p. 86-216
Graphic p. 17 from: W. Schumann, *Knaurs Buch der Erde*, Munich 1989
All other graphics: Helmut Hoffmann

Graphics in front endpaper: Gemstone cuts
Graphic in back endpaper: Gemstone occurrence

Photo p. 1: gold beryl
Photo p. 2-3: tourmaline, 6 times enlarged
Photo p. 4-5: tourmaline, 4 times enlarged
Photo p. 6: tourmaline, slightly enlarged

ISBN 978-1-4549-0953-8

Distributed in Canada by Sterling Publishing c/o Canadian Manda Group, 664 Annette Street
Toronto, Ontario, Canada M6S 2C8

For information about custom editions, special sales, and premium and corporate purchases, please contact Sterling Special Sales at 800-805-5489 or specialsales@sterlingpublishing.com.

Manufactured in Germany

8 10 9

www.sterlingpublishing.com

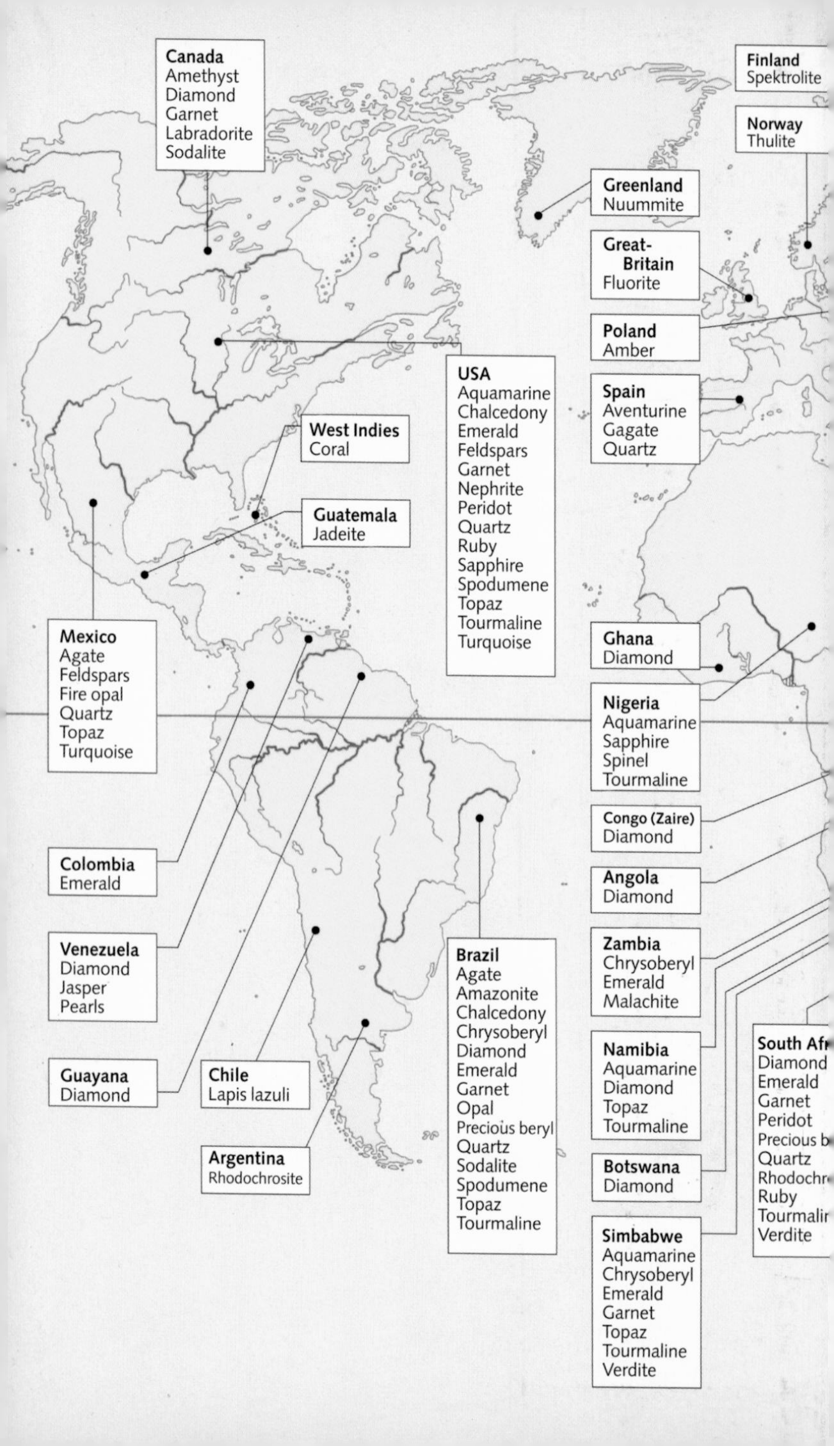

Canada
Amethyst
Diamond
Garnet
Labradorite
Sodalite
Finland
Spektrolite
Norway
Thulite
Greenland
Nuummite
Great-Britain
Fluorite
Poland
Amber
USA
Aquamarine
Chalcedony
Emerald
Feldspars
Garnet
Nephrite
Peridot
Quartz
Ruby
Sapphire
Spodumene
Topaz
Tourmaline
Turquoise
Spain
Aventurine
Gagate
Quartz
West Indies
Coral
Guatemala
Jadeite
Mexico
Agate
Feldspars
Fire opal
Quartz
Topaz
Turquoise
Ghana
Diamond
Nigeria
Aquamarine
Sapphire
Spinel
Tourmaline
Congo (Zaire)
Diamond
Angola
Diamond
Colombia
Emerald
Zambia
Chrysoberyl
Emerald
Malachite
Venezuela
Diamond
Jasper
Pearls
Brazil
Agate
Amazonite
Chalcedony
Chrysoberyl
Diamond
Emerald
Garnet
Opal
Precious beryl
Quartz
Sodalite
Spodumene
Topaz
Tourmaline
Namibia
Aquamarine
Diamond
Topaz
Tourmaline
Guayana
Diamond
Chile
Lapis lazuli
Argentina
Rhodochrosite
Botswana
Diamond
Simbabwe
Aquamarine
Chrysoberyl
Emerald
Garnet
Topaz
Tourmaline
Verdite
Diamond
Emerald
Garnet
Peridot
Quartz
Ruby
Verdite